The Cosmic 21-cm Revolution

Charting the first billion years of our universe

AAS Editor in Chief

Ethan Vishniac, John Hopkins University, Maryland, US

About the program:

AAS-IOP Astronomy ebooks is the official book program of the American Astronomical Society (AAS), and aims to share in depth the most fascinating areas of astronomy, astrophysics, solar physics and planetary science. The program includes publications in the following topics:

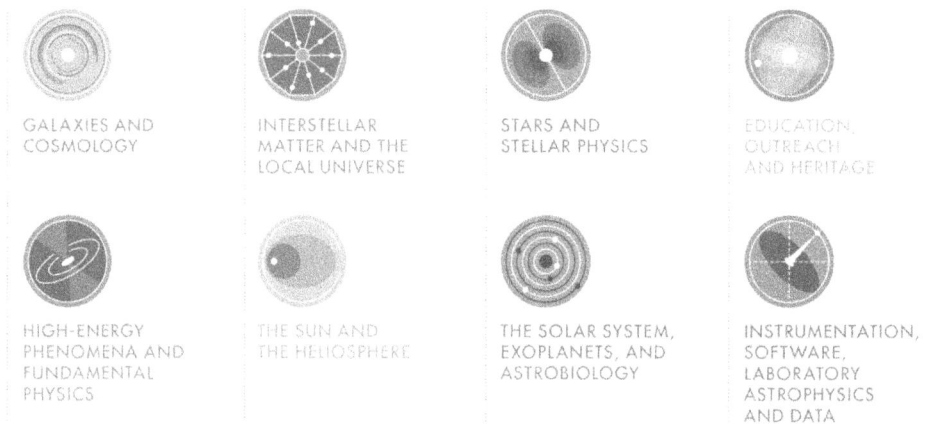

GALAXIES AND COSMOLOGY

INTERSTELLAR MATTER AND THE LOCAL UNIVERSE

STARS AND STELLAR PHYSICS

EDUCATION, OUTREACH AND HERITAGE

HIGH-ENERGY PHENOMENA AND FUNDAMENTAL PHYSICS

THE SUN AND THE HELIOSPHERE

THE SOLAR SYSTEM, EXOPLANETS, AND ASTROBIOLOGY

INSTRUMENTATION, SOFTWARE, LABORATORY ASTROPHYSICS AND DATA

Books in the program range in level from short introductory texts on fast-moving areas, graduate and upper-level undergraduate textbooks, research monographs, and practical handbooks.

For a complete list of published and forthcoming titles, please visit iopscience.org/books/aas.

About the American Astronomical Society

The American Astronomical Society (aas.org), established 1899, is the major organization of professional astronomers in North America. The membership (~7,000) also includes physicists, mathematicians, geologists, engineers and others whose research interests lie within the broad spectrum of subjects now comprising the contemporary astronomical sciences. The mission of the Society is to enhance and share humanity's scientific understanding of the universe.

The Cosmic 21-cm Revolution

Charting the first billion years of our universe

Edited by
Andrei Mesinger

Faculty of Science and Mathematics, Scuola Normale Superiore, Piazza dei Cavalieri 7, 56126 Pisa, PI, Italy

IOP Publishing, Bristol, UK

ISBN   978-0-7503-2236-2 (ebook)
ISBN   978-0-7503-2234-8 (print)
ISBN   978-0-7503-2237-9 (myPrint)
ISBN   978-0-7503-2235-5 (mobi)

DOI   10.1088/2514-3433/ab4a73

Version: 20191201

AAS–IOP Astronomy
ISSN 2514-3433 (online)
ISSN 2515-141X (print)

British Library Cataloguing-in-Publication Data: A catalogue record for this book is available from the British Library.

Published by IOP Publishing, wholly owned by The Institute of Physics, London

IOP Publishing, Temple Circus, Temple Way, Bristol, BS1 6HG, UK

US Office: IOP Publishing, Inc., 190 North Independence Mall West, Suite 601, Philadelphia, PA 19106, USA

Cover image: Galaxies during the era of reionisation in the early universe (simulation). Image credit: M. Alvarez (http://www.cita.utoronto.ca/~malvarez), R. Kaehler, and T. Abel/ESO.

Contents

Preface

The cosmic microwave background (CMB) gives us a remarkable image of the universe when it was just ~400,000 years old (less than 3% of its current age). At that time, the normal (atomic) matter in our universe recombined, dynamically separating from the CMB, and began to collapse under gravity.

However, the billion years that followed this recombination epoch are still mostly shrouded in darkness. Although observations remain sparse, we know that they must have witnessed the birth of the very first stars, black holes, and galaxies. The light from these nascent objects spread out, heating and ionizing virtually all of the atoms in existence. This epoch of reionization was the final major phase transition of our universe, and the last interesting thing to happen to most of the atoms.

As a graduate student in 2003, I remember the palpable excitement in anticipation of the first measurement of the optical depth to the CMB from the *Wilkinson Microwave Anisotropy Probe* (*WMAP*). Prior to this, we only had evidence that the universe was largely ionized up to $z < 5$–6, but had almost no clue about when this reionization actually happened. The optical depth estimate turned out to be large, stimulating a flurry of research papers on early reionization by exotic, unseen sources. Although subsequent measurements brought down the optical depth, the following decade and a half witnessed a surge of activity in inferring the ionization state of the universe from observations of high-z quasars, galaxies, and the CMB. Thanks to sophisticated observational and analysis techniques, astronomers were able to squeeze out estimates on the timing of reionization from a fairly modest amount of data. The emerging picture is that the bulk of reionization occurred around $z \sim 7$–8, driven by galaxies too faint to be observed directly.

But what else can we learn about the first billion years? They comprise the bulk of our past light cone. The number of independent modes in this light cone is orders of magnitude larger than that in the CMB. If we could tap into this vast resource, we could unlock the mysteries of how the first stars and galaxies formed, how they interacted with each other, and open up a new window for physical cosmology.

Thankfully, we have a tool to do just that: the cosmic 21 cm signal. Corresponding to the spin-flip transition of neutral hydrogen, the 21 cm line is sensitive to the temperature and ionization state of the cosmic gas, as well as to cosmological parameters. It is a line transition, so different observed frequencies correspond to different redshifts. Therefore, upcoming interferometers will allow us to map out the first billion years of our universe! The patterns of this map will tell us about the properties of the unseen first generations of galaxies, provided we know how to interpret them. Cosmic Dawn and reionization will move from being observationally starved epochs to being at the frontier of Big Data analysis.

We are truly at the cusp of a revolution—thankfully not a violent one, but one that can transform our understanding of the universe in which we live. I hope that this book can help convince you to join the revolution!

Editor biography

Andrei Mesinger

Andrei Mesinger obtained his PhD from Columbia University in 2006. After a postdoc at Yale University and a Hubble fellowship at Princeton University, he moved to Scuola Normale Superiore in Pisa in 2011 as junior faculty. He has authored over 100 publications on early structure formation and the epoch of reionization, as well as creating the widely used public simulation code 21cmFAST. In 2015, his research was recognized with a prestigious 1.5 million euro Starting Grant award from the European Research Council.

Contributors

Gianni Bernardi
INAF—Istituto di Radio Astronomia
via Gobetti 101, 40129, Bologna, Italy &
Department of Physics and Electronics
Rhodes University
PO Box 94, Grahamstown, 6140, South Africa

Emma Chapman
Blackett Laboratory
Imperial College
London, UK

Steven R. Furlanetto
UCLA Physics & Astronomy
Box 951547, PAB 3-720
Los Angeles, CA 90095-1547, USA

Lincoln J. Greenhill
Harvard-Smithsonian Center for Astrophysics
60 Garden St, Mail Stop 42,
Cambridge, MA 02138, USA

Bradley Greig
School of Physics
The University of Melbourne
Parkville, VIC 3010, Australia

Vibor Jelić
Ruđjer Bošković Institute
Division of Experimental Physics
Laboratory for Astroparticle Physics and Astrophysics
Zagreb, Croatia

Leon Koopmans
Faculty of Science and Engineering
Kapteyn Astronomical Institute
Landleven 12
9747 AD Groningen, The Netherlands

Jordan Mirocha
Department of Physics
McGill University
Montréal, Quebec, Canada

Jonathan Pober
Department of Physics
Brown University
Providence, RI, USA

Ravi Subrahmanyan
Raman Research Institute
C V Raman Avenue, Sadashivanagar,
Bangalore 560080, India

Cathryn M. Trott
International Centre for Radio Astronomy Research
Curtin University, Bentley WA, Australia

A|S IOP Astronomy

The Cosmic 21-cm Revolution

Charting the first billion years of our universe

Andrei Mesinger

Chapter 1

Theoretical Framework: The Fundamentals of the 21 cm Line

Steven R Furlanetto

We review some of the fundamental physics necessary for computing the highly-redshifted spin-flip background. We first discuss the radiative transfer of the 21 cm line and define the crucial quantities of interest. We then review the processes that set the spin temperature of the transition, with a particular focus on Wouthuysen–Field coupling, which is likely to be the most important process during and after the Cosmic Dawn. Finally, we discuss processes that heat the intergalactic medium during the Cosmic Dawn, including the scattering of Lyα, cosmic microwave background (CMB), and X-ray photons.

1.1 Radiative Transfer of the 21 cm Line

Consider a spectral line labeled 0 (the lower level) and 1 (the upper level). The radiative transfer equation for the specific intensity I_ν of photons at the relevant frequency is

$$\frac{dI_\nu}{d\ell} = \frac{\phi(\nu)h\nu}{4\pi}[n_1 A_{10} - (n_0 B_{01} - n_1 B_{10})I_\nu], \tag{1.1}$$

where $d\ell$ is a proper path length element, $\phi(\nu)$ is the line profile function, n_i denotes the number density of atoms at the different levels, and A_{ij} and B_{ij} are the Einstein coefficients for the relevant transition (here, i and j are the initial and final states, respectively). For the 21 cm line, the line frequency is $\nu_{21} = 1420.4057$ MHz. The Einstein relations associate the radiative transition rates via $B_{10} = (g_0/g_1)B_{01}$ and $B_{10} = A_{10}(c^2/2h\nu^3)$, where g is the spin degeneracy factor of each state. For the 21 cm transition, $A_{10} = 2.85 \times 10^{-15}$ s^{-1} and $g_1/g_0 = 3$.

The relative populations of hydrogen atoms in the two spin states determine the spin temperature, T_S, through the relation

doi:10.1088/2514-3433/ab4a73ch1

$$\left(\frac{n_1}{n_0}\right) = \left(\frac{g_1}{g_0}\right) \exp\left\{\frac{-T_*}{T_S}\right\}, \tag{1.2}$$

where $T_* \equiv E_{10}/k_B = 68$ mK is equivalent to the transition energy E_{10}. In almost all physically plausible situations, T_* is much smaller than any other temperature, including T_S, so all exponentials in temperature can be Taylor-expanded to leading order with high accuracy. Note, however, that T_S implicitly assumes that the level populations can be described by a single temperature—independent of each atom's velocity. In detail, velocity-dependent effects must be considered in certain circumstances (Hirata & Sigurdson 2007).

It is conventional to replace I_ν by the equivalent brightness temperature, $T_b(\nu)$, required of a blackbody radiator (with spectrum B_ν) such that $I_\nu = B_\nu(T_b)$. In the low-frequency regime relevant to the 21 cm line, the Rayleigh–Jeans formula is an excellent approximation to the Planck curve, so $T_b(\nu) \approx I_\nu c^2/2k_B\nu^2$.

In this limit, the equation of radiative transfer along a line of sight through a cloud of uniform excitation temperature T_S becomes

$$T_b'(\nu) = T_S(1 - e^{-\tau_\nu}) + T_R'(\nu)e^{-\tau_\nu}, \tag{1.3}$$

where $T_b'(\nu)$ is the emergent brightness measured at the cloud and at redshift z, the optical depth $\tau_\nu \equiv \int ds\, \alpha_\nu$ is the integral of the absorption coefficient (α_ν) along the ray through the cloud, T_R' is the brightness of the background radiation field incident on the cloud along the ray, and s is the proper distance. Because of the cosmological redshift, for the 21 cm transition an observer will measure an apparent brightness at Earth of $T_b(\nu) = T_b'(\nu_{21})/(1 + z)$, where the observed frequency is $\nu = \nu_{21}/(1 + z)$. Henceforth we will work in terms of these observed quantities.

The absorption coefficient is related to the Einstein coefficients via

$$\alpha = \phi(\nu)\frac{h\nu}{4\pi}(n_0 B_{01} - n_1 B_{10}). \tag{1.4}$$

Because all astrophysical applications have $T_S \gg T_*$, approximately three of four atoms find themselves in the excited state $(n_0 \approx n_1/3)$. As a result, the stimulated emission correction represented by the first term is significant.

The fundamental observable quantity is the change in brightness temperature induced by the 21 cm line by a patch of the intergalactic medium (IGM), relative to the incident radiation field. In most models, that incident field is simply the CMB, although if other sources create a low-frequency radio background at very high redshifts, or if there is a particular source behind the IGM patch along the line of sight from the observer, a larger radio background may exist.

Consider photons incident on the patch from this background. If any redshift into resonance with the 21 cm line, they can interact with the cloud—but only for a short time, as they will redshift out of resonance as the universe continues to expand. Thus, the Hubble expansion rate sets an effective path length through the cloud, simply equal to the distance the photon travels while it remains within the line profile. The total absorption can be calculated by integrating the IGM density across this interval, in an exactly analogous procedure to the calculation of the Gunn–Peterson Lyα optical depth (Field 1959; Gunn & Peterson 1965; Scheuer 1965). The result is

$$\tau_{10} = \frac{3}{32\pi} \frac{hc^3 A_{10}}{k_B T_S \nu_{10}^2} \frac{x_{HI} n_H}{(1 + z)(dv_{\parallel}/d\eta_{\parallel})} \tag{1.5}$$

$$\approx 0.0092 \, (1 + \delta)(1 + z)^{3/2} \frac{x_{HI}}{T_S} \left[\frac{H(z)/(1 + z)}{dv_{\parallel}/d\eta_{\parallel}} \right], \tag{1.6}$$

where n_H is the hydrogen number density, x_{HI} is the neutral fraction, and $dv_{\parallel}/d\eta_{\parallel}$ is the velocity gradient along the line of sight (here scaled to the Hubble flow). In the second part, T_S is in Kelvins, and we have scaled the density to the mean value by writing $n_H = \bar{n}_H^0 (1 + z)^3 (1 + \delta)$, where $\bar{n}_H^0$ is the mean comoving density today. Note that this expression assumes a delta-function line profile, an assumption that breaks down in regimes where the peculiar velocity gradient is large. A more careful approach is required in those cases, though note that such regions are rare in most scenarios (Mao et al. 2012).

In most circumstances, the CMB provides the background radiation source, the temperature for which is $T_\gamma(z)$. Then, $T_R' = T_\gamma(z)$, so that we are observing the contrast between high-redshift hydrogen clouds and the CMB. Because the optical depth is so small, we can then expand the exponentials in Equation (1.3), and

$$T_b(\nu) \approx \frac{T_S - T_\gamma(z)}{1 + z} \tau_{\nu_0} \tag{1.7}$$

$$\approx 9 x_{HI} (1 + \delta)(1 + z)^{1/2} \left[1 - \frac{T_\gamma(z)}{T_S} \right] \left[\frac{H(z)/(1 + z)}{dv_{\parallel}/d\eta_{\parallel}} \right] \text{mK}. \tag{1.8}$$

Thus, $T_b < 0$ if $T_S < T_\gamma$, yielding an absorption signal; otherwise, it appears in emission relative to the CMB. Both regimes are likely important for the high-z universe. Note that T_b saturates if $T_S \gg T_\gamma$, but the absorption can become arbitrarily large if $T_S \ll T_\gamma$. The observability of the 21 cm transition therefore hinges on the spin temperature; in the next section, we will describe the mechanisms that control that factor.

Of course, the other factors—the density, velocity, and ionization fields—are also important to understanding the 21 cm signal. The density field evolves through cosmological structure formation, and that same evolution drives the velocity field—both of which we will describe briefly in Chapter 3. The ionization field depends, in most scenarios, on astrophysical sources, and it will be described in detail in Chapter 2. For now, we will simply note that so long as stars drive reionization, the "two-phase" approximation is very accurate: the mean free path of ionizing photons is so short that regions around ionizing sources are essentially fully ionized, while those outside of those H II regions are nearly fully neutral. Thus to a good approximation, we can take $x_{HI} = 0$ or 1.

1.2 The Spin Temperature

Three competing processes determine T_S: (i) absorption of CMB photons (as well as stimulated emission), (ii) collisions with other particles, and (iii) scattering of UV photons. In the presence of the CMB alone, the spin states reach thermal equilibrium $(T_S = T_\gamma)$ on a timescale of $\sim T_*/(T_\gamma A_{10}) = 3 \times 10^5 (1 + z)^{-1}$ yr—much shorter than the age of the universe at all redshifts after cosmological recombination, indicating that CMB coupling establishes itself rapidly. Indeed, all the relevant processes adjust on very short timescales (compared to the Hubble time), so equilibrium is an excellent approximation.

However, the other two processes break this coupling. We let C_{10} and P_{10} be the de-excitation rates (per atom) from collisions and UV scattering, respectively. We also let C_{01} and P_{01} be the corresponding excitation rates. In equilibrium, the spin temperature is then determined by

$$n_1(C_{10} + P_{10} + A_{10} + B_{10}I_{CMB}) = n_0(C_{01} + P_{01} + B_{01}I_{CMB}), \qquad (1.9)$$

where I_{CMB} is the specific intensity of CMB photons at ν_{21}. With the Rayleigh–Jeans approximation, Equation (1.9) can be rewritten as

$$T_S^{-1} = \frac{T_\gamma^{-1} + x_c T_K^{-1} + x_\alpha T_c^{-1}}{1 + x_c + x_\alpha}, \qquad (1.10)$$

where x_c and x_α are coupling coefficients for collisions and UV scattering, respectively, and T_K is the gas kinetic temperature. Here we have used the principle of detailed balance through the relation

$$\frac{C_{01}}{C_{10}} = \frac{g_1}{g_0} e^{-T_*/T_K} \approx 3\left(1 - \frac{T_*}{T_K}\right). \qquad (1.11)$$

We have also defined the effective color temperature of the UV radiation field T_c via

$$\frac{P_{01}}{P_{10}} \equiv 3\left(1 - \frac{T_*}{T_c}\right). \qquad (1.12)$$

In the limit in which $T_c \to T_K$ (usually a good approximation), Equation (1.10) may be written

$$1 - \frac{T_\gamma}{T_S} = \frac{x_c + x_\alpha}{1 + x_c + x_\alpha}\left(1 - \frac{T_\gamma}{T_K}\right). \qquad (1.13)$$

We must now calculate x_c, x_α, and T_c, which we shall do in the next subsections.

1.2.1 Collisional Coupling

We will first consider collisional excitation and de-excitation of the hyperfine levels, which become important in dense gas. The coupling coefficient for collisions with species i is

$$x_c^i \equiv \frac{C_{10}^i}{A_{10}} \frac{T_\star}{T_\gamma} = \frac{n_i \kappa_{10}^i}{A_{10}} \frac{T_\star}{T_\gamma}, \tag{1.14}$$

where κ_{10}^i is the rate coefficient for spin de-excitation in collisions (with units of cm^3 s^{-1}). The total x_c is the sum over all relevant species i, including collisions with (1) neutral hydrogen atoms, (2) free electrons, and (3) protons.

These rate coefficients can be calculated by the quantum mechanical cross sections of the relevant processes (Zygelman 2005; Furlanetto & Furlanetto 2007a, 2007b). We will not list them in detail but show the rates in Figure 1.1. Although the atomic cross section is small, in the unperturbed IGM collisions between neutral hydrogen, atoms nearly always dominate these rates because the ionized fraction is small. Free electrons can be important in partially ionized gas; collisions with protons are only important at the lowest temperatures.

Given the densities relevant to the IGM, collisional coupling is quite weak in a nearly neutral, cold medium. Thus, the local density must be large in order for this process to effectively fix T_S. A convenient estimate of their importance is the critical overdensity, δ_{coll}, at which $x_c = 1$ for H–H collisions:

$$1 + \delta_{coll} = 0.99 \left[\frac{\kappa_{10}(88\ K)}{\kappa_{10}(T_K)}\right]\left(\frac{0.023}{\Omega_b h^2}\right)\left(\frac{70}{1+z}\right)^2, \tag{1.15}$$

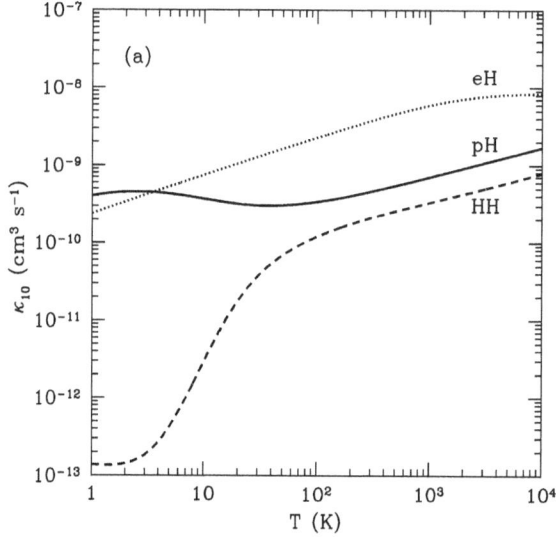

Figure 1.1. De-excitation rate coefficients for H–H collisions (dashed line), H–e$^-$ collisions (dotted line), and H–p collisions (solid line). Note that the net rates are also proportional to the densities of the individual species, so H–H collisions still dominate in a weakly ionized medium. Reproduced from Furlanetto & Furlanetto (2007b), by permission of Oxford University press on behalf of the Royal Astronomical Society.

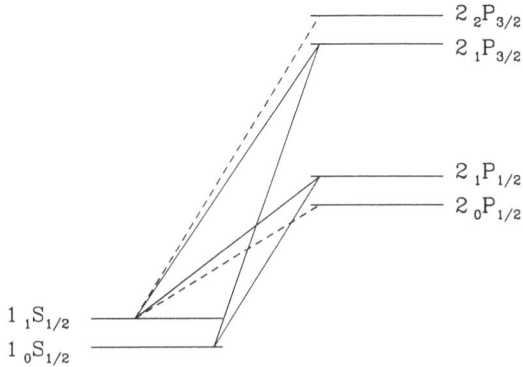

Figure 1.2. Level diagram illustrating the Wouthuysen–Field effect. We show the hyperfine splittings of the $1S$ and $2P$ levels. The solid lines label transitions that can mix the ground-state hyperfine levels, while the dashed lines label complementary allowed transitions that do not participate in mixing. Reproduced from Pritchard & Furlanetto (2006), by permission of Oxford University press on behalf of the Royal Astronomical Society.

where 88 K is the expected IGM temperature at $1 + z = 70$.[1] In the standard picture, at redshifts $z < 70$, $x_c \ll 1$ and $T_S \to T_\gamma$; by $z \sim 30$ the IGM essentially becomes invisible. However, κ_{10} is extremely sensitive to T_K in this low-temperature regime. If the universe is somehow heated above the fiducial value, the threshold density can remain modest: $\delta_{coll} \approx 1$ at $z = 40$ if $T_K = 300$ K.

1.2.2 The Wouthuysen–Field Effect

We must therefore appeal to a different mechanism to render the 21 cm transition visible during the era of the first galaxies. This is known as the Wouthuysen–Field mechanism (named after the Dutch physicist Siegfried Wouthuysen and Harvard astrophysicist George Field, who first explored it; Wouthuysen 1952; Field 1958). Figure 1.2 illustrates the effect. This shows the hyperfine sublevels of the $1S$ and $2P$ states of H I and the permitted transitions between them. Suppose a hydrogen atom in the hyperfine singlet state absorbs a Lyα photon. The electric dipole selection rules allow $\Delta F = 0, 1$, except that $F = 0 \to 0$ is prohibited (here F is the total angular momentum of the atom). Thus, the atom must jump to either of the central $2P$ states. However, these same rules now allow electrons in either of these excited states to decay to the $_1S_{1/2}$ triplet level.[2] Thus, atoms can change hyperfine states through the absorption and spontaneous re-emission of a Lyα photon (or indeed, any Lyman-series photon; see below).

The Wouthuysen–Field coupling rate depends ultimately on the total rate (per atom) at which Lyα photons scattered through the gas,

---

[1] Note that this is smaller than the CMB temperature at this time, because the IGM gas cools faster (due to adiabatic expansion) once Compton scattering becomes inefficient at $z \sim 150$.
[2] Here we use the notation $_F L_J$, where L and J are the orbital and total angular momentum of the electron.

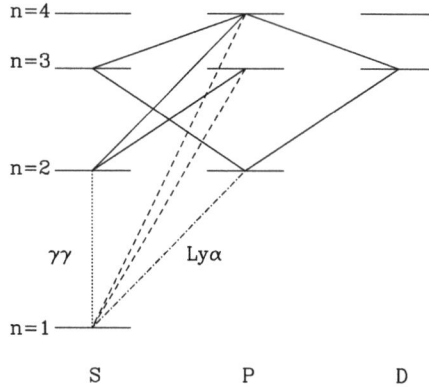

Figure 1.3. Decay chains for Lyβ and Lyγ excitations. We show Lyman-n transitions by dashed curves, Lyα by the dotted–dashed curve, cascades by solid curves, and the forbidden $2S \rightarrow 1S$ transition by the dotted curve. Reproduced from Pritchard & Furlanetto (2006), by permission of Oxford University press on behalf of the Royal Astronomical Society.

$$P_\alpha = 4\pi\sigma_0 \int d\nu \, J_\nu(\nu)\phi_\alpha(\nu), \qquad (1.16)$$

where $\sigma_\nu \equiv \sigma_0\phi_\alpha(\nu)$ is the local Lyα absorption cross section, $\sigma_0 \equiv (\pi e^2/m_e c)f_\alpha$, $f_\alpha = 0.4162$ is the oscillator strength of the Lyα transition, $\phi_\alpha(\nu)$ is the Lyα absorption profile, and J_ν is the angle-averaged specific intensity of the background radiation field.[3]

Transitions to higher Lyman-n levels have similar effects (Hirata 2006; Pritchard & Furlanetto 2006). Suppose that a UV photon redshifts into the Lyman-n resonance as it travels through the IGM. After absorption, it can either scatter (by the electron decaying directly to the ground state) or cascade through a series of intermediate levels and produce a sequence of photons. The direct decay probability for any level is ~0.8, so a Lyman-n photon will typically scatter $N_{\text{scatt}} \approx (1 - P_{nP \rightarrow 1S})^{-1} \sim 5$ times before instead initiating a decay cascade. In contrast, Lyα photons scatter hundreds of thousands of times before being destroyed, usually by redshifting all the way across the (very wide) Lyα profile. As a result, coupling from the direct scattering of Lyman-n photons is suppressed compared to Lyα by a large factor.

However, Lyman-n photons can still be important because of their cascade products, as shown in Figure 1.3. Following Lyβ absorption, the only permitted decays are to the ground state (regenerating a Lyβ photon and starting the process again) or to the $2S$ level. The Hα photon produced in the $3P \rightarrow 2S$ transition (and indeed any photon produced in a decay to an excited state) escapes to infinity. Thus, the atom will eventually find itself in the $2S$ state, which decays to the ground state

---

[3] By convention, we use the specific intensity in units of photons cm^{-2} Hz^{-1} s^{-1} sr^{-1} here, which is conserved during the expansion of the universe (whereas a definition in terms of energy instead of photon number is subject to redshifting).

via a forbidden two-photon process with $A_{2S \to 1S} = 8.2$ s^{-1}. These photons will also escape to infinity, so coupling from Lyβ photons can be completely ignored.[4]

But now consider excitation by Lyγ, also shown in Figure 1.3. This can cascade (through $3S$ or $3D$) to the $2P$ level, in which case the original Lyman-n photon is "recycled" into a Lyα photon, which then scatters many times through the IGM. Thus, the key quantity for determining the coupling induced by Lyman-n photons is the fraction $f_{\rm rec}(n)$ of cascades that terminate in Lyα photons. Our discussion in the previous paragraph shows that $f_{\rm rec}(n = 3)$ vanishes, but detailed quantum mechanical calculations show that the higher states all have $f_{\rm rec} \sim 1/3$ (Hirata 2006; Pritchard & Furlanetto 2006).

Focusing again on the Lyα photons themselves, we must relate the total scattering rate P_α to the indirect de-excitation rate P_{10} (Field 1958; Meiksin 2000). Let us first label the $1S$ and $2P$ hyperfine levels a–f, in order of increasing energy, and let A_{ij} and B_{ij} be the spontaneous emission and absorption coefficients for transitions between these levels. We write the background intensity at the frequency corresponding to the $i \to j$ transition as J_{ij}. Then,

$$P_{01} \propto B_{\rm ad} J_{\rm ad} \frac{A_{\rm db}}{A_{\rm da} + A_{\rm db}} + B_{\rm ae} J_{\rm ae} \frac{A_{\rm eb}}{A_{\rm ea} + A_{\rm eb}}. \tag{1.17}$$

The first term contains the probability for an a $\to$ d transition ($B_{\rm ad} J_{\rm ad}$), together with the probability for the subsequent decay to terminate in state b; the second term is the same for transitions to and from state e (see Figure 1.2). Next, we need to relate each A_{ij} to the total spontaneous decay rate from the $2P$ level, $A_\alpha = 6.25 \times 10^8$ Hz, the total Lyα spontaneous emission rate. This can be accomplished using a sum rule stating that the sum of decay intensities ($g_i A_{ij}$) for transitions from a given nFJ to all the $n'J'$ levels (summed over F') is proportional to $2F + 1$, which implies that the relative strengths of the permitted transitions are then (1, 1, 2, 2, 1, 5), where we have ordered the lines by (initial, final) states (bc, ad, bd, ae, be, bf). With our assumption that the background radiation field is constant across the individual hyperfine lines, we find $P_{10} = (4/27)P_\alpha$ (Meiksin 2000).

The coupling coefficient x_α is then

$$x_\alpha = \frac{4 P_\alpha}{27 A_{10}} \frac{T_\star}{T_\gamma} \equiv S_\alpha \frac{J_\alpha}{J_\nu^c}. \tag{1.18}$$

The second part evaluates J_ν "near" line center and sets $J_\nu^c \equiv 1.165 \times 10^{-10}$ $[(1 + z)/20]$ photons cm^{-2} sr^{-1} Hz^{-1} s^{-1}. S_α is a correction factor that accounts for (complicated) radiative transfer effects in the intensity near the line center (see below). The coupling threshold J_ν^c for $x_\alpha = S_\alpha$ can also be written in terms of the number of Lyα photons per hydrogen atom in the universe, which we denote $\tilde{J}_\nu^c = 0.0767 \, [(1 + z)/20]^{-2}$. This threshold is relatively easy to achieve in practice.

---

[4] In a medium with very high number density, atomic collisions can mix the two angular momentum states, but that process is unimportant in the IGM.

To complete the coupling calculation, we must determine T_c and the correction factor S_α. The former is the effective temperature of the UV radiation field, defined in Equation (1.12), and is determined by the shape of the photon spectrum at the Lyα resonance. The effective temperature of the radiation field must matter, because the energy deficit between the different hyperfine splittings of the Lyα transition (labeled bc, ad, etc. above) implies that the mixing process is sensitive to the gradient of the radiation spectrum near the Lyα resonance. More precisely, the procedure described after Equation (1.17) yields

$$\frac{P_{01}}{P_{10}} = \frac{g_1}{g_0} \frac{n_{\mathrm{ad}} + n_{\mathrm{ae}}}{n_{\mathrm{bd}} + n_{\mathrm{be}}} \approx 3\left(1 + \nu_0 \frac{d \ln n_\nu}{d\nu}\right), \tag{1.19}$$

where $n_\nu = c^2 J_\nu / 2\nu^2$ is the photon occupation number. Thus, by comparison to Equation (1.12), we find

$$\frac{h}{k_B T_c} = -\frac{d \ln n_\nu}{d\nu}. \tag{1.20}$$

A simple argument shows that $T_c \approx T_K$ (Field 1959): so long as the medium is extremely optically thick, the enormous number of Lyα scatterings forces the Lyα profile to be a blackbody of temperature T_K near the line center. This condition is easily fulfilled in the high-redshift IGM, where $\tau_\alpha \gg 1$. In detail, atomic recoils during scattering tilt the spectrum to the red and are primarily responsible for establishing this equilibrium (Field 1959).

The physics of the Wouthuysen–Field effect are actually much more complicated than naively expected because scattering itself modifies the shape of J_ν near the Lyα resonance (Chen 2004). In essence, the spectrum must develop an absorption feature because of the increased scattering rate near the Lyα resonance. Photons lose energy at a fixed rate by redshifting, but each time they scatter, they also lose a small amount of energy through recoil. Momentum conservation during each scattering slightly decreases the frequency of the photon. The strongly enhanced scattering rate near line center means that photons "flow" through that region more rapidly than elsewhere (where only the cosmological redshift applies), so the amplitude of the spectrum must be smaller. Meanwhile, the scattering in such an optically thick medium also causes photons to diffuse away from line center, broadening the feature well beyond the nominal line width.

If the fractional frequency drift rate is denoted by $\mathcal{A}$, continuity requires $n_\nu \mathcal{A} =$ constant. Because $\mathcal{A}$ increases near resonance, the number density must fall. On average, the energy loss (or gain) per scattering is (Chen 2004)

$$\frac{\Delta E_{\mathrm{recoil}}}{E} = \frac{h\nu}{m_p c^2}\left(1 - \frac{T_K}{T_c}\right), \tag{1.21}$$

where the first factor comes from recoil off an isolated atom and the second factor corrects for the distribution of initial photon energies; the energy loss vanishes when $T_c = T_K$, and when $T_c < T_K$, the gas is heated by the scattering process.

To compute S_α, we must calculate the photon spectrum near Lyα. We begin with the radiative transfer equation in an expanding universe (written in comoving coordinates), and again using units of photons cm^{-2} sr^{-1} Hz^{-1} s^{-1} for J_ν:

$$\frac{1}{cn_{\mathrm{H}}\sigma_0}\frac{\partial J_\nu}{\partial t} = -\phi_\alpha(\nu)\,J_\nu + H\nu_\alpha\,\frac{\partial J_\nu}{\partial \nu}$$
$$+ \int d\nu'\,R(\nu,\nu')\,J_{\nu'} + C(t)\psi(\nu). \tag{1.22}$$

The first term on the right-hand side describes absorption, the second describes redshifting due to the Hubble flow, and the third accounts for re-emission following absorption. $R(\nu,\nu')$ is the "redistribution function" that specifies the frequency of an emitted photon, which depends on the relative momenta of the absorbed and emitted photons as well as the absorbing atom. The last term accounts for the injection of new photons (via, e.g., radiative cascades that result in Lyα photons): C is the rate at which they are produced and $\psi(\nu)$ is their frequency distribution.

The redistribution function R is the difficult aspect of the problem, but it can be simplified if the frequency change per scattering (typically of order the absorption line width) is "small." In that case, we can expand $J_{\nu'}$ to second order in $(\nu - \nu')$ and rewrite Equation (1.22) as a diffusion problem in frequency. The steady-state version of Equation (1.22) becomes, in this so-called Fokker–Planck approximation (Chen 2004),

$$\frac{d}{dx}\left(-\mathcal{A}\,J + \mathcal{D}\,\frac{dJ}{dx}\right) + C\psi(x) = 0, \tag{1.23}$$

where $x \equiv (\nu - \nu_\alpha)/\Delta\nu_{\mathrm{D}}$, $\Delta\nu_{\mathrm{D}}$ is the Doppler width of the absorption profile, $\mathcal{A}$ is the frequency drift rate, and $\mathcal{D}$ is the diffusivity. The Fokker–Planck approximation is valid so long as (i) the frequency change per scattering ($\sim\Delta\nu_{\mathrm{D}}$) is smaller than the width of any spectral features, and either (iia) the photons are outside the line core where the Lyα line profile is slowly changing, or (iib) the atoms are in equilibrium with $T_{\mathrm{c}} \approx T_{\mathrm{K}}$.

Solving for the spectrum including scattering thus reduces to specifying $\mathcal{A}$ and $\mathcal{D}$. The drift involves the Hubble flow, which sets $\mathcal{A}_{\mathrm{H}} = -\tau_\alpha^{-1}$, where τ_α is the Gunn–Peterson optical depth for the Lyα line (Gunn & Peterson 1965; Scheuer 1965):

$$\tau_\alpha = \frac{\chi_\alpha\,n_{\mathrm{HI}}(z)\,c}{H(z)\nu_\alpha} \approx 3\times 10^5\,x_{\mathrm{HI}}\left(\frac{1+z}{7}\right)^{3/2}. \tag{1.24}$$

Because it is uniform, the Hubble flow does not introduce any diffusion. The remaining terms come from R and incorporate all the physical processes relevant to energy exchange in scattering. The drift from recoil causes (Hirata 2006)

$$\mathcal{D}_{\mathrm{scatt}} = \phi_\alpha(x)/2, \tag{1.25}$$

$$\mathcal{A}_{\mathrm{scatt}} = -(\eta - x_0^{-1})\phi_\alpha(x), \tag{1.26}$$

where $x_0 \equiv \nu_\alpha/\Delta\nu_D$ and $\eta \equiv (h\nu_\alpha^2)/(m_p c^2 \Delta\nu_D)$. The latter is the recoil parameter measuring the average loss per scattering in units of the Doppler width. The small energy defect between the hyperfine levels provides another source of slow energy exchange (Hirata 2006) and can be incorporated into the scattering in nearly the same way as recoil.

We can now solve Equation (1.23) once we choose the boundary conditions, which essentially correspond to the input photon spectrum (ignoring scattering) and the source function. Because the frequency range of interest is so narrow, two cases suffice: a flat input spectrum (which approximately describes photons that redshift through the Lyα resonance, regardless of the initial source spectrum) and a step function, where photons are "injected" at line center (through cascades or recombinations) and redshift away. In either case, the first integral over x in Equation (1.23) is trivial. At high temperatures, where spin flips are unimportant to the overall energy exchange, we can write

$$\phi\frac{dJ}{dx} + 2\left\{[\eta - (x + x_0)^{-1}]\phi + \tau_\alpha^{-1}\right\}J = 2K/\tau_\alpha. \tag{1.27}$$

The integration constant K equals J_∞, the flux far from resonance, both for photons that redshift into the line and for injected photons at $x < 0$ (i.e., redward of line center); it is zero for injected photons at $x > 0$.

The formal analytic solution, when $K \neq 0$, is most compactly written in terms of $\delta_J \equiv (J_\infty - J)/J_\infty$ (Chen 2004):[5]

$$\delta_J(x) = 2\eta \int_0^\infty dy \exp\left[-2\{\eta - (x + x_0)^{-1}\}y - \frac{2}{\tau_\alpha}\int_{x-y}^x \frac{dx'}{\phi_\alpha(x')}\right]. \tag{1.28}$$

(An analogous form also exists for photons injected at line center.) The full problem, including the intrinsic Voigt profile of the Lyα line, must be solved numerically, but including only the Lorentzian wings from natural broadening allows a simpler solution (Furlanetto & Pritchard 2006). Fortunately, this assumption is quite accurate in the most interesting regime of $T_K < 1000$ K.

The crucial aspect of Equation (1.28) is that (as expected from the qualitative argument) an absorption feature appears near the line center, with its depth roughly proportional to η, our recoil parameter. The feature is more significant when T_K is small (because in that case the average effect of recoil is large). Figure 1.4 shows some example spectra (both for a continuous background and for photons injected at line center).

Usually, the most important consequence is the suppression of the radiation spectrum at line center compared to the assumed initial condition. This decreases the total scattering rate of Lyα photons (and hence the Wouthuysen–Field coupling), with the suppression factor (defined in Equation (2.3)) as (Chen 2004)

---

[5] Here we assume the gas has a sufficiently high temperature that the different hyperfine subtransitions can be treated as one (Hirata 2006).

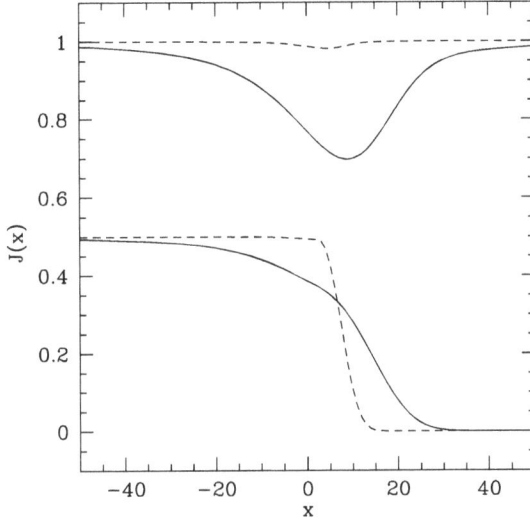

Figure 1.4. Background radiation field near the Lyα resonance at $z = 10$; $x \equiv (\nu - \nu_\alpha)/\Delta\nu_D$ is the normalized deviation from line center, in units of the Doppler width. The upper and lower sets are for continuous photons and photons injected at line center, respectively. (The former are normalized to J_∞; the latter have arbitrary normalization.) The solid and dashed curves take $T_K = 10$ and 1000 K, respectively. Reproduced from Furlanetto & Pritchard (2006), by permission of Oxford University press on behalf of the Royal Astronomical Society.

$$S_\alpha = \int_{-\infty}^{\infty} dx \, \phi_\alpha(x) \, J(x) \approx [1 - \delta_J(0)] \leqslant 1, \tag{1.29}$$

where the second equality follows from the narrowness of the line profile. Again, the Lorentzian wing approximation turns out to be an excellent one; when $T_K \gg T_\star$, the suppression is (Furlanetto & Pritchard 2006)

$$S_\alpha \sim \exp\left[-0.803\left(\frac{T_K}{1 \text{ K}}\right)^{-2/3}\left(\frac{\tau_\alpha}{10^6}\right)^{1/3}\right]. \tag{1.30}$$

Note that this form applies to both photons injected at line center as well as those that redshift in from infinity. As we can see in Figure 1.4, the suppression is most significant in cool gas.

1.3 Heating of the Intergalactic Medium

We have seen that both collisions and the Wouthuysen–Field effect couple the spin temperature to the kinetic temperature of the gas. The 21 cm brightness temperature therefore depends on processes that heat the neutral IGM. (Note that photo-ionization heating is likely the most important mechanism in setting the IGM temperature, because that process typically heats the gas to $T \sim 10^4$ K. However, by definition, that process only occurs when ionization is significant—and, in standard reionization scenarios, where $x_{HI} \approx 0$ so that the 21 cm signal vanishes.) We will review several such mechanisms in this section.

1.3.1 The Lyα Background

The photons that trigger Lyα coupling exchange energy with the IGM through the recoil in each scattering event. The typical energy exchange per scattering is small (see Equation (1.21)), but the scattering rate is extremely large. If the net heating rate per atom followed the naive expectation, $\sim P_\alpha \times (h\nu_\alpha)^2/m_p c^2$, the kinetic temperature would surpass T_γ soon after Wouthuysen–Field coupling becomes efficient.

However, the details of radiative transfer radically change these expectations (Chen 2004). In a static medium, the energy exchange must vanish in equilibrium even though scattering continues at nearly the same rate. Scattering induces an asymmetric absorption feature near ν_α (Figure 1.4), whose shape depends on the combined effects of atomic recoils and the scattering diffusivity. In equilibrium, the latter exactly counterbalances the former.

If we removed scattering, the absorption feature would redshift away as the universe expands. Thus, the energy exchange rate from scattering must simply be that required to maintain the feature in place. For photons redshifting into resonance, the absorption trough has the total energy

$$\Delta u_\alpha = (4\pi/c) \int (J_\infty - J_\nu) h\nu \, d\nu, \tag{1.31}$$

where J_∞ is the input spectrum, and we note that the $h\nu$ factor converts from our definition of specific intensity (which counts photons) to energy. The radiation background loses $\varepsilon_\alpha = H \Delta u_\alpha$ per unit time through redshifting; this energy goes into heating the gas. Relative to adiabatic cooling by the Hubble expansion, the fractional heating amplitude is

$$\frac{2}{3} \frac{\varepsilon_\alpha}{k_B T_K n_H H(z)} = \frac{8\pi}{3} \frac{h\nu_\alpha}{k_B T_K} \frac{J_\infty \Delta \nu_D}{c n_H} \int_{-\infty}^{\infty} dx \delta_J(x) \tag{1.32}$$

$$\approx \frac{0.80}{T_K^{4/3}} \frac{x_\alpha}{S_\alpha} \left(\frac{10}{1+z} \right). \tag{1.33}$$

Here we have evaluated the integral for the continuum photons that redshift into the Lyα resonance; the "injected" photons actually cool the gas slightly. The net energy exchange when Wouthuysen–Field coupling becomes important (at $x_\alpha \sim S_\alpha$) is therefore just a fraction of a degree, and in practice, gas heating through Lyα scattering is generally unimportant (Chen 2004; Furlanetto & Pritchard 2006).

Fundamentally, Lyα heating is inefficient because scattering diffusivity cancels the effects of recoil. From Figure 1.4, we see that the background spectrum is weaker on the blue side of the line than on the red. The scattering process tends to move the photon toward line center, with the extra energy deposited in or extracted from the gas. Because more scattering occurs on the red side, this tends to transfer energy from the gas back to the photons, mostly canceling the energy obtained through recoil.

1.3.2 The Cosmic Microwave Background

The previous section shows that, when considered as a two-level process that acts in isolation, Lyα scattering has only a slight effect on the gas temperature. However, in reality, this Lyα scattering always occurs in conjunction with the scattering of CMB photons within the 21 cm transition. The combination leads to an enhanced heating rate (Venumadhav et al. 2018).

In essence, the process works as follows. The CMB photons scatter through the hyperfine levels of H I to heat those atoms above their expected temperature (determined in this simple case by adiabatic cooling). Meanwhile, Lyα photons scatter through the gas as well. As they do so, they mix the hyperfine levels of the H I ground state, as depicted in Figure 1.2—this is the Wouthuysen–Field effect. CMB scattering continues to heat the hyperfine level populations during the Lyα scattering, which then sweeps up this extra energy and ultimately deposits it as thermal energy through the net recoil effect.

We can estimate the energy available to this heating mechanism by considering the CMB energy reservoir (Venumadhav et al. 2018). The CMB energy density at the 21 cm transition is $u_\nu = (4\pi/c)B_\nu \approx 8\pi(\nu_{21}^2/c^3)k_B T_\gamma$. Over a redshift interval $\Delta z = 1$, the total energy that redshifts through the line is $u_\nu \Delta\nu \approx 8\pi(\nu_{21}/c)^3 k_B T_\gamma/(1+z)^2$. However, only a fraction τ_{10} actually interacts with the line. If all of this energy is used for heating, the temperature change per H atom would be

$$\Delta T_{\text{CMB-Ly}\alpha} \approx \tau_{10}\frac{u_\nu \Delta_\nu}{(3/2)n_H} \approx 5x_{HI}\left(\frac{1+z}{20}\right)^{-1/2}\left(\frac{10\text{ K}}{T_S}\right)\text{K}. \tag{1.34}$$

A more detailed calculation of the heating rate shows that it is somewhat slower, but it does amplify the effect of the Lyα heating alone by a factor of several (Venumadhav et al. 2018). In standard models of the early radiation backgrounds, the correction is still relatively modest, but it is not negligible. For example, in the fiducial model considered by Venumadhav et al. (2018), the Lyα heating on its own modifies T_K by ~1%–5%, but with the CMB scattering included the effect is ~9%–15%. Additionally, the CMB scattering can be enhanced in some exotic physics models that decrease the spin temperature substantially.

1.3.3 The X-Ray Background

Because they have relatively long mean free paths, X-rays from galaxies and quasars are likely to be the most important heating agent for the low-density IGM (Madau et al. 1997). In particular, photons with $E > 1.5x_{HI}^{1/3}[(1+z)/10]^{1/2}$ keV have mean free paths exceeding the Hubble length (Oh 2001). Lower-energy X-rays will be absorbed in the IGM, depositing much of their energy as heat, as will a fraction of higher-energy X-rays.

X-rays heat the IGM gas by first photoionizing a hydrogen or helium atom. The resulting "primary" electron retains most of the photon energy (aside from that required to ionize it) as kinetic energy, which it must then distribute to the general IGM through three main channels: (1) collisional ionizations, which produce more secondary electrons that themselves scatter through the IGM, (2) collisional excitations of He I (which produce photons capable of ionizing H I) and H I (which produces a Lyα background), and (3) Coulomb collisions with free electrons (which distributes the kinetic energy). The relative cross sections of these processes determines what fraction of the X-ray energy goes to heating (f_{heat}), ionization (f_{ion}), and excitation (f_{excite}); clearly, it depends on both the ionized fraction x_i and the input photon energy. Through these scatterings, the primary photoelectrons, with $T \sim 10^6$ K, rapidly cool to energies just below the Lyα threshold, <10 eV, and thus equilibrate with the other IGM electrons. After that, the electrons and neutrals equilibrate through elastic scattering on a timescale $t_{eq} \sim 5[10/(1 + z)]^3$ Myr. Because $t_{eq} \ll H(z)^{-1}$, the assumption of a single-temperature fluid is an excellent one.

The details of this process have been examined numerically (Shull & van Steenberg 1985; Valdés & Ferrara 2008; Furlanetto & Johnson Stoever 2010), and Figure 1.5 shows some example results.[6] Note that the deposition fractions are smooth functions at high electron energies but, at low energies—where the atomic energy levels become relevant—can be quite complex. A number of approximate fits have been presented for the high-energy regime (Ricotti et al. 2002; Volonteri & Gnedin 2009), but they are not accurate over the full energy range. A crude but useful approximation to the high-energy limit often suffices (Chen & Kamionkowski 2004):

$$f_{heat} \sim (1 + 2x_i)/3$$
$$f_{ion} \sim f_{excite} \sim (1 - x_i)/3, \tag{1.35}$$

where x_i is the ionized fraction. In highly ionized gas, collisions with free electrons dominate and $f_{heat} \to 1$; in the opposite limit, the energy is split roughly equally between these three processes. However, the complexity of the behavior at low electron energies—together with the increasing optical thickness of the IGM in that regime, and the fact that most sources are brighter in this soft X-ray regime—suggests that a more careful treatment is needed for accurate work. Furlanetto & Johnson Stoever (2010) recommended interpolating the exact results.

1.3.4 Other Potential Heating Mechanisms

We close this section by noting that other heating mechanisms have been considered in the literature. One possibility is the heating that accompanies structure formation. When regions collapse gravitationally, they are heated by adiabatic compression (which we will discuss in Chapter 3), which is a minor effect. But, if the resulting gas flows converge at velocities above the (very small) sound speed, they can also trigger shocks, which convert a large fraction of that kinetic energy into heat. Analytic models and simulations suggest, however, that structure formation is

---

[6] Note that these results are relative to the initial X-ray energy; some others in the literature instead use present results relative to the primary electron's energy.

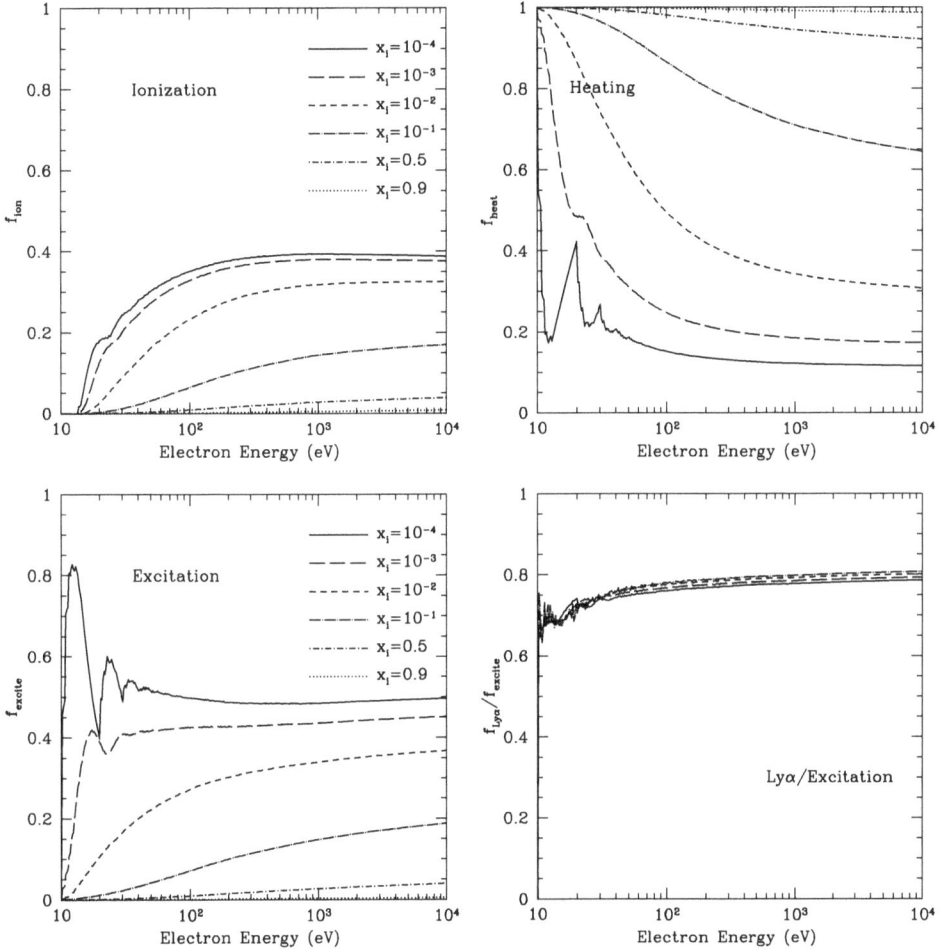

Figure 1.5. Energy deposition from fast electrons. We show the fraction of the initial X-ray energy deposited in ionization (upper left), heating (upper right), and collisional excitation (lower left), as a function of electron energy and for several different ionized fractions x_i. The lower right shows the fraction of the collisional excitation energy deposited in the H I Lyα transition, $f_{Ly\alpha}$. Reproduced from Furlanetto & Johnson Stoever (2010), by permission of Oxford University press on behalf of the Royal Astronomical Society.

still sufficiently gentle during the Cosmic Dawn that these shocks will have little effect on the 21 cm signal (Furlanetto & Loeb 2004; Kuhlen et al. 2006; McQuinn & O'Leary 2012).

Finally, exotic mechanisms like dark matter annihilation or decay, primordial black hole emission, and other speculative processes can also affect the thermal evolution of the IGM during the Dark Ages. We will discuss such possibilities further in Chapter 3.

References

Chen, X., & Kamionkowski, M. 2004, PhRvD, 70, 043502
Chen, X., & Miralda-Escudé, J. 2004, ApJ, 602, 1
Field, G. B. 1958, PIRE, 46, 240
Field, G. B. 1959, ApJ, 129, 525
Field, G. B. 1959, ApJ, 129, 536
Field, G. B. 1959, ApJ, 129, 551
Furlanetto, S. R., & Furlanetto, M. R. 2007a, MNRAS, 374, 547
Furlanetto, S. R., & Furlanetto, M. R. 2007b, MNRAS, 379, 130
Furlanetto, S. R., & Johnson Stoever, S. 2010, MNRAS, 404, 1869
Furlanetto, S. R., & Loeb, A. 2004, ApJ, 611, 642
Furlanetto, S. R., & Pritchard, J. R. 2006, MNRAS, 372, 1093
Gunn, J. E., & Peterson, B. A. 1965, ApJ, 142, 1633
Hirata, C. M. 2006, MNRAS, 367, 259
Hirata, C. M., & Sigurdson, K. 2007, MNRAS, 375, 1241
Kuhlen, M., Madau, P., & Montgomery, R. 2006, ApJ, 637, L1
Madau, P., Meiksin, A., & Rees, M. J. 1997, ApJ, 475, 429
Mao, Y., Shapiro, P. R., Mellema, G., et al. 2012, MNRAS, 422, 926
McQuinn, M., & O'Leary, R. M. 2012, ApJ, 760, 3
Meiksin, A. 2000, in Perspectives on Radio Astronomy: Science with Large Antenna Arrays, ed.
    M. P. van Haarlem (Dwingeloo: Astron), 37
Oh, S. P. 2001, ApJ, 553, 499
Pritchard, J. R., & Furlanetto, S. R. 2006, MNRAS, 367, 1057
Ricotti, M., Gnedin, N. Y., & Shull, J. M. 2002, ApJ, 575, 49
Scheuer, P. A. G. 1965, Natur, 207, 963
Shull, J. M., & van Steenberg, M. E. 1985, ApJ, 298, 268
Valdés, M., & Ferrara, A. 2008, MNRAS, 387, L8
Venumadhav, T., Dai, L., Kaurov, A., & Zaldarriaga, M. 2018, PhRvD, 98, 103513
Volonteri, M., & Gnedin, N. Y. 2009, ApJ, 703, 2113
Wouthuysen, S. A. 1952, AJ, 57, 31
Zygelman, B. 2005, ApJ, 622, 1356

The Cosmic 21-cm Revolution
Charting the first billion years of our universe
Andrei Mesinger

Chapter 2

Astrophysics from the 21 cm Background

Jordan Mirocha

The goal of this chapter is to describe the astrophysics encoded by the 21 cm background. We will begin in Section 2.1 with a brief introduction to the radiative transfer and ionization chemistry relevant to the high-z intergalactic medium (IGM). Then, in Section 2.2, we will provide a review of the most plausible sources of ionization and heating in the early universe. In Section 2.3, we will explore the variety of current 21 cm predictions and illustrate the dependencies of the global 21 cm signal and power spectrum on parameters of interest.

2.1 Properties of the High-z Intergalactic Medium

In this section, we provide a general introduction to the IGM and how its properties are expected to evolve with time. We will start with a brief recap of the 21 cm brightness temperature (Section 2.1.1), then turn our attention to its primary dependencies, the ionization state and temperature of the IGM (Section 2.1.2). In Sections 2.1.3–2.1.5, we briefly review the radiative transfer (RT) relevant to modeling ionization, heating, and Lyα coupling. Readers familiar with the basic physics may skip ahead to Section 2.2, in which we focus on the astrophysical sources most likely to heat and ionize the IGM at early times.

2.1.1 The Brightness Temperature

The differential brightness temperature of a patch of the IGM at redshift z and position $\mathbf{x}$ is given by[1]

$$\delta T_b(z,\mathbf{x}) \simeq 27(1+\delta)(1-x_i)\left(\frac{\Omega_{b,0}h^2}{0.023}\right)\left(\frac{0.15}{\Omega_{m,0}h^2}\frac{1+z}{10}\right)^{1/2}\left(1-\frac{T_R}{T_S}\right), \qquad (2.1)$$

---

[1] Refer back to Chapter 1 for a more detailed introduction.

where δ is the baryonic overdensity relative to the cosmic mean, x_i is the ionized fraction, T_R is the radiation background temperature (generally the CMB, $T_R = T_{CMB}$), and

$$T_S^{-1} \approx \frac{T_R^{-1} + x_c T_K^{-1} + x_\alpha T_\alpha^{-1}}{1 + x_c + x_\alpha} \tag{2.2}$$

is the spin temperature, which quantifies the level populations in the ground state of the hydrogen atom and itself depends on the kinetic temperature, T_K, and "color temperature" of the Lyα radiation background, T_α. Because the IGM is optically thick to Lyα photons, the approximation $T_\alpha \approx T_K$ is generally very accurate.

The collisional coupling coefficients, x_c, themselves depend on the gas density, ionization state, and kinetic temperature (see Zygelman 2005 for details). The radiative coupling coefficient, x_α, depends on the Lyα intensity, J_α, via

$$x_\alpha = \frac{S_\alpha}{1+z}\frac{J_\alpha}{J_{\alpha,0}}, \tag{2.3}$$

where

$$J_{\alpha,0} \equiv \frac{16\pi^2 T_\star e^2 f_\alpha}{27 A_{10} T_{\gamma,0} m_e c}. \tag{2.4}$$

J_α is the angle-averaged intensity of Lyα photons in units of $s^{-1}\,cm^{-2}\,Hz^{-1}\,sr^{-1}$, S_α is a correction factor that accounts for variations in the background intensity near line center (Chen & Miralda-Escudé 2004; Furlanetto & Pritchard 2006; Hirata 2006), m_e and e are the electron mass and charge, respectively, f_α is the Lyα oscillator strength, $T_{\gamma,0}$ is the CMB temperature today, and A_{10} is the Einstein A coefficient for the 21 cm transition.

A more detailed introduction to collisional and radiative coupling can be found in Chapter 1. For the purposes of this chapter, the key takeaway from Equations (2.1)–(2.3) is simply that the 21 cm background probes the ionization field, kinetic temperature field, and Lyα background intensity. We quickly review the basics of nonequilibrium ionization chemistry in the next subsection (Section 2.1.2) before moving on to sources of heating, ionization, and the Lyα background in Section 2.2.

2.1.2 Basics of Nonequilibrium Ionization Chemistry

As described in the previous section, the 21 cm brightness temperature of a patch of the IGM depends on the ionization and thermal state of the gas, as well as the incident Lyα intensity.[2] The evolution of the ionization and temperature is coupled and so they must be evolved self-consistently. The number density of hydrogen and helium ions in a static medium can be written as the following set of coupled differential equations:

---

[2] Note that Lyα photons can transfer energy to the gas (see, e.g., Venumadhav et al. 2018), though we omit this dependence from the current discussion (see Section 1.3.1).

$$\frac{dn_{\text{H\,\textsc{ii}}}}{dt} = (\Gamma_{\text{H\,\textsc{i}}} + \gamma_{\text{H\,\textsc{i}}} + \beta_{\text{H\,\textsc{i}}} n_{\text{e}}) n_{\text{H\,\textsc{i}}} - \alpha_{\text{H\,\textsc{ii}}} n_{\text{e}} n_{\text{H\,\textsc{ii}}}, \tag{2.5}$$

$$\frac{dn_{\text{He\,\textsc{ii}}}}{dt} = (\Gamma_{\text{He\,\textsc{i}}} + \gamma_{\text{He\,\textsc{i}}} + \beta_{\text{He\,\textsc{i}}} n_{\text{e}}) n_{\text{He\,\textsc{i}}} + \alpha_{\text{He\,\textsc{iii}}} n_{\text{e}} n_{\text{He\,\textsc{iii}}}$$
$$- (\beta_{\text{He\,\textsc{ii}}} + \alpha_{\text{He\,\textsc{ii}}} + \xi_{\text{He\,\textsc{ii}}}) n_{\text{e}} n_{\text{He\,\textsc{ii}}} - (\Gamma_{\text{He\,\textsc{ii}}} + \gamma_{\text{He\,\textsc{ii}}}) n_{\text{He\,\textsc{ii}}}, \tag{2.6}$$

$$\frac{dn_{\text{He\,\textsc{iii}}}}{dt} = (\Gamma_{\text{He\,\textsc{ii}}} + \gamma_{\text{He\,\textsc{ii}}} + \beta_{\text{He\,\textsc{ii}}} n_{\text{e}}) n_{\text{He\,\textsc{ii}}} - \alpha_{\text{He\,\textsc{iii}}} n_{\text{e}} n_{\text{He\,\textsc{iii}}}. \tag{2.7}$$

Each of these equations represents the balance between ionizations of species H I, He I, and He II, and recombinations of H II, He II, and He III. Associating the index i with absorbing species, $i = $ H I, He I, He II, and the index i' with ions, $i' = $ H II, He II, He III, we define Γ_i as the photoionization rate coefficient, γ_i as the rate coefficient for ionization by photoelectrons (Shull & van Steenberg 1985; Furlanetto & Stoever 2010; see Section 1.3.3), $\alpha_{i'}$ ($\xi_{i'}$) as the case B (dielectric) recombination rate coefficients, β_i as the collisional ionization rate coefficients, and $n_{\text{e}} = n_{\text{H\,\textsc{ii}}} + n_{\text{He\,\textsc{ii}}} + 2 n_{\text{He\,\textsc{iii}}}$ as the number density of electrons.

While the coefficients α, β, and ξ only depend on the gas temperature, the photo- and secondary-ionization coefficients, Γ and γ, depend on input from astrophysical sources (see Section 2.2).

The final equation necessary in a primordial chemical network is that governing the kinetic temperature evolution, which we can write as a sum of various heating and cooling processes, i.e.,

$$\frac{3}{2} \frac{d}{dt} \left(\frac{k_{\text{B}} T_{\text{K}} n_{\text{tot}}}{\mu} \right) = f^{\text{heat}} \sum_i n_i \Lambda_i - \sum_i \zeta_i n_{\text{e}} n_i - \sum_{i'} \eta_i n_{\text{e}} n_{i'}$$
$$- \sum_i \psi_i n_{\text{e}} n_i - \omega_{\text{He\,\textsc{ii}}} n_{\text{e}} n_{\text{He\,\textsc{ii}}}. \tag{2.8}$$

Here, Λ_i is the photoelectric heating rate coefficient (due to electrons previously bound to species i), $\omega_{\text{He\,\textsc{ii}}}$ is the dielectric recombination cooling coefficient, and ζ_i, $\eta_{i'}$, and ψ_i are the collisional ionization, recombination, and collisional excitation cooling coefficients, respectively, where primed indices i' indicate ions H II, He II, and He III, and unprimed indices i indicate neutrals H I, He I, and He II. The constants in Equation (2.8) are the total number density of baryons, $n_{\text{tot}} = n_{\text{H}} + n_{\text{He}} + n_{\text{e}}$; the mean molecular weight, μ; Boltzmann's constant, k_{B}; and the fraction of photoelectron energy deposited as heat, f^{heat} (sometimes denoted f_{abs}; Shull & van Steenberg 1985; Furlanetto & Stoever 2010). Formulae to compute the values of α_i, β_i, ξ_i, ζ_i, $\eta_{i'}$, ψ_i, and $\omega_{\text{He\,\textsc{ii}}}$, are compiled in, e.g., Fukugita & Kawasaki (1994) and Hui & Gnedin (1997). Terms involving helium become increasingly important in a medium irradiated by X-rays.

These equations do not yet explicitly take into account the cosmic expansion, which dilutes the density and adds an adiabatic cooling term to Equation (2.8); however, these generalizations are straightforward to implement in practice. For the duration of this chapter, we will operate within this simple chemical network, ignoring, e.g., molecular species like H_2 and HD, whose cooling channels are important in primordial gases. Though an interesting topic in their own right, molecular processes reside in the "subgrid" component of most 21 cm models, given that they influence how, when, and where stars are able to form (see Section 2.2), but do not directly affect the bulk properties of the IGM on large scales to which 21 cm measurements are sensitive.

2.1.3 Ionization and Heating around Point Sources

In order to build intuition for the progression of ionization and heating in the IGM, it is instructive to consider the impact of a single point source of UV and X-ray photons on its surroundings. Many early works focused on such 1D radiative transfer problems (Zaroubi et al. 2007; Thomas & Zaroubi 2008). In principle, this is the ideal way to simulation reionization—iterating over all sources in a cosmological volume and for each one applying 1D radiative transfer techniques over the surrounding 4π steradians. In practice, such approaches are computationally expensive, and while they provide detailed predictions (O'Shea et al. 2015; Ocvirk et al. 2016; Gnedin 2014), more approximate techniques are required to survey the parameter space and perform inference (see Chapter 4).

In 1D, the change in the intensity of a ray of photons, I_ν, is a function of the path length s, the emissivity of sources along the path j_ν, and the absorption coefficient α_ν,

$$dI_\nu = j_\nu - \alpha_\nu I_\nu. \tag{2.9}$$

If considering a point source, $j_\nu = 0$, we can integrate this radiative transfer equation (RTE) to obtain

$$I_\nu(s) = I_{\nu,0} \exp\left[-\int_0^s \alpha_\nu(s')ds'\right], \tag{2.10}$$

i.e., the intensity of photons declines exponentially along the ray. It is customary to define the optical depth

$$d\tau_\nu \equiv \alpha_\nu ds, \tag{2.11}$$

in which case we can write

$$I_\nu(s) = I_{\nu,0}e^{-\tau_\nu}. \tag{2.12}$$

In the reionization context, the optical depth of interest is that of the IGM, which is composed of (almost) entirely hydrogen and helium,[3] in which case the optical depth is

$$\tau_\nu = \sum_i \sigma_{\nu,i} N_i, \tag{2.13}$$

---

[3] Note that there will be small-scale absorption as well, though in most models this is unresolved, and parameters governing photon escape are used to quantify this additional opacity (see Sections 2.2.4 and 2.2.7).

where $i = $ HI, HeI, HeII, and $N_i = \int_0^s ds' n_i(s')$ is the column density of each species along the ray.

With a solution for $I_\nu(s)$ in hand, one can determine the photoionization and heating rates by integrating over all photon frequencies and weighting by the bound–free absorption cross section for each species. For example, the photo-ionization rate coefficient for hydrogen can be written as

$$\Gamma_{\text{HI}}(s) = \int_{\nu_{\text{HI}}}^{\infty} \sigma_{\text{HI}} I_\nu(s) \frac{d\nu}{h\nu}, \tag{2.14}$$

where ν_{HI} is the frequency of the hydrogen ionization threshold, $h\nu = 13.6$ eV.

Note that in practice the RTE is solved on a grid, in which case it may be difficult to achieve high-enough spatial resolution to ensure photon conservation. For example, a discretized version of Equation (2.14) implies that the intensity of radiation incident upon the face of a resolution element determines the photo-ionization rate within that element. However, the radiation incident on the subsequent resolution element is not guaranteed to correctly reflect the attenuation within the preceding element. As a result, in order to guarantee photon conservation, it is common to slightly reframe the calculation as follows (Abel et al. 1999):

$$\Gamma_i = A_i \int_{\nu_i}^{\infty} I_\nu e^{-\tau_\nu} (1 - e^{-\Delta\tau_{i,\nu}}) \frac{d\nu}{h\nu}, \tag{2.15}$$

$$\gamma_{ij} = A_j \int_{\nu_j}^{\infty} \left(\frac{\nu - \nu_j}{\nu_i}\right) I_\nu e^{-\tau_\nu} (1 - e^{-\Delta\tau_{j,\nu}}) \frac{d\nu}{h\nu}, \tag{2.16}$$

$$\Lambda_i = A_i \int_{\nu_i}^{\infty} (\nu - \nu_i) I_\nu e^{-\tau_\nu} (1 - e^{-\Delta\tau_{i,\nu}}) \frac{d\nu}{\nu}. \tag{2.17}$$

The normalization constant in each expression is defined as $A_i \equiv L_{\text{bol}}/n_i V_{\text{sh}}(r)$, where V_{sh} is the volume of a shell in this 1D grid of concentric spherical shells, each having thickness Δr and volume $V_{\text{sh}}(r) = 4\pi[(r + \Delta r)^3 - r^3]/3$, where r is the distance between the origin and the inner interface of each shell. We denote the ionization threshold energy for species i as $h\nu_i$. I_ν represents the SED of radiation sources and satisfies $\int_\nu I_\nu d\nu = 1$, such that $L_{\text{bol}} I_\nu = L_\nu$. Note that the total secondary-ionization rate for a given species is the sum of ionizations due to the secondary electrons from all species, i.e., $\gamma_i = f_{\text{ion}} \sum_j \gamma_{ij} n_j/n_i$.

These expressions preserve photon number by inferring the number of photo-ionizations of species i in a shell from the radiation incident upon it and its optical depth (Abel et al. 1999),

$$\Delta\tau_{i,\nu} = n_i \sigma_{i,\nu} \Delta r. \tag{2.18}$$

This quantity is not to be confused with the total optical depth between source and shell, $\tau_\nu = \tau_\nu(r)$, which sets the incident radiation field upon each shell, i.e.,

$$\tau_\nu(r) = \sum_i \int_0^r \sigma_{i,\nu} n_i(r') dr'$$
$$= \sum_i \sigma_{i,\nu} N_i(r),$$

(2.19)

where N_i is the column density of species i at distance r from the source.

In words, Equations (2.15)–(2.17) are photons propagating from a source at the origin, with bolometric luminosity $L_{\rm bol}$, and tracking the attenuation suffered between the source and some volume element of interest at radius r, $e^{-\tau}$, and the attenuation within that volume element, $\Delta\tau$, which results in ionization and heating. In each case, we integrate over the contribution from photons at all frequencies above the ionization threshold, additionally modifying the integrands for γ_{ij} and Λ_i with $(\nu - \nu_i)$-like factors to account for the fact that both the number of photo-electrons (proportional to $(\nu - \nu_j)/\nu_i$) and their energy (proportional to $\nu - \nu_i$) determine the extent of secondary ionization and photoelectric heating. Equations (2.15)–(2.17) can be solved once a source luminosity $L_{\rm bol}$, spectral shape I_ν, and density profile of the surrounding medium $n(r)$, have been specified.[4]

Figure 2.1 shows an example 1D radiative transfer model including sources of UV and X-ray photons. Because the mean free paths of UV photons are short, they generate sharp features in radial profiles of the neutral fraction of hydrogen (top) and helium (second row). The addition of X-ray sources largely influences the abundance of singly ionized helium (third row) and the extended temperature structure beyond the fully ionized region (fourth row) without dramatically modifying the sharp structures of the hydrogen and helium fraction.

Calculations like those shown in Figure 2.1 motivate two-phase models of the IGM (e.g., Furlanetto 2006; Pritchard & Loeb 2010; Mirocha et al. 2015), in which UV photons carve out relatively distinct regions of fully ionized hydrogen gas, while the hydrogen beyond these bubbles remains largely neutral.[5] The mostly neutral "bulk IGM" outside of bubbles is affected predominantly by X-rays, which have mean free paths long enough to escape the environments in which they are generated (though see Section 2.2.7). This also implies that the properties of a small patch of H I gas in the bulk IGM may be affected by many sources at cosmological distances. We focus on this limit in the next subsection.

2.1.4 Ionization and Heating on Large Scales

While the procedure outlined in the previous section is relevant to small-scale ionization and heating, i.e., that which is driven by a single (or perhaps a few) source(s) close to a volume element of interest, it is also instructive to consider the

---

[4] In practice, to avoid performing these integrals on each step of an ODE solver (for Equations (2.5)–(2.8)), the results can be tabulated as a function of τ or column density, N_i, where $\tau_{i,\nu} = \sigma_{i,\nu} N_i$ (Thomas & Zaroubi 2008; Mirocha et al. 2012; Knevitt et al. 2014).
[5] In practice, one then solves two sets of equations like Equations (2.5)–(2.8)—one for each phase of the IGM. In the fully ionized phase, the ionized fraction represents a volume-filling fraction, while in the "bulk IGM" phase, it retains its usual meaning.

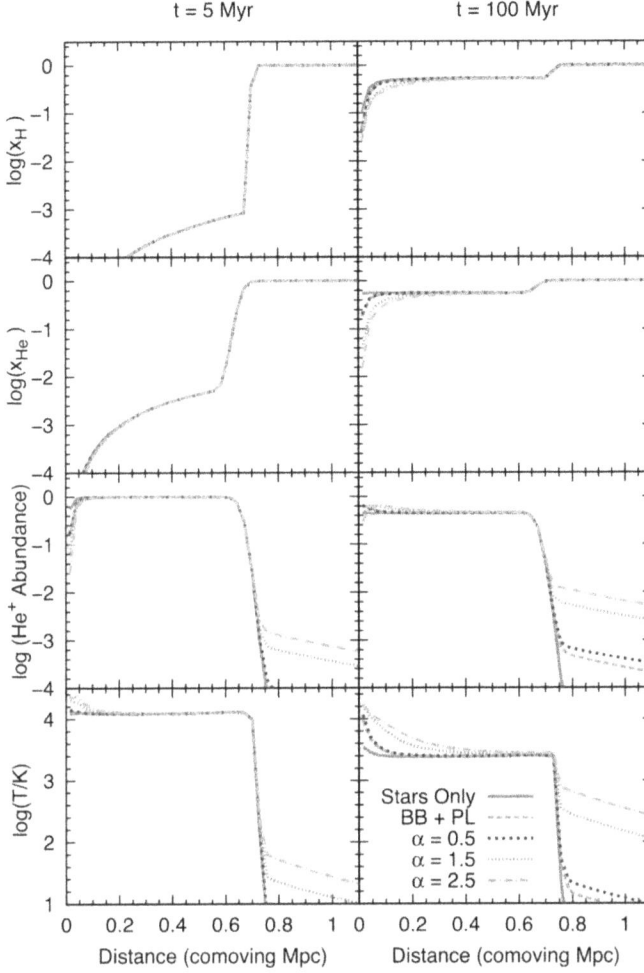

Figure 2.1. Ionization and temperature profile around stellar UV and X-ray sources in a one-dimensional radiative transfer model (reproduced from Knevitt et al. 2014, by permission of Oxford University press on behalf of the Royal Astronomical Society). From top to bottom, this includes the hydrogen neutral fraction, neutral helium fraction, singly ionized helium fraction, and kinetic temperature, while the left and right columns indicate different time snapshots after sources are first turned on. Different lines adopt different source models, from a "stars only" model (solid red), to hybrid models with stars and X-ray sources with different power-law spectra (dashed and dotted curves).

ionization and heating caused by a population of sources separated by great distances. In this limit, rather than considering the luminosity of a single source at the origin of a 1D grid, we treat the volume-averaged emissivity of sources, ε_ν, in a large "chunk" of the universe and solve for the evolution of the mean intensity in this volume, J_ν.

The transfer equation now takes its cosmological form, i.e.,

$$\left(\frac{\partial}{\partial t} - \nu H(z)\frac{\partial}{\partial \nu}\right)J_\nu(z) + 3H(z)J_\nu(z) = \frac{c}{4\pi}\varepsilon_\nu(z)(1+z)^3 - c\alpha_\nu J_\nu(z), \qquad (2.20)$$

where J_ν is the mean intensity in units of erg s^{-1} cm^{-2} Hz^{-1} sr^{-1}, ν is the observed frequency of a photon at redshift z, related to the emission frequency, ν', of a photon emitted at redshift z' as

$$\nu' = \nu\left(\frac{1+z'}{1+z}\right), \tag{2.21}$$

$\alpha_\nu = n\sigma_\nu$ is the absorption coefficient, not to be confused with recombination rate coefficient $\alpha_{\rm HII}$, and ε_ν is the comoving emissivity of sources.

The optical depth, $d\tau = \alpha_\nu ds$, experienced by a photon at redshift z and emitted at z' is an integral along a cosmological line element, summed over all absorbing species,[6] i.e.,

$$\bar{\tau}_\nu(z,z') = \sum_j \int_z^{z'} n_j(z'')\sigma_{j,\nu''}\frac{dl}{dz''}dz''. \tag{2.22}$$

The solution to Equation (2.20) is

$$\hat{J}_\nu(z) = \frac{c}{4\pi}(1+z)^2 \int_z^{z_f} \frac{\varepsilon'_\nu(z')}{H(z')}e^{-\bar{\tau}_\nu}dz', \tag{2.23}$$

where z_f is the "first light redshift" when astrophysical sources first turn on, H is the Hubble parameter, and the other variables take on their usual meaning.[7]

With the background intensity in hand, one can compute the rate coefficients for ionization and heating. These coefficients are equivalent to those for the 1D problem (Equations (2.15)–(2.17)), though the intensity of radiation at some distance R from the source has been replaced by the mean background intensity,

$$\Gamma_i(z) = 4\pi n_i(z)\int_{\nu_{\min}}^{\nu_{\max}} \hat{J}_\nu\sigma_{\nu,i}d\nu, \tag{2.24}$$

$$\gamma_{ij}(z) = 4\pi \sum_j n_j \int_{\nu_{\min}}^{\nu_{\max}} \hat{J}_\nu\sigma_{\nu,j}(h\nu - h\nu_j)\frac{d\nu}{h\nu}, \tag{2.25}$$

$$\varepsilon_{\rm X}(z) = 4\pi \sum_j n_j \int_{\nu_{\min}}^{\nu_{\max}} \hat{J}_\nu\sigma_{\nu,j}(h\nu - h\nu_j)d\nu. \tag{2.26}$$

Then, the ionization state and temperature of the gas can be updated accordingly via Equations (2.5)–(2.8).

Figure 2.2 shows predictions for the evolution of the mean ionized fraction and kinetic temperature of the IGM (Mason et al. 2018; Pober et al. 2015) using the two-

[6] In general, one must iteratively solve for $\bar{\tau}_\nu$ and J_ν. However, in many models, the bulk of cosmic reheating precedes reionization, in which case $\bar{\tau}_\nu$ can be tabulated assuming a fully neutral IGM. This approach provides a considerable speed-up computationally and remains accurate even when reionization and reheating partially overlap (Mirocha 2014).
[7] This equation can be solved efficiently on a logarithmic grid in $x \equiv 1 + z$ (Haardt & Madau 1996; Mirocha 2014), in which case photons redshift seamlessly between frequency bins over time.

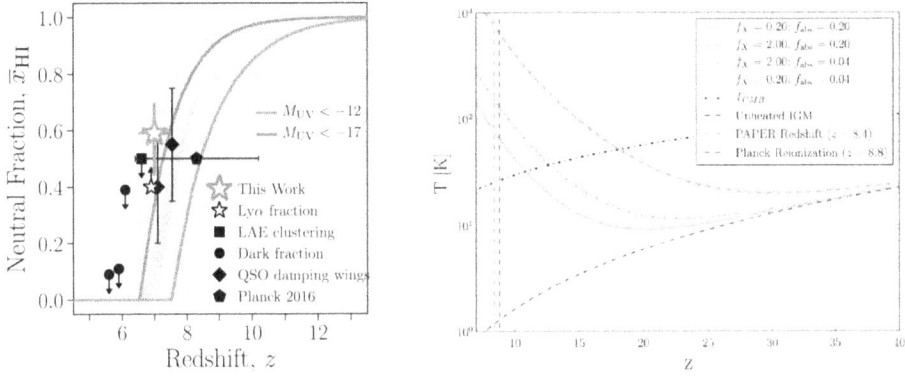

Figure 2.2. Predictions for the evolution of the mean properties of the IGM. Left: predictions for the mean neutral fraction of the IGM as a function of redshift compared to several observational constraints (Reproduced from Mason et al. 2018 © 2018. The American Astronomical Society. All rights reserved). The magenta curve includes all galaxies brighter than UV magnitude $M_{UV} < -12$, while green curve includes only brighter galaxies with $M_{UV} < -17$. Right: predictions for the mean kinetic temperature of the IGM (reproduced from Pober et al. 2015 © 2015. The American Astronomical Society. All rights reserved) for different assumptions about how efficiently galaxies produce X-rays (parameterized via f_X and the fraction of X-ray energy absorbed in the IGM, f_{abs}; see Section 2.2), compared to an early power-spectrum limit from PAPER (Parsons et al. 2014; Pober et al. 2015). Note that the PAPER limit has since been revised (Cheng et al. 2018; Kolopanis et al. 2019).

phase IGM picture described above. While current observations are consistent with reionization occurring relatively rapidly at $z \lesssim 10$, heating is generally more gradual and so far unconstrained. The strongest 21 cm signals occur when the IGM remains cold during reionization, so upper limits on the amplitude of 21 cm signals translate to lower limits on the efficiency of X-ray heating in the early universe.

2.1.5 Lyα Coupling

On scales large and small, the 21 cm background will only probe the kinetic temperature of the gas if the Lyα background intensity is strong enough to couple the spin temperature to the kinetic temperature. Determining the Lyα background intensity, J_α, requires a special solution to the cosmological radiative transfer equation (Equation (2.20)). Two effects separate this problem from the generic transfer problem outlined in Section 2.1.4: (i) the Lyman series forms a set of horizons for photons in the $10.2 < h\nu/\mathrm{eV} < 13.6$ interval, giving rise to the so-called "sawtooth modulation" of the soft UV background (Haiman et al. 1997), and (ii) the Lyα background is sourced both by photons redshifting into the line resonance as well as those produced in cascades downward from higher n transitions (Pritchard & Furlanetto 2006).

As a result, it is customary to solve the RTE in each Lyn frequency interval separately. Within each interval, bounded by the Lyn line on its red edge and Lyn + 1 on its blue edge, the optical depth is small in a primordial medium because no photon redward of the Lyman edge can ionize hydrogen or helium.[8] As a result,

---

[8] There is in principle a small opacity contribution from H_2, though we ignore this in what follows as the H_2 fraction in the IGM is expected to be small.

any photon starting its journey just redward of Lyβ will travel freely until it redshifts into the Lyα resonance, while photons originating at bluer wavelengths will encounter Lyn resonances (with $n > 2$), only a fraction of which will ultimately result in Lyα photons.

We can thus write the mean Lyα background intensity as

$$\hat{J}_\alpha(z) = \frac{c}{4\pi}(1 + z)^2 \sum_{n=2}^{n_{\max}} f^n_{\text{rec}} \int_z^{z^{(n)}_{\max}} \frac{\epsilon'_\nu(z')}{H(z')} dz', \tag{2.27}$$

where f^n_{rec} is the "recycling fraction," that is, the fraction of photons that redshift into a Lyn resonance that ultimately cascades through the Lyα resonance (Pritchard & Furlanetto 2006). The upper bound of the definite integral,

$$1 + z^{(n)}_{\max} = (1 + z)\frac{[1 - (n + 1)^{-2}]}{1 - n^{-2}}, \tag{2.28}$$

is set by the horizon of Lyn photons—a photon redshifting through the Lyn resonance at z could only have been emitted at $z' < z^{(n)}_{\max}$, as emission at slightly higher redshift would mean the photon redshifted through the Ly$(n + 1)$ resonance. The sum over Lyn levels in Equation (2.27) is generally truncated at $n_{\max} = 23$ (Barkana & Loeb 2005) because the horizon for such photons is smaller than the typical ionized bubble sourced by an individual galaxy. As a result, any Lyα photons generated by such high-n cascades are "wasted" as far as the spin temperature is concerned, as they will most likely have redshifted out of resonance before reaching any neutral gas. Lyα emission produced by recombinations in galactic H II regions is generally ignored for the same reason. Though there are some assumptions built into the $n_{\max}$ estimate, the total Lyα photon budget is relatively insensitive to the exact value of $n_{\max}$ (Barkana & Loeb 2005; Pritchard & Furlanetto 2006).

Note that in general the mean free path of photons between Lyman series resonances is very long, which makes tracking them in numerical simulations very expensive. For example, a photon emitted just redward of Lyβ and observed at the Lyα frequency at redshift z has traveled a distance

$$d_{\beta \to \alpha} \simeq 200 \, h_{70}^{-1} \left(\frac{\Omega_{m,0}}{0.3} \frac{1 + z}{20} \right)^{-1/2} \text{cMpc}, \tag{2.29}$$

where we have assumed the high-z approximation $\Omega_\Lambda \ll \Omega_m$. This exceeds a Hubble length at high z, meaning most of the Wouthuysen–Field coupling at very early times must come from photons originating just blueward of their nearest Lyn resonance. Despite their long mean free paths, fluctuations in the Lyα background inevitably arise (Barkana & Loeb 2005; Ahn et al. 2009; Holzbauer & Furlanetto 2012). However, this background is expected to become uniform (and strong) relatively quickly, meaning in general the 21 cm background is only sensitive to J_α at the earliest epochs (see Section 2.3.3).

2.2 Sources of the UV and X-Ray Background

In the previous section, we outlined the basic equations governing the ionization and temperature evolution of the IGM without actually specifying the sources of ionization and heating.[9] Instead, we used a placeholder emissivity, ε_ν, to encode the integrated emissions of sources at a frequency ν within some region R. We will now write this emissivity as an integral over the differential luminosity function (LF) of sources, dn/dL_ν, i.e.,

$$\varepsilon_\nu(z,R) = \int_0^\infty dL_\nu \frac{dn}{dL_\nu}, \tag{2.30}$$

where ν refers to the rest frequency of emission at redshift z.

It is common to rewrite the emissivity as an integral over the dark matter (DM) halo mass function (HMF), dn/dm, multiplied by a conversion factor between halo mass and galaxy light, dm/dL_ν, i.e.,

$$\varepsilon_\nu(z,R) = \int_{m_{\min}}^\infty dm \frac{dn}{dm} \frac{dm}{dL_\nu}, \tag{2.31}$$

where $m_{\min}$ is the minimum mass of DM halos capable of hosting galaxies. Because dn/dm is reasonably well determined from large N-body simulations of structure formation (Press & Schechter 1974; Sheth et al. 2001; Tinker et al. 2010), much of the modeling focus is on the mass-to-light ratio, dm/dL_ν, which encodes the efficiency with which galaxies form in halos and the relative luminosities of different kinds of sources within galaxies (e.g., stars, compact objects, diffuse gas) that emit at different frequencies.[10]

The main strength of the 21 cm background as a probe of high-z galaxies is now apparent: though 21 cm measurements cannot constrain the properties of individual galaxies, they can constrain the properties of all galaxies, in aggregate, even those too faint to be detected directly. As a result, it is common to forego detailed modeling of the mass-to-light ratio and instead relate the emissivity to the fraction of mass in the universe in collapsed halos,

$$\varepsilon_\nu(z,R) = \rho_b f_{\text{coll}}(z,R)\zeta_\nu, \tag{2.32}$$

where ρ_b is the baryon mass density, the collapsed fraction is an integral over the HMF,

$$f_{\text{coll}} = \rho_m^{-1} \int_{m_{\min}}^\infty dm\, m \frac{dn}{dm} \tag{2.33}$$

---

[9] Note that some (at least roughly) model-independent constraints on the properties of the IGM should be attainable with future 21 cm measurements (Cohen et al. 2017, 2018; Mirocha et al. 2013).

[10] Most models consider regions R that are sufficiently large that one can assume a well-populated HMF, though at very early times this approximation may break down, rendering stochasticity due to poor HMF sampling an important effect.

and ζ_ν is an efficiency factor that quantifies the number of photons emitted at frequency ν per baryon of collapsed mass in the universe. It is generally modeled as

$$\zeta_\nu = f_* N_\nu f_{\mathrm{esc},\nu}, \tag{2.34}$$

where f_* is the star formation efficiency (SFE), in this case defined to be the fraction of baryons that form stars, N_ν is the number of photons emitted per stellar baryon at some frequency ν, and $f_{\mathrm{esc},\nu}$ is the fraction of those photons that escape into the IGM. One could define additional ζ factors to represent, e.g., emission from black holes or exotic particles, in which case f_* and N_ν would be replaced by some black hole or exotic particle production efficiencies. In practice, most often three ζ factors are defined: $\zeta = \zeta_{\mathrm{ion}}$, ζ_X, and ζ_α, i.e., one efficiency factor for each radiation background that influences the 21 cm signal. A minimal model for the 21 cm background thus contains four parameters: m_{min}, ζ, ζ_X, and ζ_α. Note that ζ_X and ζ_α are often replaced by the parameters f_X and f_α, where the latter are defined such that $f_X = 1$ and $f_\alpha = 1$ correspond to fiducial values of ζ_X and ζ_α.

Because the factors constituting ζ are degenerate with each other, at least as far as 21 cm measurements are concerned, they generally are not treated separately as free parameters. However, it is still useful to consider each individually in order to determine a fiducial value of ζ and explore deviations from that fiducial model. In addition, inclusion of ancillary measurements may eventually allow ζ to be decomposed into its constituent parts (Mirocha et al. 2017; Park et al. 2019; Greig et al. 2020). For the remainder of this section, we focus on plausible values of f_*, N_ν, and $f_{\mathrm{esc},\nu}$, and the extent to which these quantities are currently understood.

2.2.1 Star Formation

Though a first-principles understanding of star formation remains elusive, the bulk properties of the star-forming galaxy population appear to obey simple scaling relationships. In this section, we outline the basic strategies used to infer the relationships between star formation and DM halos, and how such relationships can be used to inform 21 cm models.

The simplest description of the Galaxy population follows from the assumption that each DM halo hosts a single galaxy. With a model for the abundance of DM halos, i.e., the HMF, many of which are readily available (Press & Schechter 1974; Sheth et al. 2001), one can then "abundance match" halos with measured galaxy abundances (Bouwens et al. 2015; Finkelstein et al. 2015), i.e.,

$$\begin{aligned}
n(>L_h) &= \int_L^\infty \frac{dn}{dL'} dL' \\
&= n(>m_h) \\
&= \int_{m_h}^\infty \frac{dn}{dm_h'} dm_h'.
\end{aligned} \tag{2.35}$$

This procedure reveals the mapping between mass and light, dL/dm_h, upon repeated integration over a grid of L_h values, solving for the M_h value needed for abundances to match.

Results of this simple procedure show that galaxy luminosity is a function of both halo mass and cosmic time (Trenti & Stiavelli 2009; Moster et al. 2010; Behroozi et al. 2013; Tacchella et al. 2018; Mashian et al. 2016; Sun & Furlanetto 2016; Mason et al. 2015). While the HMF can be readily used to predict the abundances of halos out to arbitrary redshifts, one shortcoming of this approach is that any evolution in dL/dM must be modeled via extrapolation. As a result, predictions for deeper and/or higher redshift galaxy surveys may not be physically motivated.

To avoid the possibility of unphysical extrapolations of dL/dm_h, it is becoming more common to parameterize the Galaxy–halo connection from the outset, effectively resulting in forward models for galaxy formation that link galaxy star formation rate (SFR), $\dot{m}_*$, to halo mass, m_h, or mass accretion rate (MAR), $\dot{m}_h$, e.g.,

$$\dot{m}_*(z,m_h) = \tilde{f}_*(z,m_h)m_h(z,m_h) \tag{2.36}$$

or

$$\dot{m}_*(z,m_h) = f_*(z,m_h)\dot{m}_h(z,m_h). \tag{2.37}$$

The star formation efficiency (SFE), here indicated with $\tilde{f}_*$ and f_* to explicitly indicate whether it is tied to m_h or $\dot{m}_h$, is left as a flexible function to be calibrated empirically.[11] In the MAR-based model, one of course requires a model for the halo MAR as well as the HMF, though such models are readily available from the results of numerical simulations (McBride et al. 2009; Trac et al. 2015), or modeled approximately from the HMF itself (Furlanetto et al. 2017). Both approaches are used in the literature, and while inferred SFRs are largely in agreement, there is some difference in the interpretation of the models, which we revisit below in Section 2.2.1.1.

Finally, to complete the link between halos and galaxies, one must adopt a conversion factor between SFR and galaxy luminosity in some band. High-z measurements mostly probe the rest-UV spectrum of galaxies, so it is customary to link the SFR with the rest 1600 Å luminosity of galaxies,

$$L_{1600}(z,m_h) = l_{1600}\dot{m}_*(z,m_h), \tag{2.38}$$

where l_{1600} is of order 10^{28} erg s^{-1} Hz^{-1} $(M_\odot/\text{yr})^{-1}$ according to commonly used stellar population synthesis models, assuming constant star formation (Leitherer et al. 1999; Eldridge & Stanway 2009; Conroy et al. 2009). The precise value depends on stellar metallicity, binarity, and initial mass function (IMF), and varies from model to model. We will revisit the details of stellar spectra in Section 2.2.2, as there is a clear degeneracy between the assumed UV properties of galaxies and the inferred SFE.

The end result of this exercise is a calibrated SFE curve, which can then be used to make predictions for galaxy properties too faint or too distant to have been detected

---

[11] Note that $\tilde{f}_*$ in Equation (2.36) necessarily has units of time^{-1}, whereas f_* in Equation (2.37) is dimensionless.

by current surveys. A representative example (Tacchella et al. 2018) is shown in Figure 2.3 (which includes a common dust correction—see Section 2.2.3). The rise and fall in the SFE is a generic result of semiempirical models (Mason et al. 2015; Sun & Furlanetto 2016; Mashian et al. 2016; Tacchella et al. 2018; Behroozi et al. 2019), indicating a change in how galaxies form stars in halos above and below $\sim 10^{12}$ $M_\odot$. Such models generally agree that star formation is inefficient, with peak values $f_* \lesssim 0.1$, and $f_* \lesssim 0.01$ in the less massive (but numerous) population of galaxies residing in halos $M_h \lesssim 10^{11}$ $M_\odot$. As a result, if using the standard ζ modeling approach (see Equation (2.34)), reasonable fiducial f_* values are $f_* \sim 0.01$–0.1. Of course, evolution of the HMF implies that representative values of f_* will also evolve with time, though discerning such effects will only be possible in 21 cm analyses focused on a broad frequency range.

2.2.1.1 Physical Arguments for the Inferred Behavior of f*

Current high-z measurements support a relatively simple picture of star formation in early galaxies, in which galaxies maintain a rough equilibrium between inflow and outflow through stellar feedback (Bouché et al. 2010; Davé et al. 2012; Dekel & Mandelker 2014), the efficiency of which is a strong function of halo mass and perhaps time. Though there are quantitative differences among studies in the literature, which could arise due to different assumptions about dust, stellar populations, and/or different definitions of f_*, there does appear to be a consensus that star formation is maximally efficient (and feedback correspondingly inefficient) in $M_h \simeq 10^{11.5}$– 10^{12} $M_\odot$ halos (Mason et al. 2015; Sun & Furlanetto 2016; Moster et al. 2010; Mashian et al. 2016; Tacchella et al. 2018; Behroozi et al. 2019). The decline in the SFE below the peak is widely thought to be a signature of stellar feedback, while the decline in massive systems is likely due to shock heating and/or

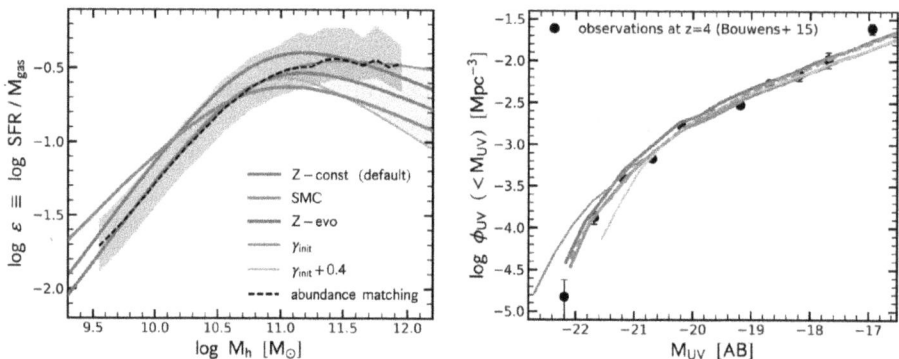

Figure 2.3. Relationship between halos and star formation recovered via semiempirical modeling (reproduced from Tacchella et al. 2018 © 2018. The American Astronomical Society. All rights reserved). Left: star formation efficiency as a function of halo mass for a variety of different approaches, including constant metallicity "Z-const," an evolving metallicity model "Z-evo," a model with Small Magellanic Cloud (SMC) dust instead of the standard relation from Meurer et al. (1999; see Section 2.2.3), and the pure abundance matching approach. Right: modeled luminosity function at $z = 4$ compared to measurements from Bouwens et al. (2015).

AGN feedback, which reduces the availability of cold gas (Faucher-Giguère et al. 2011; Croton et al. 2006; Somerville et al. 2008).

This general trend can be explained—but perhaps not understood—from relatively simple arguments. The basic idea is that star formation is fueled by the inflow of gas from the IGM, but that the overall rate of star formation in galaxies is self-regulated by feedback from stellar winds and supernovae explosions, both of which expel gas that could otherwise form stars (Dayal et al. 2014; Furlanetto et al. 2017). As a result, galaxies forming in shallow gravitational potential wells are at a disadvantage simply because the escape velocity is lower, making it easier for supernovae and winds to drive material out of the Galaxy. However, the escape velocity depends on both the mass and size of an object—if low-mass halos are sufficiently compact, they may be able to retain enough gas to continue forming stars.

To build some intuition for possible outcomes, it is common to model star formation as a balance between inflow and outflow (Bouché et al. 2010; Davé et al. 2012; Dekel & Mandelker 2014), i.e.,

$$\dot{m}_* = \dot{m}_b - \dot{m}_w, \tag{2.39}$$

where $\dot{m}_b$ is the accretion rate of baryons onto a halo and $\dot{m}_w$ is the mass-loss rate through winds (and/or supernovae). If we relate mass loss to star formation via the "mass loading factor" η, $\dot{m}_w \equiv \eta \dot{m}_*$, then we can write

$$f_* = \frac{\dot{m}_*}{\dot{m}_b} = \frac{1}{1 + \eta}. \tag{2.40}$$

One can show that, for energy-conserving winds, $\eta \propto m_h^{2/3}(1 + z)^{-1}$, while for momentum-conserving winds, $\eta \propto m_h^{1/3}(1 + z)^{-1/2}$ (Dayal et al. 2014; Furlanetto et al. 2017). Though simple, these models provide some physically motivated guidance for extrapolating models to higher redshifts and/or fainter objects than are probed by current surveys. Current measurements can still accommodate either scenario, largely due to (i) the small time baseline over which measurements are available and (ii) uncertainties in correcting for dust reddening (see Section 2.2.3).

There are numerous other techniques that are commonly employed to model star formation in high-z galaxies. For example, one need not require that star formation operate in an equilibrium with inflow and outflow, in which case Equation (2.36) may be a more sensible choice than Equation (2.37). Many efforts are now underway to simulate galaxy formation using *ab initio* cosmological simulations (Vogelsberger et al. 2014; Schaye et al. 2015; Hopkins et al. 2014; O'Shea et al. 2015; Gnedin 2014), rather than using analytic or semianalytic models. However, doing so self-consistently in statistically representative volumes is exceedingly computationally challenging; as a result, semianalytic and semiempirical prescriptions for star formation in reionization modeling remain the norm in 21 cm modeling codes (Mirocha et al. 2017; Park et al. 2019; Mutch et al. 2016). Though such approaches lack the spatial resolution to model individual galaxies or even groups of galaxies, including some information about the Galaxy population permits joint modeling of 21 cm observables as well as high-z galaxy luminosity functions, stellar-mass functions, and so on, and thus opens up the possibility of tightening constraints on the properties of galaxies using a multiwavelength approach.

2.2.1.2 Pop III Star Formation

The very first generations of stars to form in the universe did so under very different conditions than stars today, so it is not clear that the star formation models outlined in the previous section apply. The first stars, by definition, formed from chemically pristine material, as no previous generations of stars had existed to enrich the medium with heavy elements. This has long been recognized as a reason that the first stars are likely different from stars today (Abel et al. 2000; Bromm et al. 1999; O'Shea & Norman 2007; Yoshida et al. 2003). Without the energetically low-lying electronic transitions common in heavy elements, hydrogen-only gas clouds cannot cool efficiently, as collisions energetic enough to excite atoms from $n = 1$ to $n = 2$ (which subsequently cool via spontaneous emission of Lyα photons) imply temperatures of $\sim 10^4$ K, corresponding to virial masses of order $\sim 10^8\ M_\odot$.[12] Such halos are increasingly rare at $z \gtrsim 10$.

Even in the absence of metals, there are cooling channels available in halos too small to support atomic (hydrogen) line cooling, i.e., with masses $M_\odot \lesssim 10^8\ M_\odot$. Hydrogen molecules, H_2, can form using free electrons as a catalyst,[13]

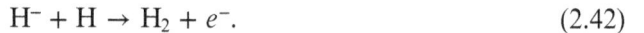

$$H + e^- \rightarrow H^- + h\nu, \tag{2.41}$$

$$H^- + H \rightarrow H_2 + e^-. \tag{2.42}$$

These reactions are limited by the availability of free electrons[14] and the survivability of H^- ions. Even in the absence of astrophysical backgrounds, the formation of H_2 is limited by the CMB, which at the high redshifts of interest can dissociate the H^- ion. Tegmark et al. (1997) found that the molecular hydrogen fraction in high-z halos scales with the virial temperature as

$$f_{H_2} \approx 3.5 \times 10^{-4} \left(\frac{T_{vir}}{10^3\ K} \right)^{1.52}. \tag{2.43}$$

Once the first stars form, the situation grows considerably more complicated. As will be detailed in the following section (Section 2.2.2), massive stars are prodigious sources of UV photons. Some of these photons originate in the Lyman–Werner band (~ 11.2–13.6 eV) and are thus capable of dissociating molecular hydrogen. This process is expected to quickly surpass H^- dissociation by the CMB as the most important mechanism capable of regulating star formation in chemically pristine halos.[15]

---

[12] Setting $T_{min} \sim 10^4$ K (or $M_{min} \sim 10^8\ M_\odot$) is thus a way to roughly exclude the effects of Pop III-hosting "mini halos" in a 21 cm model.

[13] Dust is the primary catalyst of H_2 formation in the local universe, but of course it does not exist in the first collapsing clouds.

[14] Exotic models in which an X-ray background emerges before the formation of the first stars may affect early star formation by boosting the electron fraction.

[15] If the Pop III IMF is very bottom-heavy, the resulting IR background could continue to regulate star formation via H^- photodetachment (Wolcott-Green & Haiman 2012).

A substantial literature aimed at understanding the critical Lyman Werner (LW) background intensity, J_{LW}, required to prevent star formation in high-z mini halos has emerged in the last ~20 years. For example, Visbal et al. (2014) found

$$M_{crit} = 2.5 \times 10^5 \left(\frac{1+z}{26} \right)^{-3/2} (1 + 6.96(4\pi J_{LW})^{0.47}) \, M_\odot, \qquad (2.44)$$

where J_{LW} is the LW background intensity in units of 10^{-21} erg s^{-1} cm^{-2} Hz^{-1} sr^{-1}. In principle, M_{crit} varies across the universe from region to region as a function of the local LW intensity, but it is common to use the mean LW background intensity for simplicity. Note finally that sufficiently dense clouds can self-shield themselves against LW radiation, which is an important (and still uncertain) aspect of modeling LW feedback (Wolcott-Green & Haiman 2011).

While the LW background is responsible for setting the minimum halo mass required to host star formation, the maximum mass of Pop III halos, i.e., the mass at which halos transition from Pop III to Pop II star formation, depends on the interplay of many complex processes. For example, Pop III supernovae will inject metals into the ISM of their host galaxies, which can trigger the transition to Pop II star formation provided that at least some metals are retained and efficiently mix into protostellar clouds. The timescales involved are highly uncertain and may vary from halo to halo. Some halos may even be externally enriched (Smith et al. 2015). As a result, whereas rest-frame ultra-violet luminosity functions (UVLFs) at high-z provide some insight into the Pop II SFE, the Pop III SFE, which encodes the complex feedback processes at play, is completely unconstrained.

Figure 2.4 shows some example predictions for the Pop III star formation rate density (SFRD) in a semianalytic model of Pop III star formation. Clearly, the level of Pop III star formation is subject to many unknowns, which results in a vast array of predictions spanning ~3 orders of magnitude in peak SFRD. This range is representative of the broader literature (Trenti & Stiavelli 2009; Maio et al. 2010; Crosby et al. 2013; Visbal et al. 2018; Mebane et al. 2018; Jaacks et al. 2018; Sarmento et al. 2018), with some studies favoring even slightly higher peak Pop III SFRDs $\dot{\rho}_{*,III} \sim 10^{-3} \, M_\odot$ yr^{-1} cMpc^{-3}. Even a crude constraint on the Pop III SFRD would rule out entire classes of models and thus provide a vital constraint on star formation processes in the earliest halos.

In 21 cm models, it is common to ignore a detailed treatment of individual Pop III star-forming halos and instead parameterize the impact of Pop II and Pop III halos separately. One way to do this is to assume all atomic cooling halos form Pop II stars (with $\zeta_{LW,II}$, $\zeta_{X,II}$, etc.), and all mini halos form Pop III, with their own efficiency factors $\zeta_{X,II}$, etc., and m_{min} determined self-consistently from the emergent LW background intensity (Fialkov et al. 2014; Mirocha et al. 2018). We will revisit the predictions of these models in Section 2.3.

2.2.2 UV Emission from Stars

With some handle on the efficiency of star formation, we now turn our attention to the efficiency with which stars generate UV photons, particularly in the Lyman-

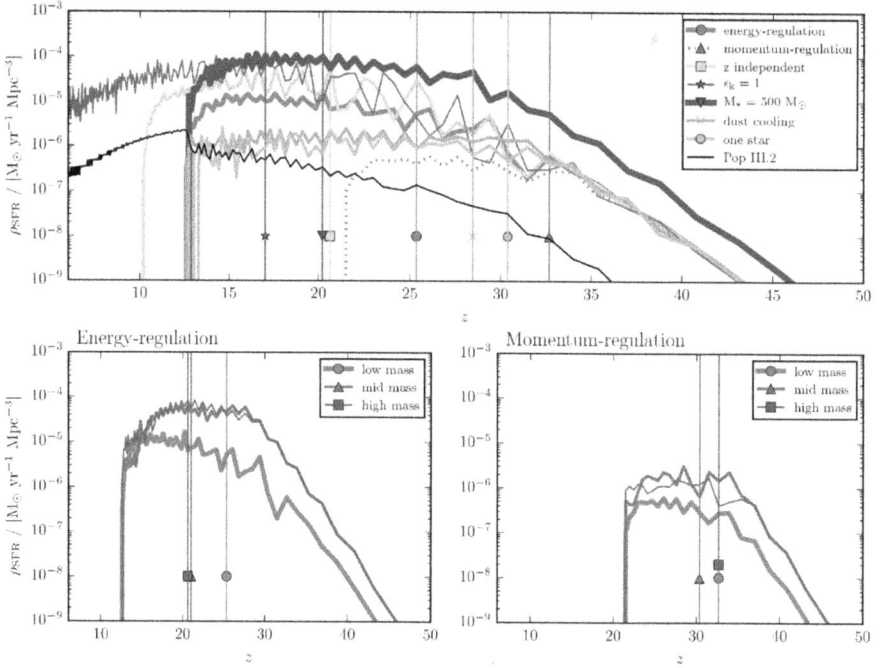

Figure 2.4. Predictions for the Pop III SFRD at high-z (reproduced from Mebane et al. 2018, by permission of Oxford University press on behalf of the Royal Astronomical Society). The upper panel shows results assuming a low-mass Salpeter-like Pop III IMF with different assumptions about the Pop II and Pop III star formation prescriptions. The lower panels study the effects of Pop III IMF between energy-regulated (left) and momentum-regulated (right) stellar feedback from Pop II stars. Note that the Pop II SFRD inferred from current UVLF measurements is of order $\sim 10^{-2}\ M_\odot\ \mathrm{yr}^{-1}\ \mathrm{cMpc}^{-3}$ at $z \sim 6$ (Bouwens et al. 2015; Finkelstein et al. 2015).

continuum and Lyman–Werner bands.[16] In principle, the 21 cm background is also sensitive to the spectrum of even harder He-ionizing photons, because photo-electrons generated from helium ionization can heat and ionize the gas, while HeII recombinations can result in H-ionizing photons. The 21 cm signal could in principle also constrain the rest-frame infrared spectrum of stars in the early universe, as IR photons can feedback on star formation at very early times through H⁻ photodetachment (Wolcott-Green & Haiman 2012). However, in this section, we focus only on the soft UV spectrum ($10.2 \lesssim E \lesssim 54.4$ eV) to which the 21 cm background is most sensitive.

The most detailed predictions for stellar spectra come from stellar population synthesis (SPS) models, which, as the name suggests, synthesize the spectra of entire stellar populations as a function of time. The key inputs to such models are

---

[16] We use this definition here loosely. Technically, the LW band is ~11.2–13.6 eV, a range which bounds photons capable of photodissociating molecular hydrogen, H_2. The Lyα background is sourced by photons in a slightly broader interval, ~10.2–13.6 eV, but it is tedious to continually indicate this distinction, and as a result, we use "LW band" to mean all photons capable of eventually generating Lyα photons.

- The stellar initial mass function (IMF), $\xi(m)$, i.e., the relative number of stars formed in different mass bins. Commonly adopted IMFs include Salpeter (1955), Chabrier (2003), Kroupa (2001), and Scalo (1998), which are all generally power laws with indices ~ -2.3, but differing in shape at the low-mass end of the distribution ($M_* < 0.5\ M_\odot$).
- Models for stellar evolution, i.e., how stars of different masses traverse the Hertzsprung–Russell diagram over time.
- Models for stellar atmospheres, i.e., predictions for the spectrum of individual stars as a function of their mass, age, and composition.

With all these ingredients, one can synthesize the spectrum of a stellar population formed some time t after a "burst,"

$$L_\nu(t) = \int_0^t dt' \int_{m_{min}}^\infty dm\xi(m)l_\nu(m,t'), \qquad (2.45)$$

where $l_\nu(m,t)$ is the specific luminosity of a star of mass m and age t, and we have assumed that ξ is normalized to the mass of the star cluster, $\int dm\xi(m) = M_*$. Equation (2.45) can be generalized to determine the spectrum of a galaxy with an arbitrary star formation history (SFH) composed of such bursts. Widely used stellar synthesis codes include STARBURST99 (Leitherer et al. 1999), BPASS (Eldridge & Stanway 2009), Flexible Stellar Population Synthesis (FSPS; Conroy et al. 2009), Stochastically Lighting up Galaxies (SLUG; da Silva et al. 2012), and the Bruzual and Charlot models (Bruzual & Charlot 2003).

Generally, 21 cm models do not operate at the level of SPS models because the 21 cm background is insensitive to the detailed spectra and SFHs of individual galaxies. Instead, because 21 cm measurements probe the relatively narrow intervals $10.2 < h\nu/eV < 13.6$ via Wouthuysen–Field coupling and $h\nu > 13.6$ eV through the ionization field, it is common to distill the predictions of SPS models into just two numbers, N_{ion} and N_α, which integrate over the age and the details of the stellar SED, i.e.,

$$N_{ion} = m_*^{-1} \int_0^\infty dt' \int_{\nu_{LL}}^\infty \frac{d\nu}{h\nu} L_\nu(t'), \qquad (2.46)$$

$$N_\alpha = m_*^{-1} \int_0^\infty dt' \int_{\nu_\alpha}^{\nu_{LL}} \frac{d\nu}{h\nu} L_\nu(t'), \qquad (2.47)$$

where ν_{LL} is the frequency of the Lyman limit (13.6 eV) and ν_α is the Lyα frequency. UV emission is dominated by massive, short-lived stars, hence the integration from $t = 0$ to $t = \infty$.

Assuming a Scalo IMF, stellar metallicity of $Z = Z_\odot/20$, and using the STARBURST99 SPS model, Barkana & Loeb (2005) reported $N_{ion} = 4000$ and $N_\alpha = 9690$ photons per baryon, the latter broken down further into subintervals between each Lyn resonance. The general expectation is for N_{ion} and N_α to increase for more top-heavy IMF and lower metallicity, meaning these values are likely to increase for Pop III stars (Bromm et al. 2001; Tumlinson & Shull 2000; Schaerer 2002). Similarly, binary evolution can effectively increase the lifetimes of massive stars, leading to a net gain in UV photon production (Stanway et al. 2016).

Note that when simultaneously modeling the UVLF, one generally assumes a constant star formation rate, in which case the UV luminosity of stellar populations asymptotes after a few hundred Myr. As a result, it is common to use the results of SPS models (with continuous star formation) at $t = 100$ Myr when converting UV luminosity to SFR, as in Equation (2.38), though in reality the detailed star formation history is (generally) unknown. In this particular case, we do not quantify the UV luminosity in photons per stellar baryon—instead, the exact values of f_*, l_{1600}, N_{ion}, and N_α should all be determined self-consistently from the model calibration. In other words, one cannot simply change N_{ion} and f_* independently, because the inferred value of f_* depends on the assumed stellar population model and thus implicitly on N_{ion} (see also Section 2.3 and Figure 2.10; Mirocha et al. 2017).

2.2.3 Attenuation of Stellar UV Emission by Dust

Even if the intrinsic stellar spectrum of galaxies were known perfectly, our ability to draw inferences about star formation is hampered by the presence of dust, which dims and reddens the "true" spectrum of galaxies. The opacity of dust is an inverse function of wavelength, meaning its impact is greatest at short wavelengths (Weingartner & Draine 2001). Unfortunately, most observations of high-z galaxies (so far) target the rest-UV spectrum of galaxies[17] and thus must be considerably "dust corrected" before star formation rates and/or efficiencies are estimated.

However, correcting for the effects of dust attenuation is not completely hopeless. If we assume that UV-heated dust grains radiate as blackbodies, we would expect to see an "infrared excess" (IRX) in galaxies with redder-than-expected UV continua, as UV reddening is suggestive of dust attenuation. If we assume for simplicity a power-law UV continuum, $f_\lambda = f_{\lambda,0} \lambda^\beta$, we would expect an excess

$$\text{IRX}_{1600} \equiv \frac{F_{\text{FIR}}}{F_{1600}} = \frac{\int_{912}^{\infty} f_{\lambda,0}(1 - e^{-\tau_\lambda})d\lambda}{f_{1600,0}\, e^{-\tau_\lambda}} \left(\frac{F_{\text{FIR}}}{F_{\text{bol}}} \right), \qquad (2.48)$$

where τ_λ is the wavelength-dependent dust opacity, equivalently written as $10^{-0.4A_{1600}} = e^{-\tau_{1600}}$, where A_{1600} is the extinction at 1600 Å in magnitudes, $f_{1600,0}$ is the intrinsic intensity at 1600 Å, and $F_{\text{FIR}}/F_{\text{bol}}$ is a correction factor that accounts for the fraction of bolometric dust luminosity emitted in the far-infrared band of observation.[18]

---

[17] This is simply due to the limited availability and sensitivity of near-infrared observations, which will soon be greatly enhanced by the *James Webb Space Telescope*.

[18] Note that the above expression assumes that all heating is done by photons redward of the Lyman limit and ignores heating by line photons.

An empirical constraint on the so-called IRX–β relation was first presented by Meurer et al. (1999), who found $A_{1600} = 4.43 - 1.99\beta_{obs}$, where β_{obs} is the logarithmic slope of the observed rest-UV spectrum. The intrinsic UV slope of young stellar populations is generally $-3 \lesssim \beta \lesssim -2$, which, coupled with the fact that observed slopes can be $-2 \lesssim \beta_{obs} \lesssim -1$ (Finkelstein et al. 2012; Bouwens et al. 2014), implies that corrections of potentially several magnitudes are likely in order. There is an ongoing debate in the field regarding the origin of the IRX–β relation, the cause of its scatter, and the possibility that it evolves with time (Narayanan et al. 2018; Salim & Boquien 2019). Early efforts with ALMA (Capak et al. 2015; Bouwens et al. 2016) are beginning to test these ideas with rest-IR observations of $z \sim 6$ star-forming galaxies, with some hints that there is evolution in IRX–β indicative of reduced dust obscuration in higher redshift galaxies. However, such inferences are currently dependent on assumptions for the dust temperature—if dust at high z is warmer than dust at low z, the data may be consistent with no evolution in A_{1600} at fixed M_{UV} or stellar-mass. These uncertainties in correcting high-z rest-UV measurements for dust reddening may affect the normalization of f_* at the factor of ~few level and could also bias the shape of the inferred f_* depending on precisely how A_{1600} scales with M_{UV} (or m_*).

2.2.4 Escape of UV Photons from Galaxies

While photons with wavelengths longer than 912 Å are most likely to be absorbed by dust grains, as described in the previous section, photons with wavelengths shortward of 912 Å will be absorbed by hydrogen and helium atoms. As a result, reionization models must also account for local attenuation, as the ionization state of intergalactic gas is of course only influenced by the ionizing photons that are able to escape galaxies. The fraction of photons that escape relative to the total number produced is quantified by the escape fraction, f_{esc}, and is the final component of the ionizing efficiency, ζ, introduced previously (see Equation (2.34)).

Current constraints on high-z galaxies and reionization suggest that f_{esc} must be ~10%–20% (Robertson et al. 2015). The result is model dependent, however, as it relies on assumptions about the UV photon production efficiency in galaxies and extrapolations to source populations beyond current detection limits. For example, if f_{esc} depends inversely on halo mass, reionization can be driven by galaxies that have yet to be detected directly (Finkelstein et al. 2019).[19]

Numerical simulations now lend credence to the idea that escape fractions of ~10%–20% are possible, with perhaps even larger f_{esc} in low-mass halos (Kimm & Cen 2014; Xu et al. 2016). The basic trend is sensible: as the depth of halo potentials declines, supernova explosions can more easily excavate clear channels through which photons escape. However, there is far from a consensus on this issue. For example, the FIRE simulations do not see evidence that f_{esc} depends on halo mass (Ma et al. 2015), and on average $f_{esc} \lesssim 5\%$, causing some tension with reionization

---

[19] This scenario is appealing because it can explain the very gradual evolution in the post-reionization ionizing background and the rarity of galaxies leaking LyC radiation at $3 \lesssim z \lesssim 6$ (Shapley et al. 2006).

constraints unless binary models (Eldridge & Stanway 2009) are employed. This result, coupled with the very high resolution in FIRE, is more suggestive of a scenario in which f_{esc} is set by very small-scale structure in the ISM, rather than the depth of the host halo potential.

Twenty-one cm measurements in principle open a new window into constraining f_{esc}. If, for example, the UV/SFR conversion factor is well known and dust can be dealt with (see Sections 2.2.2 and 2.2.3), joint fitting 21 cm power spectra and high-z galaxy LFs can isolate f_* and f_{esc} (Park et al. 2019; Greig et al. 2020). Note, however, that this is still model dependent, as f_* must be extrapolated to some limiting UV magnitude or halo mass in order to obtain the total photon production rate per unit volume.

2.2.5 X-Rays from Stellar-mass Black Holes

Though stars themselves emit few photons at energies above the He II-ionizing edge (~54.4 eV), their remnants can be strong X-ray sources and thus affect the IGM temperature. While solitary remnants will be unlikely to accrete much gas from the diffuse ISM (though see Section 2.2.5.1), remnants in binary systems may accrete gas from their companions, either via Roche-lobe overflow or stellar winds. Such systems are known as X-ray binaries (XRBs), further categorized by the mass of the donor star: "low-mass X-ray binaries" (LMXBs) are those fueled by Roche-lobe overflow from a low-mass companion, while "high-mass X-ray binaries" (HMXBs) are fed by the winds of massive companions. XRBs exhibit a rich phenomenology of time- and frequency-dependent behavior and are thus interesting in their own right. For a review, see, e.g., Remillard & McClintock (2006).

In nearby star-forming galaxies, the X-ray luminosity is generally dominated by HMXBs (Gilfanov et al. 2004; Fabbiano 2006; Mineo et al. 2012a). Furthermore, the total luminosity in HMXBs scales with the star formation rate, as expected given that the donor stars in these systems are massive, short-lived stars. An oft-used result in the 21 cm literature stems from the work of Mineo et al. (2012a), who find

$$L_X = 2.6 \times 10^{39} \left(\frac{\dot{M}_*}{M_\odot \, \mathrm{yr}^{-1}} \right) \mathrm{erg} \, \mathrm{s}^{-1}, \tag{2.49}$$

where L_X is defined here as the luminosity in the 0.5–8 keV band. This relation provides an initial guess for many 21 cm models, which add an extra factor f_X to parameterize our ignorance of how this relation evolves with cosmic time. For example, Furlanetto (2006) write

$$L_X = 3 \times 10^{40} f_X \left(\frac{\dot{M}_*}{M_\odot \, \mathrm{yr}^{-1}} \right) \mathrm{erg} \, \mathrm{s}^{-1}, \tag{2.50}$$

which is simply Equation (2.49) renormalized to a broader energy range, $0.2 < h\nu/\mathrm{keV} < 3 \times 10^4$, assuming a power-law spectrum with spectral index $\alpha_X = -1.5$, where α_X is defined by $L_E \propto E^{\alpha_X}$. Coupled with estimates for the star formation rate density at high z, the L_X–SFR relation suggests that X-ray binaries could be considerable sources of heating in the high-z IGM (Furlanetto 2006; Fragos et al. 2013; Mirocha 2014; Fialkov et al. 2014; Madau & Fragos 2017).

The normalization of these empirical L_X–SFR relations is not entirely unexpected, at least at the order-of-magnitude level. For example, if one considers a galaxy forming stars at a constant rate, a fraction $f_\bullet \simeq 10^{-3}$ of stars will be massive enough ($M_* > 20\ M_\odot$) to form a black hole assuming a Chabrier IMF. Of those, a fraction $f_{\rm bin}$ will have binary companions, with a fraction $f_{\rm surv}$ surviving the explosion of the first star for a time τ. If accretion onto these black holes occurs in an optically thin, geometrically thin disk with radiative efficiency $\varepsilon_\bullet = 0.1$, which obeys the Eddington limit, then a multicolor disk spectrum is appropriate and a fraction $f_{0.5-8} = 0.84$ of the bolometric luminosity will originate in the 0.5–8 keV band. Finally, assuming these BHs are "active" for a fraction $f_{\rm act}$ of the time, we can write (Mirabel et al. 2011; Mirocha et al. 2018)

$$L_X \sim 2 \times 10^{39}\ {\rm erg\ s^{-1}} \left(\frac{\dot{M}_*}{M_\odot\ {\rm s}^{-1}}\right)\left(\frac{\varepsilon_\bullet}{0.1}\right)\left(\frac{f_\bullet}{10^{-3}}\right)$$
$$\left(\frac{f_{\rm bin}}{0.5}\right)\left(\frac{f_{\rm surv}}{0.2}\right)\left(\frac{\tau}{20\ {\rm Myr}}\right)\left(\frac{f_{\rm act}}{0.1}\right)\left(\frac{f_{0.5-8}}{0.84}\right). \tag{2.51}$$

While several of these factors are uncertain, particularly $f_{\rm surv}$ and $f_{\rm act}$, this expression provides useful guidance in setting expectations for high redshift. For example, it has long been predicted that the first generations of stars were more massive on average than stars today owing to inefficient cooling in their birth clouds. This would boost $f_\bullet$, and thus $L_X/\dot{M}_*$, so long as most stars are not in the pair-instability supernova (PISN) mass range, in which no remnants are expected.

There are of course additional arguments not present in Equation (2.51). For example, the MCD spectrum is only a good representation of HMXB spectra in the "high-soft" state. At other times, in the so-called "low hard" state, HMXB spectra are well fit by a power law. The relative amount of time spent in each of these states is unknown. Figure 2.5 compares typical HMXB spectra with the spectrum expected from hot ISM gas (see Section 2.2.6).

In addition, physical models for the L_X–SFR relation may invoke the metallicity as a driver of changes in the relation with time and/or galaxy (stellar) mass. As the metallicity declines, one might expect the stellar IMF to change (as outlined above); however, the winds of massive stars responsible for transferring material to BHs will also grow weaker as the opacity of their atmospheres declines. As a result, increases in L_X/SFR likely saturate below some critical metallicity. Observations of nearby, metal-poor dwarf galaxies support this picture, with L_X/SFR reaching a maximum of ~10× the canonical relation quoted in Equation (2.49) (Mineo et al. 2012a).

2.2.5.1 X-Rays from Supermassive Black Holes

Though supermassive black holes (SMBHs) are exceedingly rare and thus unlikely to contribute substantially to the ionizing photon budget for reionization (Hassan et al. 2018; though see Madau & Haardt 2015), fainter—but more numerous—

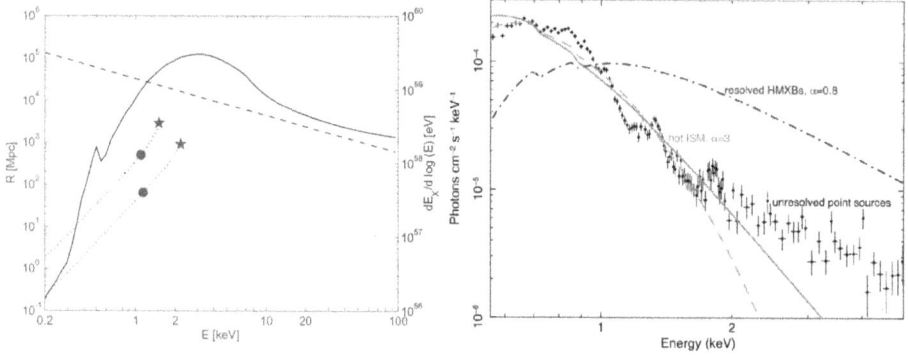

Figure 2.5. Models for the X-ray spectra of star-forming galaxies. Left: template Cygnus X-1 HMXB spectrum (solid) compared with power law (dashed), with mean free path shown on left scale and relative luminosity on right (reprinted by permission from Springer Nature: Fialkov et al. 2014). Right: typical HMXB spectrum (blue) compared to soft X-ray spectrum characteristic of bremsstrahlung emission from hot ISM gas (reproduced from Pacucci et al. 2014, by permission of Oxford University press on behalf of the Royal Astronomical Society).

intermediate-mass black holes (IMBHs) with $10^3 \lesssim M_\bullet/M_\odot \lesssim 10^6$ could have a measurable impact on the IGM thermal history (Zaroubi et al. 2007; Ripamonti et al. 2008; Tanaka et al. 2016). Growing BHs, if similar to their low-z counterparts, could also generate a strong-enough radio background to amplify 21 cm signals (Ewall-Wice et al. 2018), possibly providing an explanation for the anomalous depth of the EDGES global 21 cm measurement (Bowman et al. 2018). Such scenarios cannot yet be ruled out via independent measurements. For example, the unresolved fraction of the cosmic X-ray background still permits a substantial amount of accretion at $z \gtrsim 6$ (McQuinn 2012; Fialkov et al. 2017; Mirocha et al. 2018), while just ~10% of the radio excess reported by ARCADE-2 (Fixsen et al. 2011; Singal et al. 2018) must originate at $z \gtrsim 18$ in order to explain the EDGES signal (Feng & Holder 2018). Given the persistent challenge in explaining the existence of SMBHs at $z \gtrsim 6$, the signatures of BH growth in the 21 cm background are worth exploring in more detail.

2.2.6 X-Rays from Shocks and Hot Gas

While compact remnants of massive stars are likely the leading producer of X-rays in high-z star-forming galaxies, the supernova events in which these objects are formed may not be far behind. Supernovae inject a tremendous amount of energy into the surrounding medium, which then cools either via inverse Compton emission (in supernova remnants; Oh 2001) or eventually via bremsstrahlung radiation (in the hot interstellar medium, ISM). Because these sources are related to the deaths of massive stars, their luminosity is expected to scale with the SFR, as in the case of HMXBs. Indeed, Mineo et al. (2012b) found that diffuse X-ray emission in nearby sources follows the following relation in the 0.5–2 keV band,

$$L_X = 8.3 \times 10^{38}\left(\frac{\dot{M}_*}{M_\odot\ \mathrm{yr}^{-1}}\right)\mathrm{erg\ s}^{-1}. \tag{2.52}$$

This luminosity is that from all unresolved emission, and as a result, is not expected to trace emission from the hot ISM alone. Emission from supernova remnants will also contribute to this luminosity, as will fainter, unresolved HMXBs and LMXBs. Mineo et al. (2012b) estimated that $\sim 30\%$–40% of this emission may be due to unresolved point sources.

Though the soft X-ray luminosity from hot gas appears to be subdominant to the HMXB component in nearby galaxies, at least in total power, there are of course uncertainties in how these relations evolve. Furthermore, the bremsstrahlung emission characteristic of hot ISM gas has a much steeper $\sim \nu^{-2.5}$ spectrum than inverse Compton ($\sim \nu^{-1}$) or XRBs ($\sim \nu^{-1}$ or $\nu^{-1.5}$), and thus may heat more efficiently (owing to the $\sigma \propto \nu^{-3}$ cross section), provided soft X-rays can escape galaxies.

2.2.7 Escape of X-Rays from Galaxies

Though the mean free paths of X-rays are longer than those of UV photons, they still may not all escape from galaxies into the IGM. For example, hydrodynamical simulations suggest typical hydrogen column densities of $N_{H_1} \sim 10^{21}\,\mathrm{cm}^{-2}$ in low-mass halos (Das et al. 2017), which are substantial enough to eliminate emission below ~ 0.5 keV.

Given the many unknowns regarding X-ray emission in the early universe, 21 cm models often employ a three-parameter approach, i.e., instead of a single value of ζ_X, the specific X-ray luminosity is modeled as

$$L_{X,\nu} = L_{X,0}\left(\frac{h\nu}{1\,\mathrm{keV}}\right)^{\alpha_X} \exp\left[-\sigma_\nu N_{H_1}\right], \tag{2.53}$$

and the normalization, $L_{X,0}$, spectral index α_X, and typical column density, N_{H_1}, are left as free parameters. It is common to approximate this intrinsic attenuation with a piecewise model for L_X, i.e.,

$$L_{X,\nu} = \begin{cases} 0 & h\nu < E_{\min} \\ L_{X,0}\left(\dfrac{h\nu}{1\,\mathrm{keV}}\right)^{\alpha_X} & h\nu \geqslant E_{\min} \end{cases}. \tag{2.54}$$

Note that N_{H_1} (or $E_{\min}$) can be degenerate with the intrinsic spectrum, e.g., the SED of HMXBs in the high-soft state exhibits a turnover at energies $h\nu < 1$ keV, which could be mistaken for strong intrinsic absorption (Mirocha 2014).

2.2.8 Cosmic Rays from Supernovae

High-energy cosmic rays (CRs) produced in supernova explosions offer another potential source of ionization and heating in the bulk IGM (Nath & Biermann 1993; Sazonov & Sunyaev 2015; Leite et al. 2017), though most likely the effects are only discernible in the thermal history. Simple models suggest that CRs can raise the IGM temperature by ~ 10–200 K by $z \sim 10$ depending on the details of the CR spectrum (Leite et al. 2017). CRs are thus a potentially important, though relatively unexplored, source of heating in the high-z IGM.

2.3 Predictions for the 21 cm Background

So far we have assembled a simple physical picture of the IGM at high redshift (Section 2.1) and the sources most likely to affect its properties (Section 2.2). Here, we finally describe the generic sequence of events predicted in most 21 cm models and the sensitivity of the 21 cm background to various model parameters of interest.

Figure 2.6 shows an illustrative example using 21cmFAST (Mesinger et al. 2011), including a 2D slice of the δT_b field, the global 21 cm signal, and power spectrum on two spatial scales (Mesinger et al. 2016). Time proceeds from right to left from ~20 Myr after the big bang until the end of reionization ~1 Gyr later.

There are four distinct epochs indicated within this time period, which we describe in more detail below.

The Dark Ages: As the universe expands after cosmological recombination, Compton scattering between free electrons and photons keeps the radiation and matter temperature in equilibrium. The density is high enough that collisional coupling remains effective, and so $T_S = T_K = T_{CMB}$. Eventually, Compton scattering becomes inefficient as the CMB cools and the density continues to fall, which allows the gas to cool faster than the CMB. Collisional coupling remains effective for a short time longer and so T_K initially tracks T_S. This results in the first decoupling of T_S from T_{CMB} at

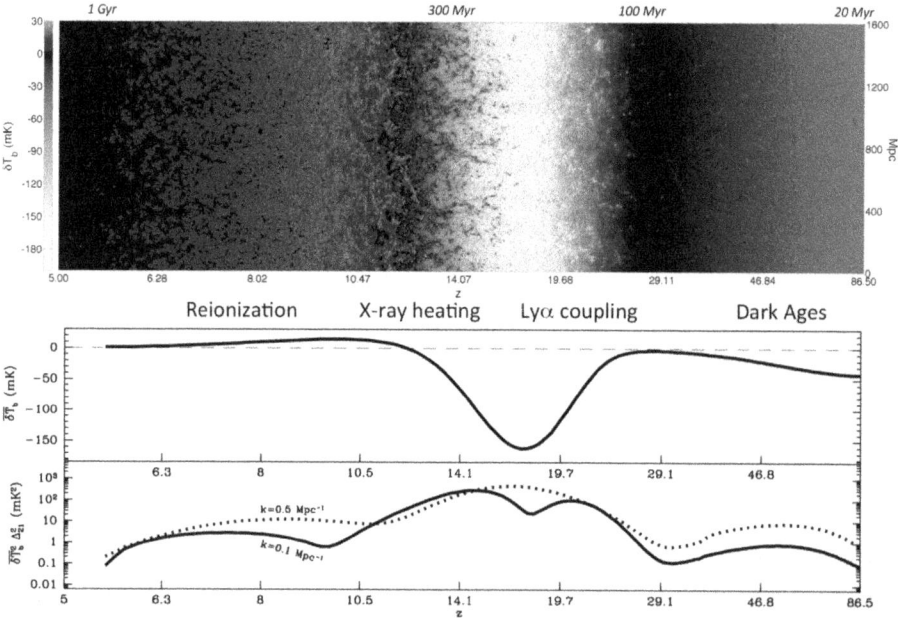

Figure 2.6. Predictions from the 21cmFAST Evolution of Structure (EoS) model suite (reproduced from Mesinger et al. 2016, by permission of Oxford University press on behalf of the Royal Astronomical Society). Top: 2D slice of the brightness temperature field, with red colors indicating a cool IGM, blue colors indicative of a heated IGM, and black representing a null signal (either due to ionization or $T_S = T_{CMB}$). Middle: global 21 cm signal, with the dashed line indicating $\delta T_b = 0$. Bottom: evolution of the dimensionless 21 cm power spectrum, $\Delta^2 = k^3 P(k)/2\pi^2$, on two different scales, $k = 0.5$ Mpc^{-1} (dotted) and $k = 0.1$ Mpc^{-1} (solid).

$z \sim 150$, resulting in an absorption signature at $z \sim 80$ ($\nu \sim 15$ MHz), which comes to an end as collisional coupling becomes inefficient, leaving T_S to reflect T_{CMB} once again.

Lyα coupling: When the first stars form, they flood the IGM with UV photons for the first time. While Lyman-continuum photons are trapped near sources, photons with energies $10.2 < h\nu/eV < 13.6$ either redshift directly through the Lyα resonance or cascade via higher Lyn levels, giving rise to a large-scale Lyα background capable of triggering Wouthuysen–Field coupling as they scatter through the medium (see also Sections 1.2.2 and 2.1.5). As a result, T_S is driven back toward T_K, which (in most models) still reflects the cold temperatures of an adiabatically cooling IGM.

X-ray Heating: The first generations of stars beget the first generations of X-ray sources, whether they be the explosions of the first stars themselves or remnant neutron stars or black holes that subsequently accrete. Though the details change depending on the identity of the first X-ray sources (see Sections 2.2.5 and 2.2.6), generally such sources provide photons energetic enough to travel great distances. Upon absorption, they heat and partially ionize the gas, eventually driving $T_S > T_{CMB}$. Once $T_S \gg T_{CMB}$, the 21 cm signal "saturates" and subsequently is sensitive only to the density and ionization fields. However, it is possible that heating is never "complete" in this sense before reionization, meaning neutral pockets of IGM gas may remain at temperatures at or below T_{CMB} until they are finally engulfed by ionized bubbles.

Reionization: As the global star formation rate density climbs, the growth of ionized regions around groups and clusters of galaxies will continue, eventually culminating in the completion of cosmic reionization. This rise in ionization corresponds to a decline in the amount of neutral hydrogen in the universe capable of producing and generating 21 cm signals. As a result, the amplitude of the 21 cm signal, both in its mean and fluctuations, falls as reionization progresses. After reionization, neutral hydrogen only remains in systems overdense enough to self-shield from the UV background.

The particular model shown in Figure 2.6 (Mesinger et al. 2016) assumes that very faint galaxies dominate the UV and X-ray emissivity, which results in relatively early features in the 21 cm background, e.g., both the power spectrum and global 21 cm signal peak in amplitude at $z \sim 18$. Reionization and reheating occur later in scenarios in which more massive halos dominate the emissivity, and may even occur simultaneously, resulting in strong 21 cm signals at $z \lesssim 12$ (Mesinger et al. 2016; Mirocha et al. 2017; Park et al. 2019).

For the remainder of this section, we focus on changes in the 21 cm signal wrought by the parameters of interest. We limit our discussion to the global 21 cm signal and power spectrum, though there are of course many other statistics one could use to constrain model parameters (see Chapter 4). We note that there is no consensus parameterization for models of galaxy formation or the 21 cm background, nor do all models incorporate the same physical processes or employ the same numerical techniques. As a result, in this section we make no effort to closely compare or homogenize results from the literature, but instead draw examples from many works in order to illustrate different aspects of the 21 cm background as a probe of galaxy formation.

2.3.1 Dependence on the Ionizing Efficiency

Generally written as ζ or ζ_{ion}, the ionizing efficiency quantifies the number of Lyman-continuum (LyC) photons that are produced in galaxies and escape into the IGM, i.e., $\zeta = f_* N_{ion} f_{esc}$ (see Sections 2.2.1–2.2.4). As a result, this parameter affects primarily the lowest redshifts $6 \lesssim z \lesssim 10$ ($\nu \gtrsim 130$ MHz), during which the bulk of reionization likely takes place.

Figure 2.7 shows predictions for the growth of ionized bubbles in the excursion set formalism (Furlanetto et al. 2004). In time, bubbles grow larger, eventually reaching typical sizes of ~tens of megaparsecs during reionization. The two-point correlation function of the ionization field (middle) grows with time as well, peaking near the midpoint of reionization (Lidz et al. 2008). This rise and fall is reflected in the 21 cm power spectrum as well (right), here modeled in the "saturated limit" $T_S \gg T_{CMB}$, in which case only fluctuations in $\psi = x_{H\,I}(1 + \delta)$ need to be considered. Larger values of ζ (thicker lines in right panel of Figure 2.7) result in stronger fluctuations on large scales and a suppression in the power on small scales.

Figure 2.8 shows results from four different numerical simulations (RT post-processed on an N-body simulation; McQuinn et al. 2007), each differing in their treatment of ζ. The key difference is how ζ depends on halo mass—here, models span the range of $\zeta \propto m_h^{-2/3}$ (S2) to $\zeta \propto m_h^{2/3}$ (S3), including the case of ζ = constant (S1). As the ionizing emissivity becomes more heavily weighted toward more massive, more rare halos (in S3 and S4), ionized structures grow larger and more spherical, while the smaller bubbles nearly vanish. This is a result of an increase in the typical bias of

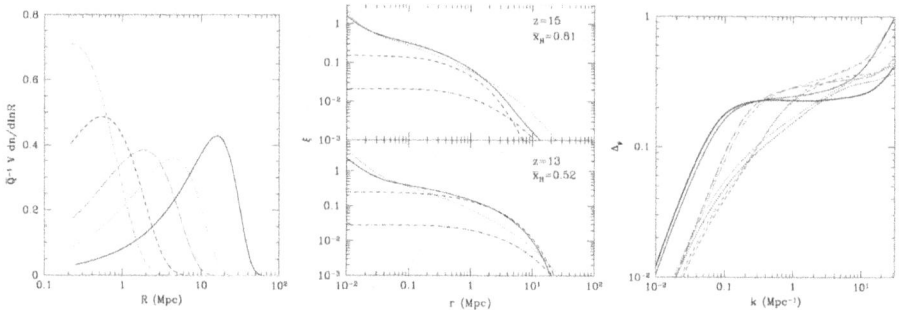

Figure 2.7. Analytic models of bubble growth during the EoR (reproduced from Furlanetto et al. 2004 © 2004. The American Astronomical Society. All rights reserved). Left: bubble size distributions at bubble filling fractions of Q = 0.037, 0.11, 0.3, 0.5, and 0.74, from left to right. Middle: correlation function of $\psi = x_{H\,I}(1 + \delta)$ (solid) at $x_{H\,I}$ = 0.81 (top) and $x_{H\,I}$ = 0.52 (bottom) as well as its constituent components, including the ionization autocorrelation function (dashed), density autocorrelation function (dotted), and cross-correlation function between ionization and density (dot-dashed). Right: dimensionless power spectrum of ψ for different values of ζ, including ζ = 12 (thin) and ζ = 40 (thick), at several neutral fractions, $x_{H\,I}$ = 0.96 (dotted), 0.8 (short-dashed), 0.5 (long-dashed), and 0.26 (solid).

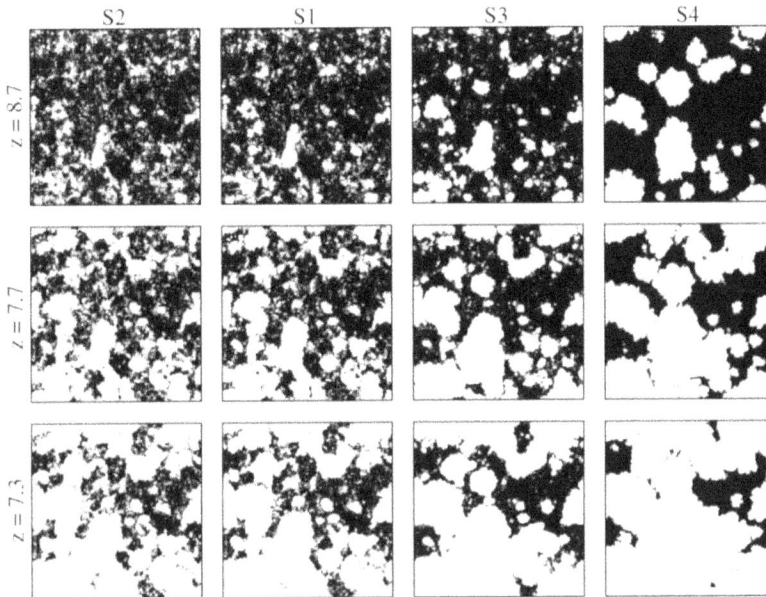

Figure 2.8. Ionization field in $(94 \text{ cMpc})^3$ box for different ionizing efficiency assumptions (reproduced from McQuinn et al. 2007, by permission of Oxford University press on behalf of the Royal Astronomical Society) as a function of redshift (top to bottom). Neutral gas is black, while ionized regions are white. Each column shows results from a different simulation, with S1–S3 using $\zeta = $ constant, $\zeta \propto m_h^{-2/3}$, and $\zeta \propto m_h^{2/3}$, respectively. S4 is the same as S1 except only halos with $m_h > 10^{10} \ M_\odot$ are included.

sources as ζ increases with m_h—because more massive halos are more clustered, ionizing photons from such halos combine to make larger ionized regions, whereas less clustered low-mass halos carve out smaller, more isolated ionized bubbles.

From Figure 2.8, it is clear that the behavior of ζ not only sets the timeline for reionization but also its topology. However, ζ is degenerate with m_{min}, because the ionizing emissivity can be enhanced both by increasing ζ directly or by increasing the number of star-forming halos by decreasing m_{min} (recall that the total number of ionizing photons emitted in a region is $N_\gamma = \zeta f_{coll}$). Despite this degeneracy, power-spectrum measurements are expected to be able to place meaningful constraints on both parameters (Greig & Mesinger 2017). The power spectrum on $k \sim 0.2 \text{ cMpch}^{-1}$ scales reliably peaks near the midpoint of reionization (Lidz et al. 2008), meaning some (relatively) model-independent constraints are expected as well.

2.3.2 Dependence on the X-Ray Efficiency and Spectrum

The progression of cosmic reheating is analogous in some respects to reionization, although driven by sources of much harder photons (see Sections 2.2.5 and 2.2.6). As a result, we must consider the total energy emitted in X-rays (per unit collapsed mass or star formation rate) in addition to parameters that control the SED of X-ray sources. This is often achieved through a three-parameter power-law model (see Section 2.2.7), including a normalization parameter (ζ_X), spectral cutoff (E_{min}), and power-law slope of X-ray emission (α_X). The combination of these parameters can capture a variety of physical models and mimic the shape of more sophisticated theoretical models (e.g., the multicolor disk spectrum; Mitsuda et al. 1984).

Holding the SED fixed, variations in ζ_X affect the thermal history much like ζ affects the ionization history: increasing ζ_X causes efficient heating to occur earlier in the universe's history, resulting in lower-frequency (and shallower) absorption troughs in the global 21 cm signal, while the peak amplitude of the power spectrum also shifts to earlier times (at fixed wavenumber). Effects of f_X on the global signal can be seen in the left panel of Figure 2.9 (see also Mirabel et al. 2011; Mirocha 2014; Fialkov et al. 2014).

Allowing the SED of X-ray sources can change the story dramatically because the mean free path of X-rays is a strong function of photon energy ($l_{\mathrm{mfp}} \propto \nu^{-3}$, due to the bound–free absorption cross section scaling $\sigma \propto \nu^{-3}$; Verner et al. 1996). This strong energy dependence means that photons with rest energies $h\nu \gtrsim 2\,\mathrm{keV}$ will not be absorbed within a Hubble length at $z \gtrsim 6$. Photons with $h\nu \lesssim 2$ keV will be absorbed on scales anywhere from ~0.1 Mpc to hundreds of megaparsecs. As a result, the "hardness" of X-ray sources will determine the spatial structure of the kinetic temperature field—soft photons will be absorbed on small scales and thus give rise to strong temperature fluctuations, while hard photons will travel great distances and heat the IGM more uniformly. The right panel of Figure 2.9 shows precisely this effect—for atomic cooling halos (solid lines), soft $\alpha_X = 3$ X-ray sources generate

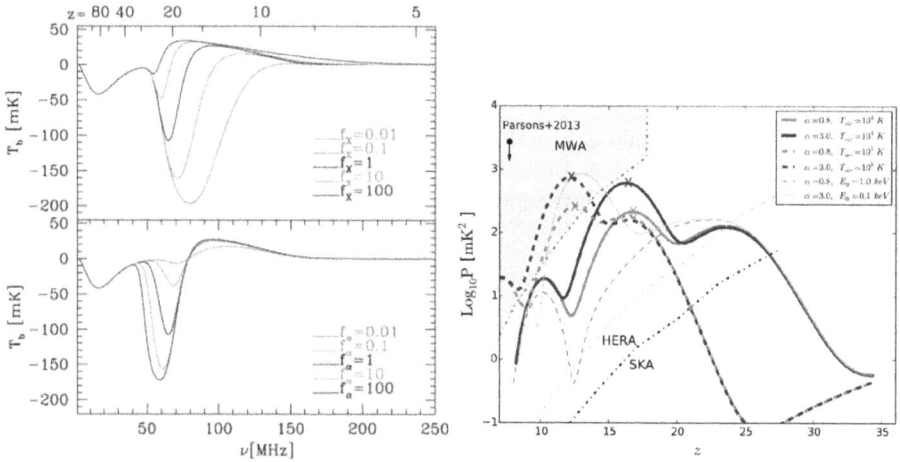

Figure 2.9. Effects of Lyα and X-ray efficiencies on the global 21 cm signal (Pritchard & Loeb 2010; Pacucci et al. 2014). Left: predictions for the global 21 cm signal showing sensitivity to the normalization of the L_X–SFR relation, f_X (top), and the production efficiency of Lyα photons, f_α (bottom; reproduced with permission from Pritchard & Loeb 2010, Copyright 2010 by the American Physical Society). Right: predictions for evolution in the 21 cm power spectrum at $k = 0.2\,\mathrm{Mpc}^{-1}$ for models with different X-ray spectra (reproduced from Pacucci et al. 2014, by permission of Oxford University press on behalf of the Royal Astronomical Society). Blue curves indicate soft power-law spectra with indices of $\alpha = 3$, while red curves are indicative of hard spectra sources with $\alpha = 0.8$. Line styles denote different minimum virial temperatures, $T_{\min}$, and lower energy cutoffs for the X-ray background, E_0.

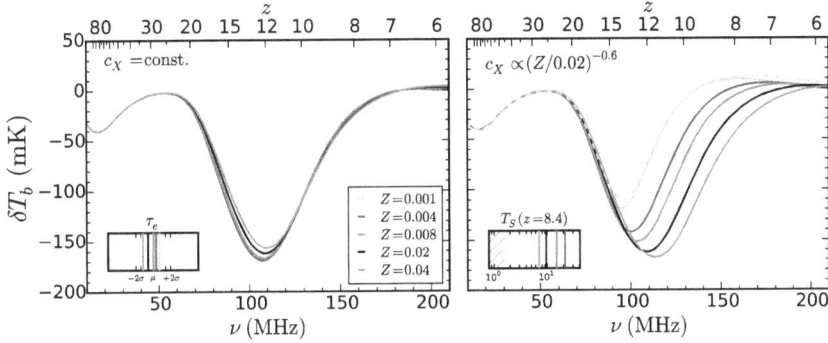

Figure 2.10. Effects of stellar metallicity on the global 21 cm signal (reproduced from Mirocha et al. 2017, by permission of Oxford University press on behalf of the Royal Astronomical Society). Left: metallicity effects assuming no link between stellar metallicity, Z_*, and X-ray luminosity. Right: metallicity effects assuming empirical relation between f_X and Z_* (Brorby et al. 2016). Insets show predictions for CMB optical depth, τ_e (left), and mean IGM spin temperature at $z = 8.4$ (right).

fluctuations ~2–3 times as strong as hard $\alpha_X = 0.8$ sources (Pacucci et al. 2014), where in this case the spectral index is defined as $L_X \propto h\nu^{-\alpha_X}$.

The hardness of X-ray emission is also controlled by the "cutoff energy," E_{min}, below which X-ray emission does not escape efficiently from galaxies (see Section 2.2.7). Alternatively, even if X-rays can escape efficiently, E_{min} could indicate an intrinsic turnover in the X-ray spectra of galaxies, e.g., that expected from a multicolor disk spectrum (Mitsuda et al. 1984). Joint constraints on ζ_X, E_{min}, and α_X are thus required to help identify the sources of X-ray emission in the early universe and the extent to which their spectra are attenuated by their host galaxies.

Finally, it is important to note that interpreting ζ_X is potentially more challenging than interpreting ζ given the additional parameters needed to describe the X-ray SED. One must be mindful of the fact that ζ_X quantifies the X-ray production efficiency in some bandpasses, generally 0.5–8 or 0.5–2 keV. As a result, changing α_X or E_{min} may be accompanied by a normalization shift so as to preserve the meaning of ζ_X. This degeneracy can be mitigated to some extent by redefining ζ_X in the ($E_{min} - 2$ keV) band so as to isolate the photons most responsible for heating (Greig & Mesinger 2017).

2.3.3 Dependence on the Lyα Efficiency

The production efficiency of Lyα photons affects when the 21 cm background first "turns on" due to Wouthuysen–Field coupling (see Sections 1.2.2 and 2.1.5). Figure 2.9 (bottom-left panel) illustrates the effect increasing f_α has on the global 21 cm signal (Pritchard & Loeb 2010). For very large values, $f_\alpha = 100$ (blue), the Dark Ages come to an end at $z \sim 30$ ($\nu \sim 45$ MHz), triggering a much deeper absorption trough than the fiducial model (with $f_\alpha = f_X = 1$; black lines). The intuition here is simple: at fixed f_X, increasing f_α drives $T_S \rightarrow T_K$ at earlier times, meaning there has been less time for sources to heat the gas.

Despite the very long mean free paths of photons that source the Lyα background, there are still fluctuations in the background intensity J_α (Barkana & Loeb 2005; Ahn et al. 2009; Holzbauer & Furlanetto 2012). As a result, there will be fluctuations in the spin temperature, as different regions transition from $T_S \approx T_{CMB}$ to $T_S \approx T_K$ at different rates. The onset of Lyα coupling is visible in the right panel of Figure 2.9, as the power (at $k = 0.2$ Mpc^{-1}) departs from its gradual descent at $z \sim 25$ ($T_{min} = 10^5$ K) and $z \sim 33$ ($T_{min} = 10^4$ K).

Because the Lyα background is sourced by photons in a relatively narrow frequency interval, $10.2 \lesssim h\nu/\text{eV} \lesssim 13.6$, the timing of Wouthuysen–Field coupling and the amplitude of fluctuations are relatively insensitive to the SED of sources. Similarly, because hydrogen gas is transparent to these photons (except at the Lyn resonances), these photons have an escape fraction $f_{esc,LW} \gtrsim 0.5$, at least in the far field limit (Schauer et al. 2015), as their only impediment is H$_2$, which is quickly dissociated by stellar Lyman-continuum emission.

2.3.4 Dependence on Stellar Metallicity

As shown in the previous sections, it is common to allow ζ, ζ_X, and ζ_α to vary independently as free parameters. However, if all features of the 21 cm background are driven by stars and their remnants, and the properties of such objects do not vary with time, then these efficiency factors will be highly correlated. For example, the number of Lyman-continuum photons produced per unit star formation is inversely proportional to stellar metallicity, Z, as is the yield in the Lyman–Werner band, so it may be more appropriate to use Z as the free parameter instead of N_{ion} and N_{LW}. It is more difficult to connect the X-ray luminosity per baryon, N_X, to Z as it depends on poorly understood details of the late stages of stellar evolution and compact binaries (Belczynski et al. 2002). However, observationally, the L_X–SFR relation (see Section 2.2.5) does appear to depend on gas-phase metallicity (Brorby et al. 2016), providing a simple empirical recipe for connecting ζ_X to Z (Mirocha et al. 2017).

Figure 2.10 shows these effects on the global 21 cm signal (Mirocha et al. 2017). In the left panel, no link between L_X/SFR and Z is assumed, while in the right panel the empirical relation with $f_X \propto Z^{-0.6}$ is adopted. In each case, though particularly in the left panel, the effect of metallicity is very small. This is because these models force a match to high-z UVLF measurements (Bouwens et al. 2015; Finkelstein et al. 2015), which means any change in Z_* also affects the 1600 Å luminosity to which UVLF measurements are sensitive. As a result, changes in metallicity make galaxies more or less bright in the UV, but to preserve UVLFs, the efficiency of star formation must compensate (see Sections 2.2.1 and 2.2.2). Once f_X depends on Z (right panel), the global 21 cm signal becomes more sensitive to changes in Z because the change in X-ray luminosity can overcome the decline in SFE as Z decreases (Mirocha et al. 2017).

In reality, the metallicity is a function of galaxy mass and time, so the simple constant Z_* models above are of course simplistic. Note also that the models in Figure 2.10 only include atomic cooling halos. As a result, observed signals peaking at lower frequencies (like the EDGES 78 MHz signal Bowman et al. 2018) likely require mini halos and/or non-standard source prescriptions (Mirocha et al. 2018; Mirocha & Furlanetto 2019).

2.3.5 Dependence on the Minimum Mass

The minimum halo mass (or equivalent virial temperature) for star formation sets the total number of halos emitting UV and X-ray photons as a function of redshift and thus influences all points in 21 cm background, unlike the ζ factors, which largely impact a single feature. Fiducial models often adopt the mass corresponding to a virial temperature of 10^4 K, as gas in halos of this mass will be able to cool atomically, i.e., there is not an obvious barrier to star formation in halos of this mass. Reducing $m_{\min}$, as is justified if star formation in mini halos is efficient, results in a larger halo population, while increasing $m_{\min}$ of course reduces the halo population. Moreover, for models in which low-mass halos are the dominant sources of emission, the typical star-forming halo is less biased than that drawn from a model in which high-mass halos dominate. As a result, changing $m_{\min}$ in principle affects both the timing of events in the 21 cm background as well as the amplitude of fluctuations.

As shown in Figure 2.11 (Mesinger et al. 2014), $m_{\min}$ indeed affects all features of the 21 cm background, both in the global signal and fluctuations (see also, e.g., Fialkov et al. 2017; Mirocha et al. 2015). With no other changes to the model, the effects are largely systematic, i.e., the timing of features in the global signal and power spectrum is shifted without a dramatic change in their amplitude. Notice also that changing $m_{\min}$ can serve to mimic the effects of including warm DM (e.g., red dotted versus magenta dashed–dotted curve), which suppresses the formation of small structures that would otherwise (presumably) host galaxies.

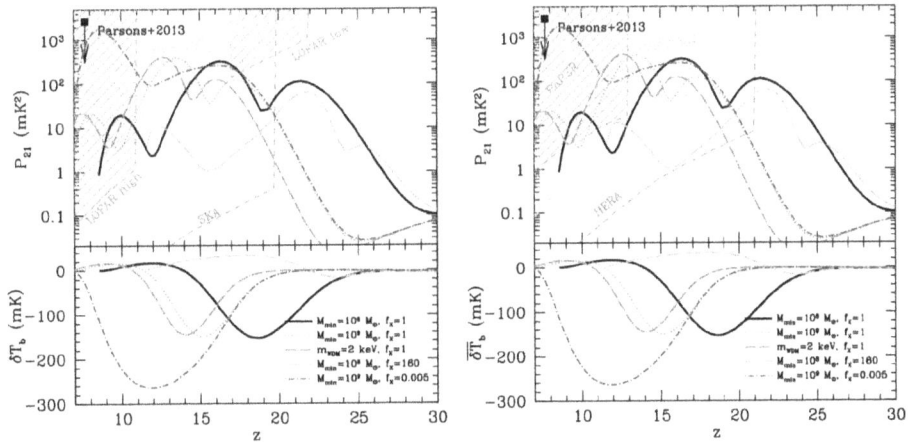

Figure 2.11. Effects of the minimum mass on the global 21 cm signal and 21 cm power spectrum on $k = 0.1$ Mpc^{-1} scales (reproduced from Mesinger et al. 2014, by permission of Oxford University press on behalf of the Royal Astronomical Society). Solid black, red dotted, and magenta dashed–dotted curves hold constant $f_X = 1$ and vary the $m_{\min}$ by hand (first two) and via a 2 keV warm dark matter particle (magenta). Left and right panels differ only in which sensitivity curves are included for comparison.

Not depicted in Figure 2.11 is the possibility that m_{min} evolves with time. Initially, only a mild redshift-dependence is expected just from linking m_{min} to a constant virial temperature of ~500 K (Tegmark et al. 1997), which is required for molecular cooling and thus star formation to initially begin (see Section 2.2.1.2). However, the ability of mini halos to form stars also depends on their ability to accrete and retain gas, which is influenced by the relative velocity between baryons and DM (Tseliakhovich & Hirata 2010; Fialkov et al. 2012). Soon after the first sources form, m_{min} will react to the LW background (Haiman et al. 1997; Machacek et al. 2001; Visbal et al. 2014) and likely rise to the atomic cooling threshold, $T_{min} \sim 10^4$ K, at $z \gtrsim 10$ (Trenti & Stiavelli 2009; Mebane et al. 2018). During reionization, this threshold may grow even higher, as ionization inhibits halos from accreting fresh gas from which to form stars (Gnedin 2000; Noh & Mcquinn 2014; Yue et al. 2016).

Figure 2.12 shows 21 cm power spectra for various models of feedback in the first star-forming halos (Fialkov et al. 2013). Both the strength of feedback and type of feedback (LW and/or baryon-velocity streaming in this particular example) change the power spectrum by a factor of ~2–3 while fundamentally altering its shape.

The signatures of mini halos and feedback in the global 21 cm signal are likely more subtle. Figure 2.13 shows predictions for the amplitude and shape of the global 21 cm absorption signal with (green) and without (gray) a model for Pop III star formation and LW feedback (Mirocha et al. 2018). While the effects of Pop III sources on the position of the absorption trough alone are difficult to distinguish from uncertainties in Pop II source models (quantified by gray contours), as shown in the left column, Pop III signals do affect the symmetry of the trough and the derivative of the signal (middle and right, respectively). As a result, any inferred skew in the global signal (to high frequencies) may be an indicator of efficient Pop III star formation in the early universe.

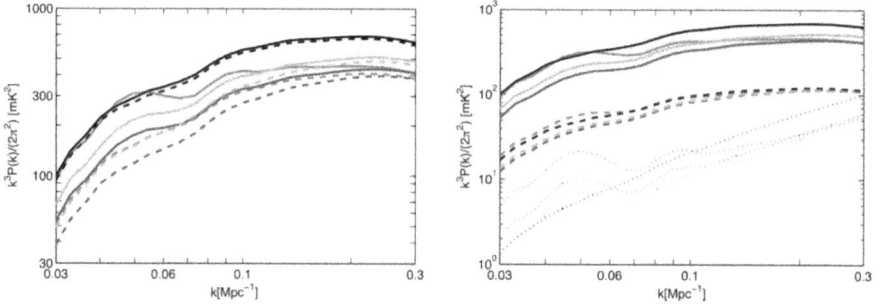

Figure 2.12. Effects of LW feedback on the 21 cm power spectrum (reproduced from Fialkov et al. 2013, by permission of Oxford University press on behalf of the Royal Astronomical Society). Left: power spectra with different feedback models, including no feedback (red), weak (blue), strong (green), and saturated (black). Dashed curves exclude the baryon–DM velocity offset effect (Tseliakhovich & Hirata 2010). Right: all models here include the baryon-velocity offset effect—line styles indicate power spectra at three different redshifts. Note that because changes to the strength of feedback shifts the timing of events, models are compared at fixed increments relative to the "heating redshift," z_0, which in these models occurs between $z \sim 15$ and $z \sim 18$ (Fialkov et al. 2013). Dashed lines are power spectra at $z = z_0$, while those at $z = z_0 + 3$ and $z = z_0 + 9$ are shown by solid and dotted lines, respectively.

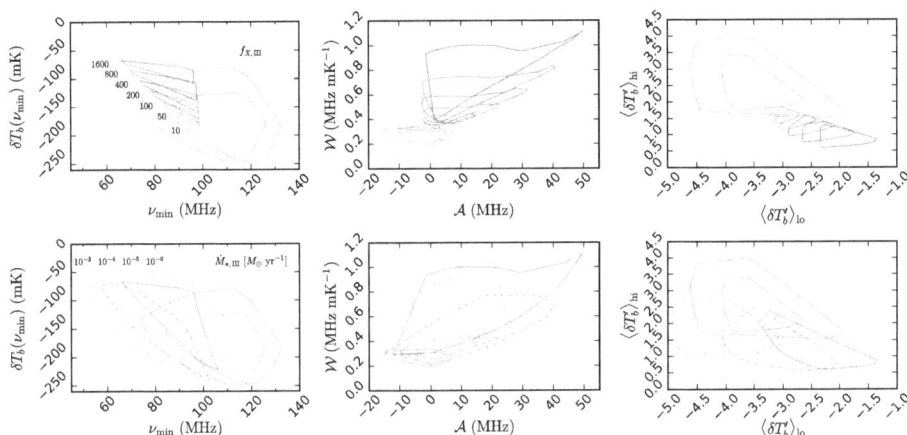

Figure 2.13. Potential Pop III signatures in the global 21 cm signal (reproduced from Mirocha et al. 2018, by permission of Oxford University press on behalf of the Royal Astronomical Society). Comparison between Pop II-only and Pop III models performed in three diagnostic spaces, including the absorption trough position (left), the prominence of its wings, $\mathcal{W}$, and its asymmetry, $\mathcal{A}$ (middle), and the mean slopes at frequencies above and below the extremum (right). Black contours enclose sets of Pop II models generated by Monte Carlo sampling a viable range of parameter space constrained by current observations, while the green polygons are slices through a 3D Pop III model grid, first assuming a fixed Pop III SFE and varying the X-ray production efficiency (top row), and then for different SFE models having "marginalized" over all $f_{X,III}$ (bottom row). Measurements falling in regions of overlap between the green and black contours would have no clear evidence of Pop III, while measurements falling only within the green contours would be suggestive of Pop III.

2.4 Summary

In this chapter, we have introduced the fundamentals of ionization and heating in the high-z IGM (Section 2.1), the sources most likely to provide the UV and X-ray photons that drive reionization and reheating (Section 2.2), and the signatures of these events in the global 21 cm signal and power spectrum (Section 2.3). However, this is by no means an all-inclusive account of work in the field. There are a variety of modeling codes (Furlanetto et al. 2004; Barkana & Loeb 2005; Pritchard & Furlanetto 2007; Thomas et al. 2009; Mesinger et al. 2011; Fialkov et al. 2014; Mirocha 2014; Santos et al. 2010; Raste & Sethi 2018), each designed with different trade-offs in mind and often with their own methods for parameterizing the effects of astrophysical sources. To date, no systematic effort has been undertaken to compare the results of these codes or to (attempt to) converge on a "concordance" parameterization of the Cosmic Dawn. As a result, we encourage readers to be aware of the assumptions underlying different models, how parameters are defined, and extent to which these choices impact the inferences drawn from current and future experiments.

References

Abel, T., Bryan, G. L., & Norman, M. L. 2000, ApJ, 540, 39
Abel, T., Norman, M. L., & Madau, P. 1999, ApJ, 523, 66
Ahn, K., Shapiro, P. R., Iliev, I. T., Mellema, G., & Pen, U.-L. 2009, ApJ, 695, 1430

Barkana, R., & Loeb, A. 2005, ApJ, 626, 1

Behroozi, P. S., Wechsler, R. H., & Conroy, C. 2013, ApL, 762, L31

Behroozi, P., Wechsler, R. H., Hearin, A. P., & Conroy, C. 2019, MNRAS, 488, 3143

Belczynski, K., Kalogera, V., & Bulik, T. 2002, ApJ, 572, 407

Bouché, N., Dekel, A., Genzel, R., et al. 2010, ApJ, 718, 1001

Bouwens, R. J., Aravena, M., Decarli, R., et al. 2016, ApJ, 833, 72

Bouwens, R. J., Illingworth, G. D., Oesch, P. A., et al. 2014, ApJ, 793, 115

Bouwens, R. J., Illingworth, G. D., Oesch, P. A., et al. 2015, ApJ, 803, 34

Bowman, J. D., Rogers, A. E. E., Monsalve, R. A., Mozdzen, T. J., & Mahesh, N. 2018, Natur, 555, 67

Bromm, V., Coppi, P. S., Coppi, P. S., Larson, R. B., & Larson, R. B. 1999, ApJ, 527, L5

Bromm, V., Kudritzki, R. P., Kudritzki, R. P., & Loeb, A. 2001, ApJ, 552, 464

Brorby, M., Kaaret, P., Prestwich, A., & Mirabel, I. F. 2016, MNRAS, 457, 4081

Bruzual, G., & Charlot, S. 2003, MNRAS, 344, 1000

Capak, P. L., Carilli, C., Jones, G., et al. 2015, Natur, 522, 455

Chabrier, G. 2003, PASP, 115, 763

Chen, X., & Miralda-Escudé, J. 2004, ApJ, 602, 1

Cheng, C., Parsons, A. R., Kolopanis, M., et al. 2018, ApJ, 868, 26

Cohen, A., Fialkov, A., & Barkana, R. 2018, MNRAS, 478, 2193

Cohen, A., Fialkov, A., Barkana, R., & Lotem, M. 2017, MNRAS, 472, 1915

Conroy, C., Gunn, J. E., & White, M. 2009, ApJ, 699, 486

Crosby, B. D., O'Shea, B. W., Smith, B. D., Turk, M. J., & Hahn, O. 2013, ApJ, 773, 108

Croton, D. J., Springel, V., White, S. D. M., et al. 2006, MNRAS, 365, 11

da Silva, R. L., Fumagalli, M., & Krumholz, M. 2012, ApJ, 745, 145

Das, A., Mesinger, A., Pallottini, A., Ferrara, A., & Wise, J. H. 2017, MNRAS, 469, 1166

Davé, R., Finlator, K., & Oppenheimer, B. D. 2012, MNRAS, 421, 98

Dayal, P., Ferrara, A., Dunlop, J. S., & Pacucci, F. 2014, MNRAS, 445, 2545

Dekel, A., & Mandelker, N. 2014, MNRAS, 444, 2071

Eldridge, J. J., & Stanway, E. R. 2009, MNRAS, 400, 1019

Ewall-Wice, A., Chang, T. C., Lazio, J., et al. 2018, ApJ, 868, 63

Fabbiano, G. 2006, ARA&A, 44, 323

Faucher-Giguère, C.-A., Kereš, D., & Ma, C.-P. 2011, MNRAS, 417, 2982

Feng, C., & Holder, G. 2018, ApJ, 858, L17

Fialkov, A., Barkana, R., Pinhas, A., & Visbal, E. 2014, MNRAS, 437, L36

Fialkov, A., Barkana, R., Tseliakhovich, D., & Hirata, C. M. 2012, MNRAS, 424, 1335

Fialkov, A., Barkana, R., & Visbal, E. 2014, Natur, 506, 197

Fialkov, A., Barkana, R., Visbal, E., Tseliakhovich, D., & Hirata, C. M. 2013, MNRAS, 432, 2909

Fialkov, A., Cohen, A., Barkana, R., & Silk, J. 2017, MNRAS, 464, 3498

Finkelstein, S. L., D'Aloisio, A., Paardekooper, J.-P., et al. 2019, ApJ, 879, 36

Finkelstein, S. L., Papovich, C., Salmon, B., et al. 2012, ApJ, 756, 164

Finkelstein, S. L., Ryan, R. E. Jr., Papovich, C., et al. 2015, ApJ, 810, 71

Fixsen, D. J., Kogut, A., Levin, S., et al. 2011, ApJ, 734, 5

Fragos, T., Lehmer, B. D., Naoz, S., Zezas, A., & Basu-Zych, A. 2013, ApJ, 776, L31

Fukugita, M., & Kawasaki, M. 1994, MNRAS, 269, 563

Furlanetto, S. R. 2006, MNRAS, 371, 867

Furlanetto, S. R., Mirocha, J., Mebane, R. H., & Sun, G. 2017, MNRAS, 472, 1576
Furlanetto, S. R., & Pritchard, J. R. 2006, MNRAS, 372, 1093
Furlanetto, S. R., & Johnson Stoever, S. 2010, MNRAS, 404, 1869
Furlanetto, S., Zaldarriaga, M., & Hernquist, L. 2004, ApJ, 613, 1
Gilfanov, M., Grimm, H. J., & Sunyaev, R. 2004, MNRAS, 347, L57
Gnedin, N. Y. 2000, ApJ, 542, 535
Gnedin, N. Y. 2014, ApJ, 793, 29
Greig, B., & Mesinger, A. 2017, MNRAS, 472, 2651
Greig, B., Mesinger, A., & Koopmans, L. V. E. 2020, MNRAS, 491, 1398
Haardt, F., & Madau, P. 1996, ApJ, 461, 20
Haiman, Z., Rees, M. J., & Loeb, A. 1997, ApJ, 476, 458
Hassan, S., Davé, R., Mitra, S., et al. 2018, MNRAS, 473, 227
Hirata, C. M. 2006, MNRAS, 367, 259
Holzbauer, L. N., & Furlanetto, S. R. 2012, MNRAS, 419, 718
Hopkins, P. F., Kereš, D., Oñorbe, J., et al. 2014, MNRAS, 445, 581
Hui, L., & Gnedin, N. Y. 1997, MNRAS, 292, 27
Jaacks, J., Thompson, R., Finkelstein, S. L., & Bromm, V. 2018, MNRAS, 475, 4396
Kimm, T., & Cen, R. 2014, ApJ, 788, 121
Knevitt, G., Wynn, G. A., Power, C., & Bolton, J. S. 2014, MNRAS, 445, 2034
Kolopanis, M., Jacobs, D. C., Cheng, C., et al. 2019, ApJ, 883, 133
Kroupa, P. 2001, MNRAS, 322, 231
Leite, N., Evoli, C., D'Angelo, M., et al. 2017, MNRAS, 469, 416
Leitherer, C., Schaerer, D., Goldader, J. D., et al. 1999, ApJS, 123, 3
Lidz, A., Zahn, O., McQuinn, M., Zaldarriaga, M., & Hernquist, L. 2008, ApJ, 680, 962
Ma, X., Kasen, D., Hopkins, P. F., et al. 2015, MNRAS, 453, 960
Machacek, M. E., Bryan, G. L., & Abel, T. 2001, ApJ, 548, 509
Madau, P., & Fragos, T. 2017, ApJ, 840, 39
Madau, P., & Haardt, F. 2015, ApJ, 813, L8
Maio, U., Ciardi, B., Dolag, K., Tornatore, L., & Khochfar, S. 2010, MNRAS, 407, 1003
Mashian, N., Oesch, P. A., & Loeb, A. 2016, MNRAS, 455, 2101
Mason, C. A., Trenti, M., & Treu, T. 2015, ApJ, 813, 21
Mason, C. A., Treu, T., Dijkstra, M., et al. 2018, ApJ, 856, 2
McBride, J., Fakhouri, O., & Ma, C.-P. 2009, MNRAS, 398, 1858
McQuinn, M. 2012, MNRAS, 426, 1349
McQuinn, M., Lidz, A., Zahn, O., et al. 2007, MNRAS, 377, 1043
Mebane, R. H., Mirocha, J., & Furlanetto, S. R. 2018, MNRAS, 479, 4544
Mesinger, A., Ewall-Wice, A., & Hewitt, J. 2014, MNRAS, 439, 3262
Mesinger, A., Furlanetto, S., & Cen, R. 2011, MNRAS, 411, 955
Mesinger, A., Greig, B., & Sobacchi, E. 2016, MNRAS, 459, 2342
Meurer, G. R., Heckman, T. M., & Calzetti, D. 1999, ApJ, 521, 64
Mineo, S., Gilfanov, M., & Sunyaev, R. 2012a, MNRAS, 419, 2095
Mineo, S., Gilfanov, M., & Sunyaev, R. 2012b, MNRAS, 426, 1870
Mirabel, I. F., Dijkstra, M., Laurent, P., Loeb, A., & Pritchard, J. R. 2011, A&A, 528, A149
Mirocha, J. 2014, MNRAS, 443, 1211
Mirocha, J., & Furlanetto, S. R. 2019, MNRAS, 483, 1980
Mirocha, J., Furlanetto, S. R., & Sun, G. 2017, MNRAS, 464, 1365

Mirocha, J., Harker, G. J. A., & Burns, J. O. 2013, ApJ, 777, 118

Mirocha, J., Harker, G. J. A., & Burns, J. O. 2015, ApJ, 813, 11

Mirocha, J., Mebane, R. H., Furlanetto, S. R., Singal, K., & Trinh, D. 2018, MNRAS, 478, 5591

Mirocha, J., Skory, S., Burns, J. O., & Wise, J. H. 2012, ApJ, 756, 94

Mitsuda, K., Inoue, H., Koyama, K., et al. 1984, PASJ, 36, 741

Moster, B. P., Somerville, R. S., Maulbetsch, C., et al. 2010, ApJ, 710, 903

Mutch, S. J., Geil, P. M., Poole, G. B., et al. 2016, MNRAS, 462, 250

Narayanan, D., Conroy, C., Davé, R., Johnson, B. D., & Popping, G. 2018, ApJ, 869, 70

Nath, B. B., & Biermann, P. L. 1993, MNRAS, 265, 241

Noh, Y., & Mcquinn, M. 2014, MNRAS, 444, 503

Ocvirk, P., Gillet, N., Shapiro, P. R., et al. 2016, MNRAS, 463, 1462

Oh, S. P. 2001, ApJ, 553, 499

O'Shea, B. W., & Norman, M. L. 2007, ApJ, 654, 66

O'Shea, B. W., Wise, J. H., Xu, H., & Norman, M. L. 2015, ApJ, 807, L12

Pacucci, F., Mesinger, A., Mineo, S., & Ferrara, A. 2014, MNRAS, 443, 678

Park, J., Mesinger, A., Greig, B., & Gillet, N. 2019, MNRAS, 484, 933

Parsons, A. R., Liu, A., Aguirre, J. E., et al. 2014, ApJ, 788, 106

Pober, J. C., Ali, Z. S., Parsons, A. R., et al. 2015, ApJ, 809, 62

Press, W. H., & Schechter, P. 1974, ApJ, 187, 425

Pritchard, J. R., & Furlanetto, S. R. 2006, MNRAS, 367, 1057

Pritchard, J. R., & Furlanetto, S. R. 2007, MNRAS, 376, 1680

Pritchard, J. R., & Loeb, A. 2010, PhRvD, 82, 023006

Raste, J., & Sethi, S. 2018, ApJ, 860, 55

Remillard, R. A., & McClintock, J. E. 2006, ARA&A, 44, 49

Ripamonti, E., Mapelli, M., & Zaroubi, S. 2008, MNRAS, 387, 158

Robertson, B. E., Ellis, R. S., Furlanetto, S. R., & Dunlop, J. S. 2015, ApJ, 802, L19

Salim, S., & Boquien, M. 2019, ApJ, 872, 23

Salpeter, E. E. 1955, ApJ, 121, 161

Santos, M. G., Ferramacho, L., Silva, M. B., Amblard, A., & Cooray, A. 2010, MNRAS, 406, 2421

Sarmento, R., Scannapieco, E., & Cohen, S. 2018, ApJ, 854, 75

Sazonov, S., & Sunyaev, R. 2015, MNRAS, 454, 3464

Scalo, J. 1998, in ASP Conf. Ser. 142, The Stellar Initial Mass Function, ed. G. Gilmore, & D. Howell (San Francisco, CA: ASP), 201

Schaerer, D. 2002, A&A, 382, 28

Schauer, A. T. P., Whalen, D. J., Glover, S. C. O., & Klessen, R. S. 2015, MNRAS, 454, 2441

Schaye, J., Crain, R. A., Bower, R. G., et al. 2015, MNRAS, 446, 521

Shapley, A. E., Steidel, C. C., Pettini, M., Adelberger, K. L., & Erb, D. K. 2006, ApJ, 651, 688

Sheth, R. K., Mo, H. J., & Tormen, G. 2001, MNRAS, 323, 1

Shull, J. M., & van Steenberg, M. E. 1985, ApJ, 298, 268

Singal, J., Haider, J., Ajello, M., et al. 2018, PASP, 130, 036001

Smith, B. D., Wise, J. H., O'Shea, B. W., Norman, M. L., & Khochfar, S. 2015, MNRAS, 452, 2822

Somerville, R. S., Hopkins, P. F., Cox, T. J., Robertson, B. E., & Hernquist, L. 2008, MNRAS, 391, 481

Stanway, E. R., Eldridge, J. J., & Becker, G. D. 2016, MNRAS, 456, 485

Sun, G., & Furlanetto, S. R. 2016, MNRAS, 460, 417

Tacchella, S., Bose, S., Conroy, C., Eisenstein, D. J., & Johnson, B. D. 2018, ApJ, 868, 92

Tanaka, T. L., O'Leary, R. M., & Perna, R. 2016, MNRAS, 455, 2619

Tegmark, M., Silk, J., Rees, M. J., et al. 1997, ApJ, 474, 1

Thomas, R. M., & Zaroubi, S. 2008, MNRAS, 384, 1080

Thomas, R. M., Zaroubi, S., Ciardi, B., et al. 2009, MMNRAS, 393, 32

Tinker, J. L., Robertson, B. E., Kravtsov, A. V., et al. 2010, ApJ, 724, 878

Trac, H., Cen, R., & Mansfield, P. 2015, ApJ, 813, 54

Trenti, M., & Stiavelli, M. 2009, ApJ, 694, 879

Tseliakhovich, D., & Hirata, C. 2010, PhRvD, 82, 083520

Tumlinson, J., & Shull, J. M. 2000, ApJ, 528, L65

Venumadhav, T., Dai, L., Kaurov, A., & Zaldarriaga, M. 2018, PhRvD, 98, 103513

Verner, D. A., Verner, D. A., Ferland, G. J., et al. 1996, ApJ, 465, 487

Visbal, E., Haiman, Z., & Bryan, G. L. 2018, MNRAS, 475, 5246

Visbal, E., Haiman, Z., Terrazas, B., Bryan, G. L., & Barkana, R. 2014, MNRAS, 445, 107

Vogelsberger, M., Genel, S., Springel, V., et al. 2014, MNRAS, 444, 1518

Weingartner, J. C., & Draine, B. T. 2001, ApJ, 548, 296

Wolcott-Green, J., & Haiman, Z. 2011, MNRAS, 418, 838

Wolcott-Green, J., & Haiman, Z. 2012, MNRAS, 425, L51

Xu, H., Wise, J. H., Norman, M. L., Ahn, K., & O'Shea, B. W. 2016, ApJ, 833, 84

Yoshida, N., Abel, T., Hernquist, L., & Sugiyama, N. 2003, ApJ, 592, 645

Yue, B., Ferrara, A., & Xu, Y. 2016, MNRAS, 463, 1968

Zaroubi, S., Thomas, R. M., Sugiyama, N., & Silk, J. 2007, MNRAS, 375, 1269

Zygelman, B. 2005, ApJ, 622, 1356

The Cosmic 21-cm Revolution
Charting the first billion years of our universe
Andrei Mesinger

Chapter 3

Physical Cosmology from the 21 cm Line

Steven R Furlanetto

We describe how the high-z 21 cm background can be used to improve both our understanding of the fundamental cosmological parameters of our universe and the exotic processes originating in the dark sector. The 21 cm background emerging during the cosmological Dark Ages, the era between hydrogen recombination and the formation of the first luminous sources (likely at $z \sim 30$), is difficult to measure but provides several powerful advantages for these purposes: in addition to the lack of astrophysical contamination, it will allow probes of very small-scale structure over a very large volume. Additionally, the 21 cm background is sensitive to the thermal state of intergalactic hydrogen and therefore probes any exotic processes (including, e.g., dark matter scattering or decay and primordial black holes) during that era. After astrophysical sources have formed, cosmological information can be separated from astrophysical effects on the 21 cm background through methods such as redshift space distortions, joint modeling, and by searching for indirect effects on the astrophysical sources themselves.

3.1 Introduction

The previous chapter has shown how the first galaxies and black holes have enormous implications for the 21 cm background. However, all of these astrophysical processes occur within the framework of cosmological structure formation—a process we would like to probe to understand the fundamental properties of our universe. This chapter will examine ways in which the 21 cm background can be used to probe the cosmology. Just as fluctuations in the cosmic microwave background (CMB) and galaxy distribution can be used as probes of cosmology, so can the H I distribution at $z > 10$. We shall see that the 21 cm background offers an unparalleled probe of the matter distribution in our universe (Tegmark & Zaldarriaga 2009): interferometric measurements can, at least in principle, map the distribution of gas over a wide range in both redshift and physical scale. Moreover, we have seen that the amplitude of the 21 cm signal depends sensitively

doi:10.1088/2514-3433/ab4a73ch3

on the thermal state of the IGM. Although the combination of adiabatic expansion and X-ray heating determines that state in the standard scenario, any "exotic" process—like dark matter decay, scattering between dark matter and baryons, X-rays from primordial black holes, etc.—that changes this energy balance will leave a signature in the 21 cm background. The low temperature of the hydrogen gas before any astrophysical X-ray background forms means that the 21 cm line is an exceptionally sensitive calorimeter for these processes.

In this chapter, we will review some of these potential cosmological probes. We begin in Section 3.2 with a discussion of cosmology in the "Dark Ages" before structure forms—an era that should be uncontaminated by astrophysics, although it is also extraordinarily hard to observe. Then, in Section 3.3, we consider how cosmological information can be extracted from the signal even in the presence of astrophysical processes. Finally, in Section 3.4, we briefly point out that 21 cm measurements can offer strong synergies with other cosmological probes.

3.2 Cosmology in the Dark Ages

3.2.1 Setting the Stage: The Standard Cosmological Paradigm

Before astrophysical sources turn on, the 21 cm background depends on the thermal evolution of the intergalactic medium (IGM) and the earliest stages of structure formation in the universe. We will therefore first describe these processes in the context of the standard cosmological paradigm.

3.2.1.1 Thermal Evolution

Let us begin with the thermal evolution. If it were thermally isolated, the IGM gas would simply cool adiabatically as the universe expands. For an ideal gas, this cooling rate can be written as $(\gamma - 1)(\dot{\rho}_b/\rho_b)\bar{T}_e$, where ρ_b is the baryon density, $\bar{T}_e$ is the electron temperature (equal to the hydrogen temperature in this regime), and $\gamma = 5/3$ is the adiabatic index of a monatomic gas. For gas at the mean density, the factor $(\dot{\rho}_b/\rho_b) = -3H$, due to the Hubble expansion.

However, the gas is not actually thermally isolated: free electrons may exchange energy with CMB photons through Compton scattering. Although cosmological recombination at $z \sim 1100$ results in a nearly neutral universe, a small fraction $\bar{x}_e \sim 10^{-4}$ of electrons are "frozen out" following the recombination process. These free electrons scatter off CMB photons and, for a long period, maintain thermal equilibrium with that radiation field. The timescale for Compton cooling is

$$t_C \equiv \left(\frac{8\sigma_T a_{rad} T_\gamma^4}{3m_e c}\right)^{-1} = 1.2 \times 10^8 \left(\frac{1+z}{10}\right)^{-4} \text{yr}, \qquad (3.1)$$

where $T_\gamma \propto (1 + z)$ is the background radiation temperature (in this case, the CMB), σ_T is the Thomson cross section, a_{rad} is the radiation constant, and m_e is the electron mass.

Including both adiabatic cooling and Compton heating, the temperature evolution of gas at the mean cosmic density is therefore described by

$$\frac{d\bar{T}_e}{dt} = \frac{\bar{x}_e}{(1 + \bar{x}_e)}\left[\frac{T_\gamma - \bar{T}_e}{t_C(z)}\right] - 2H\bar{T}_e. \tag{3.2}$$

The first term describes Compton heating. For an electron–proton gas, $\bar{x}_e = n_e/(n_e + n_H)$, where n_e and n_H are the electron and hydrogen densities; the relation is more complicated when helium is included. This prefactor appears because the electrons must share the energy they gain from Compton scattering with the other particles. The last term on the right-hand side of Equation (3.2) yields the adiabatic scaling $\bar{T}_e \propto (1 + z)^2$ in the absence of Compton scattering.

The temperature evolution therefore depends on the residual fraction of free electrons after cosmological recombination (Seager et al. 1999; Ali-Haïmoud & Hirata 2011; Chluba & Thomas 2011). It is easy to estimate this residual fraction, but in detail depends on the complex physics of hydrogen recombination. In a simple picture, the hydrogen recombination rate is

$$\frac{d\bar{x}_e}{dt} = -\alpha_B(T_e)\bar{x}_e^2\bar{n}_H, \tag{3.3}$$

where $\alpha_B \propto T_e^{-0.7}$ is the case-B recombination coefficient. (The case-B coefficient ignores recombinations to the ground state, which generate a new ionizing photon and so do not change the net ionized fraction.) In the standard cosmology, the fractional change in $\bar{x}_e$ per Hubble time is therefore

$$\frac{\dot{n}_e}{H^{-1}n_e} \approx 7x(1 + z)^{0.8}, \tag{3.4}$$

where n_e is the electron fraction. Electrons "freeze out" and cease to recombine effectively when this factor becomes of order unity; after that point, the Hubble expansion time is shorter than the recombination time. More precise numerical calculations account for the large photon density during cosmological recombination, line emission, and recombinations to higher energy levels, among other factors (Seager et al. 1999; Ali-Haïmoud & Hirata 2011; Chluba & Thomas 2011). Figure 3.1 shows the result of one such calculation, which yields $x \approx 3 \times 10^{-4}$ by $z \approx 200$.

Inserting this electron density into Equation (3.2), we find that the small fraction of residual electrons maintains thermal equilibrium between the gas and CMB down to $z \approx 200$, when Compton heating finally becomes inefficient. Figure 3.1 shows a more exact calculation: note how the gas and CMB temperatures begin to depart at $z \sim 200$, after which the gas follows the expected adiabatic cooling track.

3.2.1.2 Density Fluctuations

Measurements of fluctuations in the 21 cm background depend on variations in the density, spin temperature, and ionization fraction. We therefore briefly consider here how fluctuations in those quantities can emerge during the Dark Ages. A complete treatment of the power spectrum and structure formation is well outside the scope of this work, but we will summarize the key points for understanding the 21 cm signal.

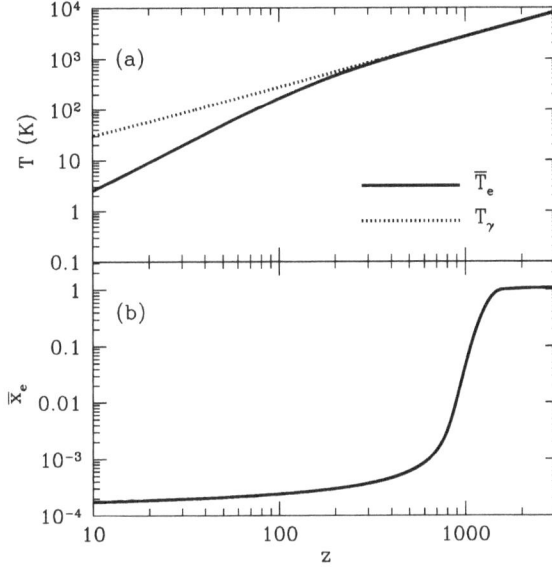

Figure 3.1. Thermal and ionization history during the Dark Ages (panels a and b, respectively). The free-electron fraction decreases rapidly after recombination at $z \sim 1100$ and then "freezes out" at later times. Meanwhile, Compton scattering keeps $\bar{T}_e \approx T_\gamma$ until $z \sim 200$, after which the declining CMB energy density and small residual ionized fraction are no longer sufficient to maintain thermal equilibrium between the gas and CMB. Past that point, $\bar{T}_e \propto (1+z)^2$ as appropriate for an adiabatically expanding nonrelativistic gas. These results were produced with the publicly available code RECFAST (http://www.astro.ubc.ca/people/scott/recfast.html).

Density fluctuations grow through gravity and, in the linear regime, their statistical properties can be calculated precisely. Combining the mass and momentum conservation equations for a perfect fluid with the Poisson equation for the gravitational potential yields an evolution equation for the Fourier transform $\delta_\mathbf{k}$ of the fractional density perturbation δ,

$$\frac{\partial^2 \delta_\mathbf{k}}{\partial t^2} + 2H \frac{\partial \delta_\mathbf{k}}{\partial t} = 4\pi G \bar{\rho} \delta - \frac{c_s^2 k^2}{a^2} \delta_\mathbf{k}, \qquad (3.5)$$

where the last term is the pressure force (which vanishes for cold dark matter) and c_s^2 is the sound speed. This linear equation has two independent solutions, one of which grows in time and eventually comes to dominate the density evolution.[1]

While the fluctuations are small—and thus very nearly Gaussian—the density field can be accurately characterized by the power spectrum,

$$P(\mathbf{k}) = \delta_\mathbf{k} \delta_{\mathbf{k}'}^* = (2\pi)^3 \delta^D(\mathbf{k} - \mathbf{k}') P(\mathbf{k}). \qquad (3.6)$$

---

[1] Note that this solution uses Eulerian perturbation theory, which breaks down when δ is still relatively small. It suffices during the Dark Ages, but greater accuracy is necessary during later eras.

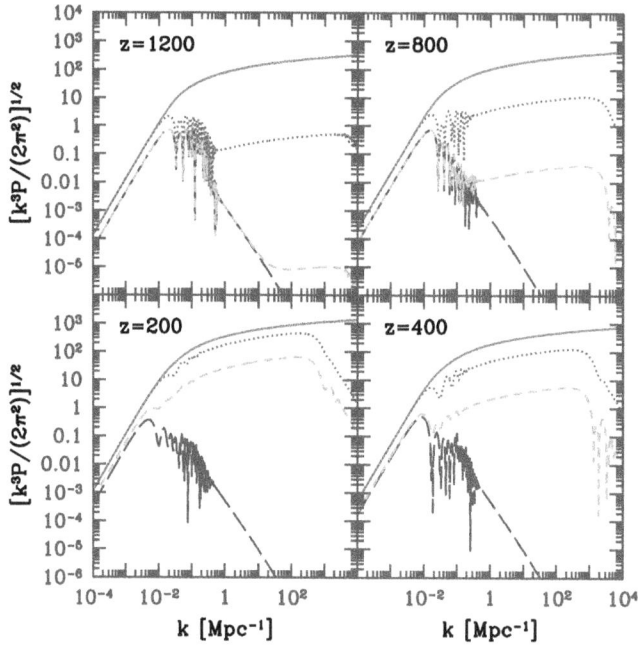

Figure 3.2. Power spectra for density and temperature fluctuations as a function of comoving wavenumber at four different redshifts. The curves show dimensionless power spectra for the CDM density (solid), baryon density (dotted), baryon temperature (short-dashed), and photon temperature (long-dashed). These curves do not include the relative streaming of the baryons and cold dark matter. Reproduced from Naoz & Barkana (2005), by permission of Oxford University press on behalf of the Royal Astronomical Society.

The power spectrum is the expectation value of the Fourier amplitude at a given wavenumber; for a homogeneous, isotropic universe it depends only on the magnitude of the wavenumber.[2] The power spectrum therefore represents the variance in the density field as a function of smoothing scale.

The solid curve in Figure 3.2 shows the cold dark matter power spectrum at several redshifts during the Dark Ages (taken from Naoz & Barkana 2005). The most obvious feature is the flattening at $k \sim 0.1$ Mpc^{-1}, which results from stagnation in the growth of small-scale structure during the radiation era. The power spectrum is otherwise quite simple. However, the 21 cm line will probe the fluctuations in the IGM gas, which we will require more physics to understand.

For example, Equation (3.2) describes the evolution of the mean IGM temperature. However, that field actually fluctuates as well, for two reasons (Naoz & Barkana 2005). First, the CMB temperature itself has fluctuations, so each electron will scatter off a different local T_γ: the power spectrum of the CMB temperature fluctuations is shown by the long-dashed curve in Figure 3.2. Additionally, the adiabatic expansion term depends on the local density, because gravity slows the

---

[2] Henceforth we will suppress the **k** subscript for notational simplicity.

expansion of overdense regions and hence decreases the cooling rate (and of course in underdense regions, the cooling accelerates). Thus, the IGM will be seeded by small temperature fluctuations reflecting its density structure.

To describe these fluctuations, we write δ_T as the fractional gas temperature fluctuation and δ_γ as the photon density fluctuation (so that $\delta_\gamma = 4\delta_{T_\gamma}$, the fractional CMB temperature fluctuation). Then, the perturbed version of Equation (3.2) is

$$\frac{d\delta_T}{dt} = \frac{2}{3}\frac{d\delta_b}{dt} + \frac{x_e(t)}{t_C(z)}\left[\delta_\gamma\left(\frac{\bar{T}_\gamma}{\bar{T}_e} - 1\right) + \frac{\bar{T}_\gamma}{\bar{T}_e}(\delta_{T_\gamma} - \delta_T)\right], \tag{3.7}$$

where the first term describes adiabatic cooling due to expansion (allowing for variations in the expansion rate) and the second accounts for variations in the rate of energy exchange through Compton scattering (which can result from variations in either the gas or photon temperatures).

Meanwhile, the fluctuations in the baryon temperature affect the baryon density evolution as well. Equation (3.5) implicitly assumed that temperature fluctuations were driven (only) by density fluctuations; allowing a more general relation, we obtain (Naoz & Barkana 2005)

$$\frac{\partial^2\delta}{\partial t^2} + 2H\frac{\partial\delta}{\partial t} = \frac{3}{2}H^2(\Omega_c\delta_c + \Omega_b\delta_b) - \frac{k^2}{a^2}\frac{k_B\bar{T}_e}{\mu m_H}(\delta_b + \delta_T). \tag{3.8}$$

This, together with Equations (3.7), (3.2), and a more precise version of (3.3) for the temperature and ionized fraction evolution, provides a complete set of equations to trace the density and temperature evolution (modulo one more effect that we will discuss next).

Figure 3.2 shows the resulting power spectra for the dark matter density, baryon density, baryon temperature, and photon temperature perturbations at four different redshifts. The photon fluctuations are not directly observable, but the others can in principle be probed through the 21 cm line. The photon perturbations are strongly suppressed on scales below the sound horizon, thanks to their large pressure. Near recombination, the baryonic perturbations are also suppressed on these scales, especially in the temperature, because they interact so strongly with the CMB. However, after recombination, the baryons fall into the dark matter potential wells, with their perturbations rapidly growing, and temperature fluctuations also grow, thanks largely to the variations in the adiabatic cooling rate. The turnover at very small scales in the baryonic power spectrum is due to the finite pressure of the gas.

3.2.1.3 Relative Streaming of Baryons and Cold Dark Matter

There is one additional effect on the baryonic power spectrum that may provide insight into cosmology during this era: "streaming" of baryonic matter relative to dark matter (Tseliakhovich & Hirata 2010). As a relativistic fluid, CMB photons have a very high pressure that drives acoustic waves throughout that component. While these photons are coupled to the baryons through Compton scattering, they can drag the baryonic component along with them—it is these acoustic waves in the photon–baryon fluid that we see in CMB fluctuations. Once recombination occurs, the radiation drag force decreases, and the baryons begin to fall into the potential wells of dark matter fluctuations (which have not participated in the acoustic waves). This transition can be seen in the dotted and short-dashed curves in Figure 3.2. Because the radiation sound speed is $\sim c/\sqrt{3}$, it is near the causal horizon at the time of recombination, corresponding to ~150 comoving Mpc today, where they can be observed as "baryon acoustic oscillations" in the matter power spectrum (and, because their physical scale is well known, used as a standard ruler to measure cosmological parameters).

There is an additional effect relevant to high redshifts, however. Even after the radiation driving becomes ineffective, the baryons are left with a relic velocity imprinted by the pressure force with a root mean square (rms) speed of $v_{bc} \approx 10^{-4}c = 30$ km s^{-1}, which decays with redshift as $1/a$ (Tseliakhovich & Hirata 2010). This streaming velocity between the baryonic and dark matter components is coherent over very large scales—comparable to the acoustic scale —and fades only slowly over time. Figure 3.3 shows the variance of the relative velocity perturbations as a function of the mode wavenumber k at $z = 15$. The power is substantial on scales up to the sound horizon at decoupling (~140 comoving Mpc), but it declines rapidly at $k > 0.5$ Mpc^{-1}, indicating that the streaming velocity is coherent over a scale of several comoving Mpc. Therefore, in the rest frame of small-scale fluctuations (such as those that will eventually collapse into galaxies), the baryons appeared to be moving coherently as a "wind," which can suppress the formation of the first galaxies (Dalal et al. 2010; Fialkov et al. 2014) and can result in a standard ruler during the Cosmic Dawn (Muñoz 2019b, 2019a).

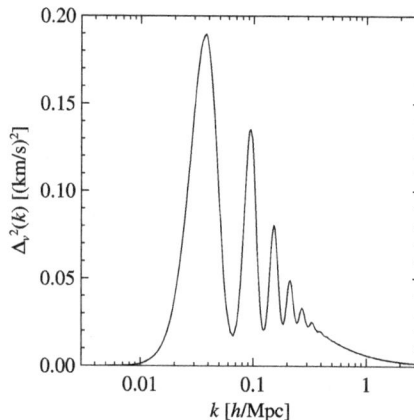

Figure 3.3. The power spectrum of the velocity difference perturbations (between baryons and dark matter) as a function of comoving wavenumber k at $z = 15$. Reproduced from Dalal et al. (2010). © 2010 IOP Publishing Ltd and SISSA Medialab srl. All rights reserved.

3.2.2 The Global 21 cm Signal during the Dark Ages

With the thermal history from Figure 3.1, we can easily calculate the global 21 cm signal throughout the Dark Ages—without any way of generating a Lyα background, the competition between the CMB and particle collisions sets the spin temperature. The left panel of Figure 3.4 shows the result: once the gas temperature begins to fall below the CMB temperature at $z \sim 150$, the 21 cm signal can be seen in absorption. At these early times, the gas is relatively dense and has a high-enough temperature for collisional coupling to be substantial, so $T_S \rightarrow T_K$. However, as the gas expands and cools, collisional coupling becomes inefficient, and eventually the spin temperature begins to return to T_γ. By the time we expect star formation to begin in earnest ($z < 30$), $T_S \approx T_\gamma$, so the 21 cm line is nearly invisible.

The global 21 cm signal is therefore sensitive to the IGM thermal history, so any process that affects the temperature evolution can also be probed by the 21 cm line. This method provides a particularly powerful probe of nonstandard physics because the low gas temperature during this period makes the 21 cm line a sensitive calorimeter of additional heating (or cooling) and/or of an excess radio background over and above the CMB (Feng & Holder 2018).

While many such processes tend to heat the IGM and therefore decrease the amplitude of the 21 cm signal (e.g., Chen & Kamionkowski 2004; Furlanetto et al. 2006; Shchekinov & Vasiliev 2007; Chuzhoy 2008), the recent claim of a detection of a 21 cm absorption feature at 78 MHz ($z \sim 17$) by the EDGES collaboration (Bowman et al. 2018) has triggered interest in nonstandard models that amplify the 21 cm signal (though note that the EDGES signal has not yet been independently confirmed, and with such a challenging analysis, systematics and foreground

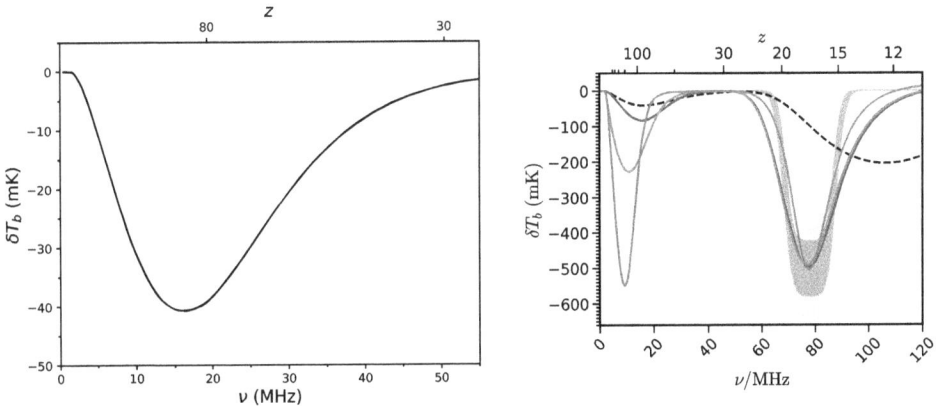

Figure 3.4. Left: the Dark Ages 21 cm absorption trough in the standard cosmology. The shape at $z > 30$ is independent of astrophysical sources. Right: the black dashed line shows the mean 21 cm brightness temperature (averaged across the sky) in the standard cosmology. The gray contours show schematically the reported EDGES absorption signal (Bowman et al. 2018). The solid curves are phenomenological models that invoke extra cooling to match the amplitude of the EDGES signal but that also dramatically affect the Dark Ages absorption trough at $z > 50$. Courtesy of J. Mirocha, based on calculations in Mirocha & Furlanetto (2019).

contamination remain a concern; Hills et al. 2018; Draine & Miralda-Escudé 2018; Spinelli et al. 2019; Bradley et al. 2019). The EDGES feature is more than twice as deep as expected if gas follows standard thermal history, even if astrophysical sources were able to drive $T_S \rightarrow T_K$ and no X-ray heating had yet occurred. It therefore requires new physics that either cools the IGM or increases the radio background against which the gas absorbs. A multitude of processes have been proposed to explain the anomalous detection, including dark matter–baryon scattering (Muñoz & Kovetz 2015; Barkana 2018; Slatyer & Wu 2018; Hirano & Bromm 2018), millicharged dark matter (Muñoz & Loeb 2018; Berlin et al. 2018; Kovetz et al. 2018), axions (Moroi et al. 2018), neutrino decay (Chianese et al. 2019), charge sequestration (Falkowski & Petraki 2018), quark nuggets (Lawson & Zhitnitsky 2019), dark photons (Jia & Liao 2019), and interacting dark energy (Costa et al. 2018), and it has been used to constrain additional exotic processes like dark matter annihilation (Cheung et al. 2019).

Although the EDGES signal at $z \sim 17$ is very likely past the end of the Dark Ages —after the first astrophysical sources formed—the same new physics would have affected the Dark Ages signal. The solid curves in Figure 3.4 illustrate how models that invoke excess cooling to explain the EDGES result (shown schematically by the gray contours) could also greatly amplify the Dark Ages signal. (Here the curves use a phenomenological parameterized cooling model as in Mirocha & Furlanetto 2019; physically motivated models will differ in the details.) (Note that the signal still vanishes at $z \sim 30$ because collisional coupling is still inefficient.) In these cases, even though the exotic physics have implications at relatively low redshifts, Dark Ages observations serve to break degeneracies between astrophysics and that new physics.

3.2.3 The Power Spectrum during the Dark Ages

As we have discussed, maps of 21 cm emission provide a sensitive probe of the power spectrum of density and temperature fluctuations (Kleban et al. 2007; Mao et al. 2008). Figure 3.5, from Pritchard & Loeb (2012), shows examples of how modes of the 21 cm fluctuation power spectrum evolve in the standard cosmology. Although the strongest fluctuations occur at $z \sim 10$, when astrophysical sources dominate, the Dark Ages signal are not too far behind: this is because, although the fractional density fluctuations are small at those times, the mean temperature can be relatively large during the absorption era—even in the standard cosmology.

The 21 cm power spectrum during the Dark Ages offers several advantages over other probes of the density field. Because they use the cosmological redshift to establish the distance to each observed patch, 21 cm measurements probe three-dimensional volumes—unlike the CMB, which probes only a narrow spherical shell around recombination. Additionally, the 21 cm line does not suffer from Silk damping (photon diffusion), which suppresses the CMB fluctuations on relatively large scales. The number of independent modes accessible through this probe is therefore (Loeb & Zaldarriaga 2004)

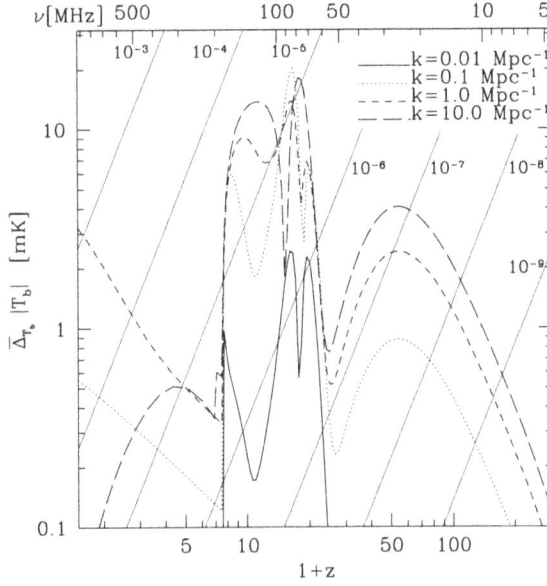

Figure 3.5. 21 cm fluctuations are substantial during the Dark Ages. The curves show the amplitude of the 21 cm brightness temperature fluctuations at several different wavenumbers from the Dark Ages to low redshifts, in the standard model of cosmology. Diagonal lines compare these fluctuations to the foreground brightness temperature (from Galactic synchrotron): each scales the foreground by the number shown. Note that exotic cooling scenarios described in Section 3.2.2 could significantly amplify these fluctuations. Reproduced from Pritchard & Loeb (2012). © IOP Publishing Ltd. All rights reserved.

$$N_{21cm} \sim 8 \times 10^{11} \left(\frac{k_{max}}{3 \text{ Mpc}^{-1}} \right)^3 \left(\frac{\Delta \nu}{\nu} \right) \left(\frac{1+z}{100} \right)^{-1/2}, \qquad (3.9)$$

where $\Delta \nu$ is the bandwidth of the observation. The choice of k_{max}—the smallest physical scale to be probed—is not obvious. The Jeans length during the Dark Ages corresponds to $k_{max} \sim 1000 \text{ Mpc}^{-1}$. Accessing these small-scale modes in three dimensions would require an enormous instrument, but our relatively conservative choice in Equation (3.9) shows that even a more modest effort provides a massive improvement over the information contained in all the measurable modes of the CMB, $N_{CMB} \sim 10^7$. Moreover, because the density fluctuations are still small at such early epochs, these modes remain in the linear or mildly nonlinear regime, allowing a straightforward interpretation of them in terms of the fundamental parameters of our universe (Lewis & Challinor 2007).

In principle, measurements of the 21 cm power spectrum during the Dark Ages will therefore enable a number of precision cosmological measurements, even within the "standard" cosmology. For example, because 21 cm fluctuations extend to such small scales, they expand the dynamic range of power spectrum measurements over several orders of magnitude. This is useful for a number of specific cosmological parameters, as illustrated by the curves in Figure 3.6, but also including the running of the spectral index of the matter power spectrum (Mao et al. 2008), a key

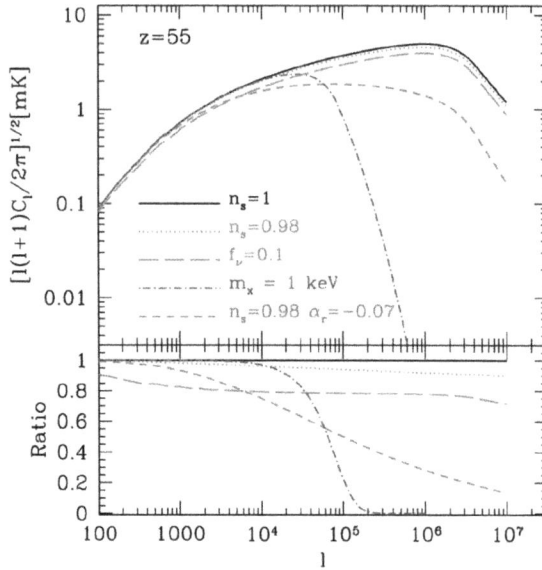

Figure 3.6. The angular power spectrum of 21 cm fluctuations at $z = 55$ in "standard" cosmologies, varying some of the parameters. The 21 cm power spectrum is a sensitive probe of cosmological parameters. The solid and dotted curves use ΛCDM with the power-law index of density fluctuations $n_s = 1$ and 0.98, respectively; the short-dashed curve adds a "running" to the spectral index. The long-dashed curve assumes 10% of the matter density is in the form of massive neutrinos (with masses 0.4 keV), while the dotted–dashed curve assumes warm dark matter with particle masses of 1 keV. Reprinted figure with permission from Loeb & Zaldarriaga (2004). Copyright 2004 by the American Physical Society.

parameter that can test the underlying assumptions of the inflationary paradigm, as well as the total spatial curvature and neutrino masses (Mao et al. 2008).

The large number of modes is also essential to constraining primordial non-Gaussianity in the cosmological density field, which is another tool for testing inflationary models. CMB measurements already offer constraints, which can best be improved by larger-volume surveys. However, low-z galaxy surveys suffer from contamination, because nonlinear structure formation also generates non-Gaussianity. The "clean," small-scale Dark Ages 21 cm field is an excellent opportunity to further constrain the non-Gaussianity (Chen et al. 2018)—in principle, 21 cm Dark Ages measurements can test the generic inflationary picture itself (Muñoz 2015).

The potential insights we have discussed so far are useful within the standard cosmology, but the exotic physics mechanisms discussed in the previous section will inevitably affect the 21 cm power spectrum as well, though the implications have only recently begun to be explored. Most obviously, the power spectrum amplitude is proportional to the square of the mean brightness temperature. But more subtle signatures that depend on the particulars of the physics can also appear. For example, if the cooling is triggered by scattering between the baryons and a fraction of the dark matter that has a modest charge, the scattering rate will be modulated by the relative streaming velocity of dark matter and baryons. Such a model therefore leaves distinct features in the 21 cm power spectrum that trace the velocity structure. The detailed implications of most of these exotic models are mostly unexplored, but any model in which (1) the energy exchange depends on local density, velocity, or temperature; or (2) in which background radio energy deposition is inhomogeneous, should generically leave signatures in the power spectrum.

3.3 Cosmology during the Era of Astrophysics

Although the Dark Ages offer a clean and powerful test bed to probe cosmology, in practice they will be difficult to explore. Not only do the astrophysical foregrounds become significantly stronger at low frequencies (as shown by the diagonal lines in Figure 3.5, which increase toward higher redshifts), but Earth's ionosphere also becomes opaque at very low frequencies (Datta et al. 2014), which may necessitate observations from space or the Moon.

Because the 21 cm background will likely be observed at lower redshifts—after the first luminous sources have appeared—we will next consider how cosmological information might be extracted from observations during the Cosmic Dawn. The effects we have already described will also affect the global 21 cm signal and the power spectrum during the later era. However, many astrophysical mechanisms (those described in Chapter 2) also affect the 21 cm background during this era, so the challenge is to separate the cosmological information from the astrophysical. In this section, we shall consider some strategies to do that.

3.3.1 Isolating the Matter Power Spectrum

Conceptually, the simplest method is simply to measure the 21 cm power spectrum and fit simultaneously for both astrophysical and cosmological parameters. We will discuss the practicalities of such fits in Chapter 4, but it will suffice for now to note that it is not a trivial exercise, and adding additional cosmological parameters will be a challenge (Clesse et al. 2012; Kern et al. 2017; Park et al. 2019). Moreover, the astrophysical effects can be very strong and hence mask any cosmological effects.

Short of modeling the astrophysics precisely enough to extract cosmology—a strategy whose potential won't really be known until the astrophysics is better understood[3]—the chief hope for isolating the power spectrum is that there exists an era in which astrophysics processes can largely be ignored. This is not an entirely unreasonable expectation. Recall from Chapter 1 that the 21 cm brightness temperature is

$$T_b(\nu) \approx 9\, x_{\text{HI}}(1+\delta)\,(1+z)^{1/2}\left[1 - \frac{T_\gamma(z)}{T_S}\right]\left[\frac{H(z)/(1+z)}{dv_\parallel/dr_\parallel}\right]\text{mK}. \qquad (3.10)$$

Astrophysical effects determine the neutral fraction x_{HI} and the spin temperature T_S, but the other factors—density and velocity—are driven by cosmological processes.

---

[3] However, note that recent work has shown that the ionization field may be constructed as a perturbative expansion around the density field in certain regimes, which suggests precision modeling may indeed be possible (McQuinn & D'Aloisio 2018; Hoffmann et al. 2019).

Thus, we can imagine that cosmological information will show up clearly in the power spectrum if, for example, there exists a period in which $T_S \gg T_\gamma$ (so that the temperature effects can be ignored) but in which ionization fluctuations are not yet significant.

Simple estimates show that such a period is far from impossible—but also not guaranteed. Reionization requires at least one ionizing photon per baryon, or of order ~10 eV of ionizing energy released by stars per baryon. Heating the IGM above the CMB temperature—so that $(1 - T_\gamma/T_S) \approx 1$—requires only ~$10^{-2}$ eV (corresponding to a temperature of ~100 K, though only a fraction of the X-ray energy would actually be used to heat the IGM; see Section 1.3.3). Thus, if early sources produce at least ~10^{-3} as much energy in X-rays as they do in ionizing photons, heating would occur before reionization—leaving open the possibility that a period exists in which astrophysics can be ignored. More complex astrophysical processes can also enable such a period as well—for example, strong photoheating feedback can delay reionization relative to X-ray heating (Mesinger et al. 2013). Whether this is more than speculation remains uncertain: calibrating the X-ray luminosity of star-forming galaxies to local measurements suggests that the reionization epoch may overlap with the X-ray heating epoch (Mirocha et al. 2017; Park et al. 2019), but if the EDGES measurement is confirmed, heating must actually occur very early in the Cosmic Dawn, well before reionization is complete.

3.3.2 Redshift Space Distortions

To this point, we have largely ignored the last factor in Equation (3.10): the velocity gradient. However, it offers another route to extracting cosmological information. Usually, we expect the fluctuations from the other terms—density, ionization fraction, Lyα flux, and temperature—to be isotropic, because the processes responsible for them have no preferred direction (e.g., $\delta(\mathbf{k}) = \delta(k)$). However, peculiar velocity gradients introduce anisotropic distortions through the "Kaiser effect" (Kaiser 1984), which emerge because 21 cm observations use the line's observed frequency as a proxy for the distance of the cloud that produced it.

Consider a spherical overdense region. Because of its enhanced gravitational potential, the region expands less quickly than an average region of the same mass. Therefore, to an observer using redshift as a distance indicator, the apparent radial size of the overdense region is smaller than that of the average region. Of course, its transverse size can be measured by its angular extent on the sky and so is unaffected by the velocity structure. Thus, to the observer, the spherical region is distorted, appearing larger along the plane of the sky than in the radial direction. An underdense region is distorted in the opposite way: because it expands faster than average, it appears larger in the radial direction than along the plane of the sky. Thus, redshift space distortions will exaggerate intrinsic density fluctuations, but they do so in an anisotropic way—making this source of fluctuations separable from others (Barkana & Loeb 2005; McQuinn et al. 2006).

To see these effects, we start by labeling the coordinates in redshift space with **s**. Assuming that the radial extent of the volume is small, so that the Hubble parameter

H is constant throughout the volume, these coordinates are related to the real space $\mathbf{r}$ by

$$\mathbf{s}(\mathbf{r}) = \mathbf{r} + \frac{U(\mathbf{r})}{H}, \tag{3.11}$$

where $U(\mathbf{r}) = \mathbf{v} \cdot \hat{\mathbf{x}}$ is the radial component of the peculiar velocity.

Now consider a set of particles with number density $n(\mathbf{r})$ that are biased with respect to the dark matter by a factor b. Number conservation demands that the fractional overdensity in redshift space be related to that in real space via $[1 + \delta_s(\mathbf{s})]d^3\mathbf{s} = [1 + \delta(\mathbf{r})]d^3\mathbf{r}$. The Jacobian of the transformation is

$$d^3\mathbf{s} = d^3\mathbf{r}\left[1 + \frac{U(\mathbf{r})}{r}\right]^2\left[1 + \frac{dU(\mathbf{r})}{dr}\right], \tag{3.12}$$

because only the radial component of the volume element, $r^2 dr$, changes from real to redshift space. Thus, the density observed in redshift space increases if the peculiar velocity gradient is smaller than the Hubble flow, while the redshift space density will be smaller if the peculiar velocity gradient is larger. Thus, assuming $|U(r)| \ll Hr$,

$$\delta_s(\mathbf{r}) = \delta(\mathbf{r}) - \left(\frac{d}{dr} + \frac{2}{r}\right)\frac{U(r)}{H}. \tag{3.13}$$

Importantly, the peculiar velocity field itself is a function of the dark matter density field, as described qualitatively above. More rigorously, in Fourier space, the components of the peculiar velocity $\mathbf{u_k}$ are directly related to those of the density field, because the latter is the source of the gravitational fluctuations that drive the velocity gradients (e.g., Kaiser 1984):

$$\mathbf{u_k} = -i\frac{aHf(\Omega)}{k}\delta_\mathbf{k}\hat{\mathbf{k}}, \tag{3.14}$$

where $f(\Omega)$ is a function of the growth rate of cosmological perturbations.

Equation (3.13) shows that there are two corrections from the redshift space conversion. To see which of these dominates, consider a plane wave perturbation, $U \propto e^{i\mathbf{k}\cdot\mathbf{r}}$. Then, the derivative term is $\sim kU/H_0$ while the last term is $\sim U/H_0 r$. But r is the median distance to the survey volume, and k corresponds to a mode entirely contained inside it, so $kr \gg 1$, and we may ignore the last term. If we further make the small-angle approximation and make a Fourier transform, so that $\hat{\mathbf{x}}$ is also approximately a constant over the relevant volume, the Fourier transform of Equation (3.13) is

$$\delta_s(\mathbf{k}) = \delta(\mathbf{k})\left[1 + \beta\mu_\mathbf{k}^2\right], \tag{3.15}$$

where $\mu_\mathbf{k} = \hat{\mathbf{k}} \cdot \hat{\mathbf{x}}$ is the cosine of the angle between the wave vector and the line of sight. Here, $\beta = f(\Omega_m)/b$ corrects for a possible bias between the tracers we are studying and the growth rate of dark matter perturbations. For the case of 21 cm fluctuations in the IGM gas, the bias factor is very close to unity except below the Jeans filtering scale.

The redshift space distortions therefore provide an anisotropic amplification to the background signal, because only modes along the line of sight are affected. Averaged over all modes, these distortions amplify the signal by a factor $\approx (1 + \mu^2)^2 \approx 1.87$ (Bharadwaj & Ali 2004).

For the purposes of extracting cosmological information, the anisotropies are helpful in that they imprint angular structure on the signal, which may allow us to separate the many contributions to the total power spectrum (Barkana & Loeb 2005). Brightness temperature fluctuations in Fourier space have the form

$$\delta_{21} = \mu^2 \beta \delta + \delta_{\text{iso}}, \tag{3.16}$$

where we have collected all of the statistically isotropic terms—including those due to astrophysics—into δ_{iso}. Ignoring "second-order" terms (see below) and setting $\beta = 1$, the total power spectrum can be written as

$$P_{21}(\mathbf{k}) = \mu^4 P_{\delta\delta} + 2\mu^2 P_{\delta_{\text{iso}}\delta} + P_{\delta_{\text{iso}}\delta_{\text{iso}}}. \tag{3.17}$$

(Here we have written the normal density power spectrum as $P_{\delta\delta}$ for clarity.) By separately measuring these three angular components (which requires, in principle, estimates at just a few values of μ), we can isolate the contribution from density fluctuations $P_{\delta\delta}$. Measuring this component, without any astrophysical contributions, will provide the desired cosmological constraints.

However, in practice, the angular dependence of the power spectrum will not be so simple. To write Equation (3.17), we must ignore "second-order" terms in the perturbation expansion of the 21 cm field, such as the density and the ionization field perturbations. But the latter is not actually a small term (because, at least in the standard reionization scenarios, $x_{\text{HI}} = 0$ or 1), so its contributions do not decrease rapidly in higher-order terms (Lidz et al. 2007). During reionization, these additional terms complicate the angular dependence and will significantly complicate attempts to separate the μ^n powers during reionization, making it more difficult (McQuinn et al. 2006; Shapiro et al. 2013). Moreover, if the first H II regions are highly biased —thus overlapping the regions with the largest peculiar velocities—the redshift space distortions can be suppressed (Mesinger et al. 2011; Mao et al. 2012). The redshift space distortions are also more complicated if the heating and reionization eras overlap (Ghara et al. 2015).

3.3.3 Indirect Effects of Cosmology on the 21 cm Background

Finally, it is also worth noting that cosmological processes can have direct effects on astrophysical sources and hence indirect effects on the 21 cm background. For example, consider warm dark matter. If dark matter has a nonzero velocity dispersion, then it can easily escape from shallow potential wells, suppressing the formation of small dark matter halos. Because the first phases of galaxy formation occur in small halos, warm dark matter delays galaxy formation, which in turn delays the formation of Lyα, X-ray, and ionizing backgrounds and changes the timing (and potentially spatial fluctuations) of the 21 cm background (e.g., Barkana et al. 2001; Yue & Chen 2012; Lopez-Honorez et al. 2017). Of course, this dark

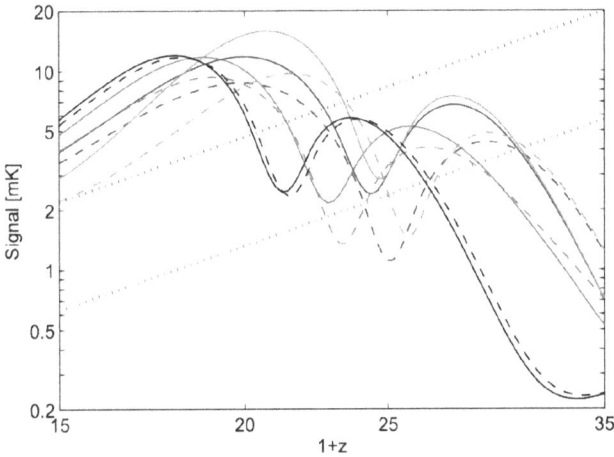

Figure 3.7. The relative streaming velocity between baryons and dark matter can impact the 21 cm power spectrum during the Cosmic Dawn. The solid curves show the time evolution of several example 21 cm power spectra, evaluated at $k \sim 0.1\, h$ Mpc^{-1}, without including streaming, while the dashed curves show the same scenarios with streaming included. The dotted lines show the expected sensitivities of current and future 21 cm radio arrays. Reproduced from Fialkov et al. (2014), by permission of Oxford University press on behalf of the Royal Astronomical Society.

matter effect is degenerate with astrophysical processes (for example, strong feedback in small galaxies may also suppress their star formation rates), and so requires careful analysis, and other cosmological changes can have qualitatively similar effects (e.g., Yoshida et al. 2003), although such effects may be distinct from large swaths of astrophysical parameter space (Sitwell et al. 2014).

Another example is the interaction between astrophysical processes and the relative streaming between baryons and dark matter. The streaming velocity also suppresses the formation of small baryonic halos, because a shallow potential well cannot accrete gas traveling by it at sufficiently large velocity. However, unlike in the case of warm dark matter, the streaming effect is spatially variable, so, at least in some circumstances, the streaming effect will imprint spatial structure on the radiation backgrounds and hence on the 21 cm power spectrum (Dalal et al. 2010; Fialkov et al. 2014; Muñoz 2019b). Figure 3.7 shows an example of this effect during the era in which the first stars appear.

Finally, dark matter annihilation offers another interesting example of the interaction between cosmology and astrophysics. The heating from dark matter annihilation occurs very uniformly. If astrophysically driven X-ray heating begins within such a preheated medium, the associated large-scale peak in the power spectrum occurs in emission rather than absorption, providing a distinct signature for (some) dark matter annihilation scenarios (Evoli et al. 2014; Lopez-Honorez et al. 2016).

3.4 21 cm Cosmology in a Larger Context

Although the focus of this book is on the 21 cm line itself, it is worth emphasizing that observations of the high-z spin-flip background will ultimately be combined

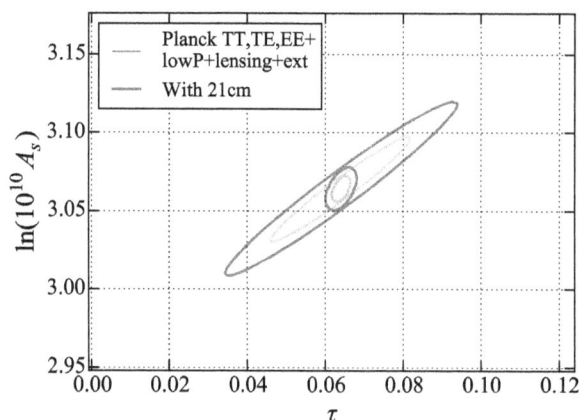

Figure 3.8. Constraints on the amplitude of the primordial power spectrum, A_s, and the CMB optical depth to electron scattering, τ. The elongated curves show the current constraints (at 1σ and 2σ) from combining the *Planck* satellite and several other probes. The tighter constraints show the combination with forecasted measurements from HERA, illustrating how the 21 cm measurement of the ionization history substantially improves constraints on other cosmological parameters. Figure based on calculations from Liu et al. (2016), provided courtesy of A. Liu.

with many other observations. For cosmological measurements, it is therefore useful to understand synergies between the 21 cm background and other probes (Liu et al. 2016; Liu & Parsons 2016). Some of the key advantages of the 21 cm line have already been described: it can probe small physical scales and high redshifts. Additionally, it can break degeneracies within other probes. Figure 3.8 shows an example. If the 21 cm line can measure the reionization history, it provides an independent estimate of the CMB optical depth to electron scattering, which is otherwise nearly degenerate with the amplitude of the initial power spectrum, and thus allows a precise measurement of that parameter.

References

Ali-Haïmoud, Y., & Hirata, C. M. 2011, PhRvD, 83, 043513

Barkana, R. 2018, Natur, 555, 71

Barkana, R., Haiman, Z., & Ostriker, J. P. 2001, ApJ, 558, 482

Barkana, R., & Loeb, A. 2005, ApJ, 624, L65

Berlin, A., Hooper, D., Krnjaic, G., & McDermott, S. D. 2018, PhRvL, 121, 011102

Bharadwaj, S., & Ali, S. S. 2004, MNRAS, 352, 142

Bowman, J. D., Rogers, A. E. E., Monsalve, R. A., Mozdzen, T. J., & Mahesh, N. 2018, Natur, 555, 67

Bradley, R. F., Tauscher, K., Rapetti, D., & Burns, J. O. 2019, ApJ, 874, 153

Chen, X., & Kamionkowski, M. 2004, PhRvD, 70, 043502

Chen, X., Palma, G. A., Hitschfeld, B. S., & Sypsas, S. 2018, PhRvL, 121, 161302

Cheung, K., Kuo, J.-L., Ng, K.-W., & Sming Tsai, Y.-L. 2019, PhLB, 789, 137

Chianese, M., Di Bari, P., Farrag, K., & Samanta, R. 2019, PhLB, 790, 64

Chluba, J., & Thomas, R. M. 2011, MNRAS, 412, 748

Chuzhoy, L. 2008, ApJ, 679, L65

Clesse, S., Lopez-Honorez, L., Ringeval, C., Tashiro, H., & Tytgat, M. H. G. 2012, PhRvD, 86, 123506

Costa, A. A., Landim, R. C. G., Wang, B., & Abdalla, E. 2018, EPJC, 78, 746

Dalal, N., Pen, U.-L., & Seljak, U. 2010, JCAP, 11, 7

Datta, A., Bradley, R., Burns, J. O., et al. 2014, arXiv:1409.0513

Draine, B. T., & Miralda-Escudé, J. 2018, ApJ, 858, L10

Evoli, C., Mesinger, A., & Ferrara, A. 2014, JCAP, 2014, 024

Falkowski, A., & Petraki, K. 2018, arXiv:1803.10096

Feng, C., & Holder, G. 2018, ApJ, 858, L17

Fialkov, A., Barkana, R., Pinhas, A., & Visbal, E. 2014, MNRAS, 437, L36

Furlanetto, S. R., Oh, S. P., & Pierpaoli, E. 2006, PhRvD, 74, 103502

Ghara, R., Roy Choudhury, T., & Datta, K. K. 2015, MNRAS, 447, 1806

Hills, R., Kulkarni, G., Daniel Meerburg, P., & Puchwein, E. 2018, Natur, 564, E32

Hirano, S., & Bromm, V. 2018, MNRAS, 480, L85

Hoffmann, K., Mao, Y., Xu, J., Mo, H., & Wand elt, B. D. 2019, MNRAS, 487, 3050

Jia, L.-B., & Liao, X. 2019, PhRvD, 100, 035012

Kaiser, N. 1984, ApJ, 282, 374

Kern, N. S., Liu, A., Parsons, A. R., Mesinger, A., & Greig, B. 2017, ApJ, 848, 23

Kleban, M., Sigurdson, K., & Swanson, I. 2007, JCAP, 2007, 009

Kovetz, E. D., Poulin, V., Gluscevic, V., et al. 2018, PhRvD, 98, 103529

Lawson, K., & Zhitnitsky, A. R. 2019, PDU, 24, 100295

Lewis, A., & Challinor, A. 2007, PhRvD, 76, 083005

Lidz, A., Zahn, O., McQuinn, M., et al. 2007, ApJ, 659, 865

Liu, A., & Parsons, A. R. 2016, MNRAS, 457, 1864

Liu, A., Pritchard, J. R., Allison, R., et al. 2016, PhRvD, 93, 043013

Loeb, A., & Zaldarriaga, M. 2004, PhRvL, 92, 211301

Lopez-Honorez, L., Mena, O., Moliné, Á., Palomares-Ruiz, S., & Vincent, A. C. 2016, JCAP, 2016, 004

Lopez-Honorez, L., Mena, O., Palomares-Ruiz, S., & Villanueva-Domingo, P. 2017, PhRvD, 96, 103539

Mao, Y., Shapiro, P. R., Mellema, G., et al. 2012, MNRAS, 422, 926

Mao, Y., Tegmark, M., McQuinn, M., Zaldarriaga, M., & Zahn, O. 2008, PhRvD, 78, 023529

McQuinn, M., & D'Aloisio, A. 2018, JCAP, 2018, 016

McQuinn, M., Zahn, O., Zaldarriaga, M., Hernquist, L., & Furlanetto, S. R. 2006, ApJ, 653, 815

Mesinger, A., Ferrara, A., & Spiegel, D. S. 2013, MNRAS, 431, 621

Mesinger, A., Furlanetto, S., & Cen, R. 2011, MNRAS, 411, 955

Mirocha, J., & Furlanetto, S. R. 2019, MNRAS, 483, 1980

Mirocha, J., Furlanetto, S. R., & Sun, G. 2017, MNRAS, 464, 1365

Moroi, T., Nakayama, K., & Tang, Y. 2018, PhLB, 783, 301

Muñoz, J. B. 2019a, arXiv:1904.07868

Muñoz, J. B. 2019b, arXiv:1904.07881

Muñoz, J. B., Ali-Haïmoud, Y., & Kamionkowski, M. 2015, PhRvD, 92, 083508

Muñoz, J. B., Kovetz, E. D., & Ali-Haïmoud, Y. 2015, PhRvD, 92, 083528

Muñoz, J. B., & Loeb, A. 2018, Natur, 557, 684

Naoz, S., & Barkana, R. 2005, MNRAS, 362, 1047

Park, J., Mesinger, A., Greig, B., & Gillet, N. 2019, MNRAS, 484, 933

Pritchard, J. R., & Loeb, A. 2012, RPPh, 75, 086901

Seager, S., Sasselov, D. D., & Scott, D. 1999, ApJ, 523, L1

Shapiro, P. R., Mao, Y., Iliev, I. T., et al. 2013, PhRvL, 110, 151301

Shchekinov, Y. A., & Vasiliev, E. O. 2007, MNRAS, 379, 1003

Sitwell, M., Mesinger, A., Ma, Y.-Z., & Sigurdson, K. 2014, MNRAS, 438, 2664

Slatyer, T. R., & Wu, C.-L. 2018, PhRvD, 98, 023013

Spinelli, M., Bernardi, G., & Santos, M. G. 2019, MNRAS, 489, 4007

Tegmark, M., & Zaldarriaga, M. 2009, PhRvD, 79, 083530

Tseliakhovich, D., & Hirata, C. 2010, PhRvD, 82, 083520

Yoshida, N., Sokasian, A., Hernquist, L., & Springel, V. 2003, ApJ, 598, 73

Yue, B., & Chen, X. 2012, ApJ, 747, 127

Chapter 4

Inference from the 21 cm Signal

Bradley Greig

In the previous chapters, we have discussed in depth the astrophysical and cosmological information that is encoded by the cosmic 21 cm signal. However, once we have a measurement, how do we extract this information from the signal? This chapter focuses on the inference of the interesting astrophysics and cosmology once we obtain a detection of the 21 cm signal.

Essentially, inference of the astrophysics can be broken down into three parts:

1. **Characterization of the observed data:** The observed 21 cm signal varies spatially as well as along the line of sight (frequency or redshift dimension) to provide a full three-dimensional movie of the intergalactic medium in the early universe. However, we cannot perform a full pixel-by-pixel comparison between theoretical models and the observed signal. Instead, we require a variety of statistical methods to average the observational data in order to be able to better characterize and compare the behavior of the faint signal.

2. **An efficient method to model the 21 cm signal:** In order to interpret the observations and understand the astrophysical processes responsible, we must be able to produce physically motivated models capable of replicating the signal. Further, these must be as computationally efficient as possible in order to be able to realistically investigate the 21 cm signal.

3. **A robust probabilistic framework to extract the physics:** The observed 21 cm signal is dependent on numerous physical processes, which within our models or simulations are described by many unknown parameters. Further, these contain approximations in order to deal with the requisite dynamic range. We must be able to characterize our ignorance in a meaningful way in order to be truly able to infer the astrophysical processes of the Epoch of Reionization (EoR) and Cosmic Dawn.

In this chapter, we will focus on each separately, discussing the current state of the art in inferring astrophysical and cosmological information from the 21 cm signal.

doi:10.1088/2514-3433/ab4a73ch4

4.1 What Do We Actually Measure?

The 21 cm signal from the neutral hydrogen in the intergalactic medium is measured by its brightness temperature, T_b. However, this cannot be measured directly; instead, it is expressed as a brightness temperature contrast, δT_b, relative to the cosmic microwave background (CMB) temperature, T_{CMB} (Furlanetto et al. 2006),

$$\delta T_b(\mathbf{x},\nu) \equiv T_b(\mathbf{x},\nu) - T_{CMB,0}. \quad (4.1)$$

As such, this brightness temperature contrast can be seen either in emission or absorption, dependent on the 21 cm brightness temperature which itself is dependent on the excitation state of the neutral hydrogen (i.e., its spin temperature, T_S; see Section 1.2). We can re-express Equation (4.1) in terms of T_S to recover

$$\delta T_b(\mathbf{x},\nu) \equiv \frac{T_S(\mathbf{x},\nu) - T_{CMB}(z)}{1 + z}(1 - e^{-\tau_{\nu_0}(\mathbf{x},\nu)}), \quad (4.2)$$

where τ_{ν_0} is the optical depth of the 21 cm line (see e.g., Section 1.1). $\delta T_b(\mathbf{x},\nu)$ varies spatially due to its two-dimensional angular position on the sky while it varies along the line-of-sight direction owing to the 21 cm line being redshifted by cosmological expansion (i.e., adding a frequency or time dependence to the signal). Thus, measuring $\delta T_b(\mathbf{x},\nu)$ can reveal a full three-dimensional movie of the neutral hydrogen in the early universe.

Unfortunately, $\delta T_b(\mathbf{x},\nu)$ is faint. Further, in reality it is buried under numerous astrophysical foregrounds, all of which are orders of magnitude brighter (see e.g., Chapter 6). In order to deal with this faint signal coupled with the astrophysical foregrounds, typically we seek to compress the data to boost the signal-to-noise or specifically tailor methods to extract the faint signal. In Section 4.2, we will discuss the numerous methods proposed in order to tease out the faint astrophysical signal from the noise.

4.2 Optimal Methods for Characterizing the 21 cm Signal

The first step in our efforts to be able to infer information about the astrophysical processes responsible for reionization and the Cosmic Dawn is to explore optimal methods to characterize the 21 cm signal. In this section, we summarize the wide variety of approaches considered in the literature, highlighting the leverage that each is able to provide with respect to the underlying astrophysical processes. Note that throughout this chapter, all investigations into detecting the 21 cm signal are generated theoretically, either analytically or numerically. Thus, we urge the reader to refer to the corresponding references in order to understand the limiting assumptions.

4.2.1 Global Signal

The simplest way to deal with such a faint signal is to average it over as large a volume as possible. Because the 21 cm signal is visible across the entire sky, one can produce a complete sky-averaged (global) 21 cm brightness temperature as a function of frequency (redshift).

Although the two-dimensional spatial information from the 21 cm signal is lost, the main advantage is that it is relatively cheap to observe, requiring comparatively simple instrumentation (see, e.g., Section 8.3). For example, a single radio dipole is capable of seeing essentially the entire sky at any one time, which has formed the basis for several single dipole experiments to measure the 21 cm signal. In Figure 4.1, we show a representative model of the global 21 cm signal, highlighting the major cosmological milestones that have been discussed in previous chapters. Thus, in each frequency bin, we measure an all-sky average of the 21 cm brightness temperature.

The global signal has been studied extensively in the literature (see, e.g., Furlanetto 2006; Pritchard & Loeb 2010, 2012; Mirocha et al. 2013; Fialkov & Barkana 2014; Mirocha 2014; Mirocha et al. 2015, 2017; Cohen et al. 2017; Fialkov et al. 2018; Mirocha et al. 2018; Fialkov & Barkana 2019). Roughly speaking, the global 21 cm signal can be broken up into five major turning points (e.g., Furlanetto et al. 2006; Pritchard & Loeb 2010) corresponding to (A) a minimum during the Dark Ages, where collisional coupling becomes ineffective, (B) a maximum at the transition from the Dark Ages to the Lyα pumping regime (Lyα pumping from the first sources becomes efficient), (C) a minimum at the commencement of X-ray heating taking the signal back toward emission, (D) a maximum once the 21 cm signal becomes saturated during the EoR and finally (E) when reionization is complete. Importantly, both the amplitude of the 21 cm signal as well as the frequency (redshift) of these transitions is strongly dependent on the underlying astrophysical processes. Thus, measuring both the amplitude and frequency of the turning points can reveal information on the underlying astrophysics.

The second turning point (end of the Dark Ages) can, under certain simple assumptions, be used to place limits on the spin temperature, T_S. Details on T_S, through Equations (2.2) and (2.3), can provide an estimate on the overall amplitude of the angle-averaged intensity of Lyα photons, J_α. The relative depth of the third turning point (heating epoch) can be used to place limits on the comoving heating rate density, that is, the amount of heating that the IGM has undergone owing to heating sources (e.g., X-rays from HMXBs, the ISM, or other more exotic scenarios; see, e.g., Sections 1.3 and 2.2 for further details,). Finally, if the spin temperature

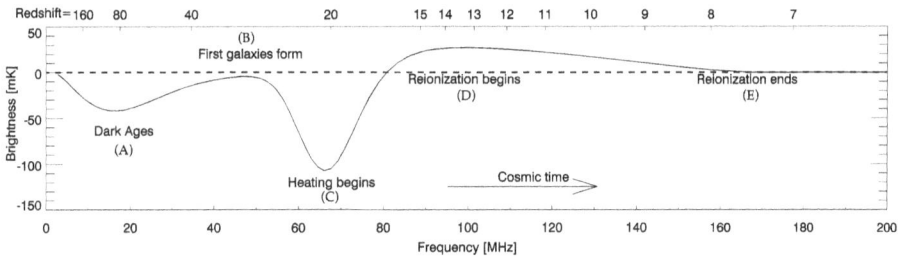

Figure 4.1. A representative example of the all-sky-averaged (global) 21 cm brightness temperature signal, demarcating the major cosmological transitions. Reproduced from Pritchard & Loeb (2012). © IOP Publishing Ltd. All rights reserved.

saturates ($T_S \gg T_{CMB}$) during the EoR, then the expression for the brightness temperature (Equation (1.8)) collapses into an approximate proportionality ($T_S \propto x_{HI}(1 + \delta_{nl})$) with the underlying ionization fraction, x_{HI}. Tracking the evolution of the ionized fraction, i.e., the reionization history, reveals the time span of reionization and the number density of ionizing photons produced.

Unfortunately, the estimates for the amplitude of the ionizing, Lyα, and X-ray backgrounds from the global signal cannot directly reveal insights into the population of sources responsible (e.g., their typical emission spectra) as these amplitudes are convolved with the underlying galaxy number density. In compressing the entirety of the signal down into these five turning points, we cannot separate out the two contributions. However, this degeneracy can be broken when further spatial information is used (e.g., the 21 cm power spectrum; Section 4.2.2).

To highlight the expected variation in the global 21 cm signal as a result of the underlying astrophysical processes, in Figure 4.2, we show ~200 theoretical models of the global 21 cm signal from Cohen et al. (2017). Here, the authors explore the maximal variation in the global 21 cm signal when varying the ionization and heating properties of the astrophysical sources. Some common features in the signal are as follows: the depth of the absorption trough deepens for lower X-ray luminosities (including some models which never appear in emission as a result of inefficient heating) or the turning points push to later times when the minimum masses of sources increases (i.e., require more massive halos in which stars can form and produce ionizing photons).

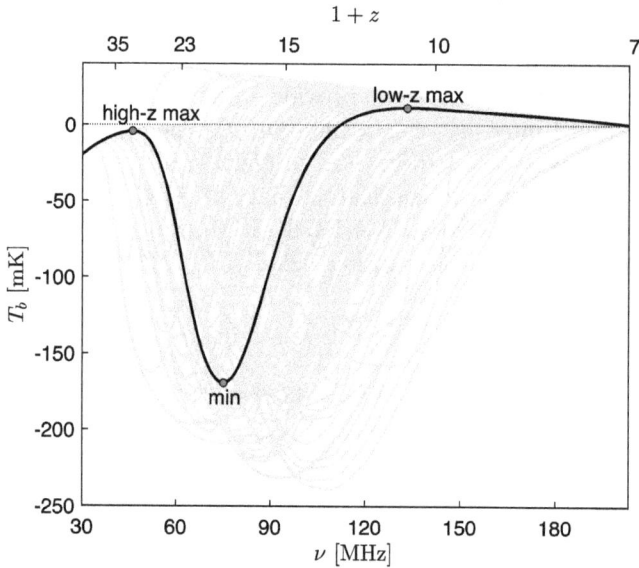

Figure 4.2. The all-sky-averaged (global) 21 cm brightness temperature signal obtained when varying the astrophysical parameters in ~200 theoretical models. Reproduced from Cohen et al. (2017), by permission of Oxford University press on behalf of the Royal Astronomical Society.

4.2.2 Power Spectrum

After the global signal, the next simplest and most straightforward approach to characterize the 21 cm signal is through the power spectrum. This is the Fourier transform of the two-point correlation function—basically, a measure of the excess signal (above random) on all possible spatial scales. The workhorse statistic for any signal containing structural information, the power spectrum is simply the number of modes (in Fourier space) as a function of physical scale (or size). It produces a distribution of modes characterizing the amount of structural information which is contained within the signal. The power spectrum is the natural method for observing the 21 cm signal from a radio interferometer, as these measure differences in the arrival times of the cosmological signal between radio dipoles or dishes of some fixed separation. Thus, a radio interferometer is sensitive to the spatial fluctuations rather than the total amplitude.

To obtain the 21 cm power spectrum, we normalize the 21 cm brightness temperature, $\delta T_b(\mathbf{x})$ to be a zero-mean quantity, $\delta_{21}(\mathbf{x}) = (\delta T_b(\mathbf{x}) - \delta \bar{T}_b)/\delta \bar{T}_b$, which amplifies the fluctuations (spatial information) in the signal. The power spectrum, $P_{21}(\mathbf{k})$, is computed by the angle-averaged sum of the Fourier transform of the 21 cm brightness temperature fluctuations via

$$\langle \delta_{21}(\mathbf{k}_1)\delta_{21}^*(\mathbf{k}_2) \rangle = (2\pi)^3 \delta_D(\mathbf{k}_1 - \mathbf{k}_2)P_{21}(\mathbf{k}_1), \quad (4.3)$$

where δ_D is the Dirac delta function, $\langle \rangle$ denotes the ensemble average, and * corresponds to the complex conjugate. Typically, the 21 cm power spectrum is converted into a dimensionless quantity through $\Delta_{21}^2(\mathbf{k}) = (k^3/2\pi^2)P_{21}(\mathbf{k})$. Typically, the Fourier modes are then averaged in spherical shells to obtain the spherically averaged power spectrum, $P_{21}(k)$, which considerably improves the overall signal-to-noise, at the cost of averaging over some spatial information. Alternatively, one can also measure the two-dimensional cylindrically averaged power spectrum, $P_{21}(k_\parallel, k_\perp)$, decomposing it into modes perpendicular to the line of sight ($k_\perp$; spatially averaging the two-dimensional angular modes on the sky in annuli) and along the line-of-sight ($k_\parallel$; in frequency) direction. The strength of the two-dimensional 21 cm power spectrum is that most of the contamination of the signal by the astrophysical foregrounds can be contained in what is referred to as the EoR "wedge" while the remaining Fourier modes can be clean tracers of the cosmological signal (see Section 6.2.1.2).

The advantage of the power spectrum over the global signal is that it provides a measure of the spatial fluctuations in the 21 cm signal. However, it does not encode all the available spatial information from the 21 cm signal. If these fluctuations were truly Gaussian, the power spectrum would contain all the information, and any higher-order n-point correlation functions would contain no additional information. The structural complexity of the large- and small-scale processes of reionization and the Cosmic Dawn results in the signal being highly non-Gaussian. As such, the power spectrum does not reveal all available information, meaning there is further constraining power from the higher-order n-point statistics. In Sections 4.2.3 and

4.2.4, we will return to this. Nevertheless, the power spectrum still contains a wealth of information, and observationally is considerably easier to measure.

The sensitivity of the 21 cm power spectrum to the underlying astrophysics can be highlighted when we decompose the 21 cm brightness temperature fluctuations through a perturbative analysis (i.e., Taylor expansion) from which we recover the following (see, e.g., Barkana & Loeb 2005; Santos et al. 2005; Mao et al. 2008),

$$\delta_{21} \propto C_b \delta_b + C_x \delta_x + C_\alpha \delta_\alpha + C_T \delta_T - \delta_{\partial v}, \qquad (4.4)$$

Simply put, fluctuations in the 21 cm brightness temperature field, δ_{21}, are driven by a sum of contributions from the underlying density field, δ_b, the ionization fraction δ_x, the Lyα coupling coefficient δ_α, the temperature of the neutral hydrogen δ_T, and the line-of-sight peculiar velocity gradient, $\delta_{\partial v}$. Computing the power spectrum then measures the combined signal from the power spectra of each field as well as the cross-power spectra of each. Thus, if we measure the 21 cm power spectrum across cosmic time, we will be sensitive to the epochs when each component dominates (similar to the global signal) and also the spatial scales on which the signal is strongest. This, similar to the global signal, is depicted in Figure 4.3.

However, rather than using one single Fourier mode, we have a range of spatial scales over which to recover astrophysical information. This provides access to both the small-scale and large-scale physical processes. For example, during the EoR, the 21 cm power spectrum is dominated by the contribution from the ionization field, which contains particular structural information on the reionization process due to the characteristic size of the H II regions as well as their clustering (e.g., Lidz et al. 2008). One can equally obtain the spectrum of the sources responsible for heating the IGM from the structural information, owing to the strong dependence of the mean free path with the energy of the X-ray sources.

In Figure 4.4 we show the variation in the three-dimensional spherically averaged 21 cm power spectrum at a single redshift ($z = 9$) when varying three different astrophysical parameters under the assumption of $T_S \gg T_{CMB}$ (see e.g., Greig & Mesinger 2015). Inset tables correspond to the parameter being varied and the resultant IGM neutral fraction (stage of reionization). In the top left panel, we vary the ionizing efficiency, ζ, a proxy for the number of ionizing photons produces by the sources. The shape of the 21 cm power spectrum differs considerably with ionizing

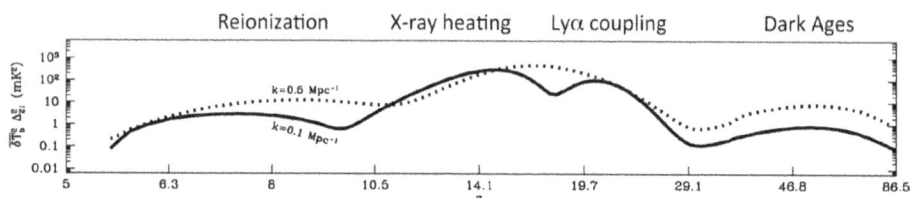

Figure 4.3. The 21 cm power spectrum amplitude for two different Fourier modes, $k = 0.1$ Mpc^{-1} (solid) and $k = 0.5$ Mpc^{-1} (dashed). Peaks in the 21 cm power spectrum amplitude correspond to the different cosmic milestones. Reproduced from Mesinger et al. (2016), by permission of Oxford University press on behalf of the Royal Astronomical Society.

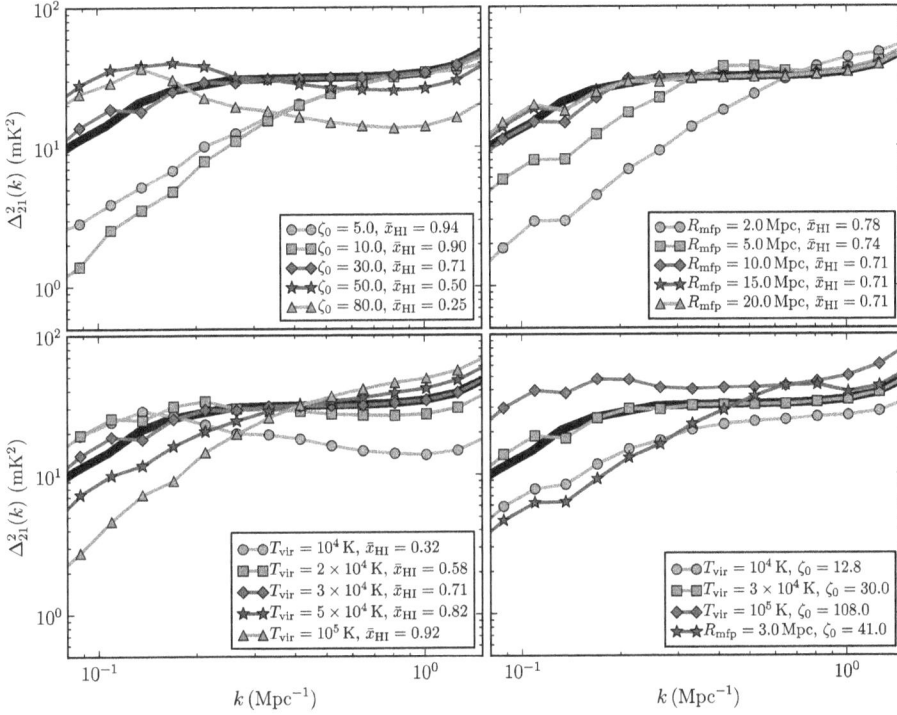

Figure 4.4. The three-dimensional spherically averaged 21 cm power spectrum at $z = 9.0$ when varying astrophysical parameters controlling different astrophysical processes, assuming $T_S \gg T_{CMB}$. Top left: the number of ionizing photons produced per baryon (ionizing efficiency, ζ), top right: maximum ionizing photon horizon (proxy for maximum allowable bubble size, R_{mfp}) and bottom left: minimum mass of halo hosting star-forming galaxy (represented here as T_{vir}). Bottom right: several models at the same ionization fraction. Reproduced from Greig & Mesinger (2015), by permission of Oxford University press on behalf of the Royal Astronomical Society.

efficiency. In the early stages, the 21 cm PS matches the density (matter) power spectrum, while in the latter stages it follows the ionization field.

Similar behavior is observed for varying T_{vir}, a proxy for the minimum mass of halos hosting star-forming galaxies. Increasing this threshold, results in fewer sources to contribute to reionization. In the top right panel, the maximum photon horizon, R_{mfp}, is varied. Essentially, in this specific work it acts as a maximum allowable bubble size. Note that in this case, the change in R_{mfp} does not alter the neutral fraction strongly, thus the changes in the 21 cm power spectrum are purely as a result in changes to the size of the ionized regions. Finally, in the bottom right we highlight astrophysical models with the same IGM neutral fraction (i.e., the same stage of reionization). Despite being at the same point in reionization, the amplitude and shape of the 21 cm power spectrum differs considerably, highlighting the sensitivity of the 21 cm power spectrum to the underlying astrophysical parameters.

While this example is only for the epoch of reionization, the same strong sensitivity of the 21 cm power spectrum to the underlying astrophysics is true for both the heating or Lyα coupling epochs. This highlights the strength and utility of the 21 cm power spectrum for recovering the astrophysical information. As such numerous authors have explored the impact of various astrophysical processes on the 21 cm power spectrum (see e.g., Bowman et al. 2006; Furlanetto et al. 2006; Iliev et al. 2006; McQuinn et al. 2006, 2007; Pritchard & Furlanetto 2007; Lidz et al. 2008; Santos et al. 2008; Baek et al. 2010; Harker et al. 2010; Mesinger et al. 2013; Fialkov & Barkana 2014; Pober et al. 2014; Greig & Mesinger 2015; Geil et al. 2016; Greig & Mesinger 2017a; Hassan et al. 2017; Cohen et al. 2018; Greig & Mesinger 2018; Park et al. 2019; Seiler et al. 2019).

4.2.3 Bispectrum

The logical extension beyond the power spectrum, the bispectrum, B, is simply the Fourier transform of the three-point correlation function,

$$\langle \delta_{21}(\mathbf{k}_1)\delta_{21}(\mathbf{k}_2)\delta_{21}(\mathbf{k}_3)\rangle = (2\pi)^3\delta_{\mathrm{D}}(\mathbf{k}_1 - \mathbf{k}_2 - \mathbf{k}_3)B(\mathbf{k}_1,\mathbf{k}_2,\mathbf{k}_3), \qquad (4.5)$$

where the δ_{D} enforces that the Fourier modes must form closed triangles. It measures the excess probability of the underlying quantity as a function of three spatial positions in real space. The bispectrum provides a scale-dependent measure of the non-Gaussianity of the 21 cm signal, and as such contains additional astrophysical information beyond that held in the power spectrum. However, it suffers from lower signal-to-noise as there are less modes to average over to boost the signal.

Whereas the power spectrum is relatively trivial to interpret as it is a measure of the power over a single length scale, k, the bispectrum is the measure of power over all possible triangle configurations that satisfy the closure condition from δ_{D}. Thus in order to simplify the interpretation of the bispectrum, it is common to consider several simplified triangle configurations. These are typically: (i) the equilateral triangle ($k_1 = k_2 = k_3$), (ii) the isosceles triangle ($k_1 > k_2 = k_3$), (iii) folded triangle ($k_1 = 2k_2 = 2k_3$), (iv) elongated triangle ($k_1 = k_2 + k_3$) and (v) the squeezed triangle ($k_1 \simeq k_2 \gg k_3$). Each, corresponds to different physical properties of the real-space field.

While a detailed discussion of the 21 cm bispectrum is beyond the scope of this chapter, it is fruitful to provide a brief explanation and example of the various configurations (see for example Lewis 2011 and Watkinson et al. 2019 for more detailed discussions). The equilateral configuration is essentially an extension of the power spectrum, in the sense that it is expressed as a single amplitude scale, k. Generally speaking, it produces the largest amplitude signal and as such is the most commonly studied configuration. It is sensitive to the spherical symmetry of the 21 cm signal such as the scale of the ionized H II regions during reionization or the hot/cold spots due to IGM heating. Typically, its amplitude grows during the EoR as the signal becomes more non-Gaussian, due to the topology of the ionization field. Shifting toward isosceles or folded triangles, these become more sensitive to planar or filamentary structures in the underlying 21 cm signal. Thus, as the topology of either the ionized or X-ray-heated regions deviate away from spherical symmetry (i.e., either multiple contributing sources or overlap of ionized regions), the signal should increase with increasing angle. The squeezed limit correlates the small-scale signal from two modes with a large-scale mode, for example capturing the impact of the large-scale environment (i.e., from X-ray heating) on the small-scale power spectrum (i.e., source clustering).

In addition to the structural information in the bispectrum amplitude, the relative sign of the bispectrum under certain triangle configurations and on certain spatial scales can equally reveal insights into the underlying processes. As discussed in Majumdar et al. (2018) and Hutter et al. (2019), the sign of the bispectrum during reionization can help distinguish between whether the non-Gaussianity is driven by the topology of the ionized regions (where the bispectrum is negative owing to the below-average contribution from the ionized regions) or is being driven by the matter and cross-bispectra (where it is positive). Thus, different reionization models are easily distinguishable by the 21 cm bispectrum.

In recent times, the 21 cm bispectrum has gained considerable traction in interpreting the astrophysics of reionization and the Cosmic Dawn (see e.g., Bharadwaj & Pandey 2005; Pillepich et al. 2007; Yoshiura et al. 2015; Shimabukuro et al. 2016, 2017; Watkinson et al. 2017; Majumdar et al. 2018; Hutter et al. 2019; Trott et al. 2019; Watkinson et al. 2019). Alternatively, rather than exploring the information from the amplitude of the bispectrum, Gorce & Pritchard (2019) introduced a three-point correlation function based solely on the phases of the Fourier modes (e.g., Obreschkow et al. 2013), termed the triangle correlation function. In focusing solely on the phases, it is sensitive to the characteristic size of the ionized regions and thus explores the topology of reionization, which places it in a similar vein to other topological-based approaches (Section 4.2.7) or the size distribution of ionized regions (Section 4.2.8). However, not all experiments are designed to measure this phase information. In fact, several experiments are specifically designed to throw away this phase information for increased sensitivity to specific spatial scales. These are referred to redundant configurations and are discussed in Chapter 7.

4.2.4 Trispectrum

Following the bispectrum, the trispectrum is the Fourier transform of the four-point correlation function. Already at the level of the bispectrum, the relative signal-to-noise of the signal is becoming weak; thus, in the foreseeable future it is unlikely a measurement of the trispectrum during the EoR or earlier will be achievable. Nevertheless, Cooray et al. (2008) explored the trispectrum of the 21 cm fluctuations, focusing on fundamental cosmology rather than the astrophysics of the reionization process. These authors find that the anisotropies from the 21 cm signal are sensitive to primordial non-Gaussianities, an important quantity in constraining inflationary models.

4.2.5 One-point Statistics

Rather than measuring the Fourier transform (e.g., power spectrum) of the 21 cm brightness temperature signal, $\delta_{21}(\mathbf{x})$, we can instead measure the one-point statistics (or moments) of the probability distribution function (PDF). In fact, we have already discussed the lowest-order one-point statistic, that is, the mean of $\delta T_b(\mathbf{x})$ given by the global signal (see Section 4.2.1). These one-point statistics of the PDF essentially measure the deviations away from a fully Gaussian PDF, thus they are by definition sensitive to the non-Gaussian nature of the 21 cm signal. Generally speaking, the one-point statistics of $\delta T_b(\mathbf{x})$ are given by

$$m_n = \frac{1}{N} \sum_{i=0}^{N} (\delta T_b(\mathbf{x}_i) - \delta \bar{T}_b)^n, \qquad (4.6)$$

where m_n is the nth-order moment and N is the number of pixels over which the signal is measured. For the 21 cm signal, these moments would be generated from the observed two-dimensional tomographic maps of the 21 cm signal.

The next lowest-order statistic of the PDF following the mean is the variance, σ^2. The variance is equivalent to the average of the power spectrum over all Fourier modes, k,

$$\sigma^2 = (\delta \bar{T}_b)^2 \int \frac{d^3k}{(2\pi)^3} P(\mathbf{k}). \qquad (4.7)$$

As it is the average over all spatial information, the variance itself is less sensitive to the underlying astrophysics than the power spectrum. However, the strength of one-point statistics shines through when using the higher-order moments in combination with the variance (or power spectrum). The next two higher-order moments are referred to as the skewness and the kurtosis. Equivalent to the variance's relation to the power spectrum, the skewness and kurtosis are the average over all Fourier modes of the bispectrum and trispectrum, respectively (the three- and four-point correlation functions). As such, whereas the power spectrum only measures the two-point correlations, the skewness and kurtosis reveal insights from the non-Gaussian properties of the 21 cm signal.

The amplitude of the variance is sensitive to differences in the 21-cm brightness temperature. For example, during the EoR, as the number of ionised regions increases (i.e. the contrast between the 21-cm signal from the neutral regions compared to zero signal from the ionised regions) the variance increases. It subsequently turns over as most of the volume is ionised. During the epoch of heating, for increasing X-ray efficiencies (i.e. increased heating) the peak of the variance decreases in amplitude while shifting to earlier times. Increasing the efficiency allows the X-ray heating to occur at earlier times, reducing the contrast between $T_{CMB/TS}$ resulting in a lower amplitude peak in the variance.

The skewness is a measure of the asymmetry of the underlying PDF. A negative skewness corresponds to a longer tail towards a lower amplitude signal and a positive skewness corresponds to a longer tail towards higher amplitude signals. For example, during the epoch of heating a decreasing X-ray efficiency results in larger skewness owing to a more asymmetric PDF of 21-cm brightness temperatures due to the increasing contrast in T_{CMB}/T_S. The kurtosis is essentially a measure of the outliers of the distribution, with increasing positive (negative) kurtosis corresponding to larger positive (negative) amplitude outliers. Clearly, these one-point statistics are capable of distinguishing between different astrophysical models and as such have been explored extensively in numerous works (e.g., Wyithe & Morales 2007; Harker et al. 2009; Patil et al. 2014; Watkinson & Pritchard 2014, 2015; Kittiwisit et al. 2016; Kubota et al. 2016; Watkinson et al. 2015; Shimabukuro et al. 2015; Ross et al. 2017).

Alternatively, the direct 21 cm PDF or the difference PDF has also been studied (e.g., Barkana & Loeb 2008; Gluscevic & Barkana 2010; Ichikawa et al. 2010; Pan & Barkana 2012). The difference PDF is the difference between the brightness temperature separated by some spatial scale, r. The advantages of the difference PDF is that it can bypass the fact that interferometric observations cannot easily determine the zero flux threshold of the 21 cm signal and that it includes more data by being dependent on spatial scales (similar to two-point correlation functions or the power spectrum). The difference PDF can be more sensitive to the ionizing sources and sizes of the ionized regions as it is a direct measure of the distribution of separated pixel pairs that are either both ionized, ionized, and neutral or both neutral.

4.2.6 Wavelets

Thus far we have only considered either real-space quantities such as the one-point statistics or the Fourier transform of the n-point correlation functions (i.e., the power spectrum and bispectrum). The Fourier transform measures the amplitude of the fluctuations of a given spatial scale, and in order to increase the signal-to-noise, we must average the signal over all line-of-sight modes within some observed bandwidth. As a result, we average over modes containing different redshift evolutions and thus increase the bias of the signal. This can be minimized somewhat, for the case of the power spectrum, by averaging the signal over relatively narrow observing bandwidths. However, it still results in some loss in fidelity of the signal.

Instead, Trott (2016) explored the potential usage of wavelets, which provide multiple alternatives to the Fourier basis set. Specifically, they explored the application of the Morlet transform. This provides a family of curves, which provide the ability to localize the 21 cm signal both spatially and in frequency. The equivalent to the power spectrum, the Morlet power spectrum is capable of providing an unbiased estimator which maximizes the three-dimensional nature of the 21 cm signal. Preliminary analysis shows that the Morlet power spectrum performs more optimally than the Fourier power spectrum. A physical interpretation of the Morlet power spectrum in the context of the evolution of the 21 cm signal has yet to be explored.

4.2.7 Topological Measurements of the 21 cm Signal

Up until this point, we have only discussed methods of characterizing the 21 cm signal using just the amplitude of the spatial (e.g., Fourier) information. This is primarily driven by the difficulty in measuring the 21 cm signal and the low signal-to-noise of the first-generation experiments. However, the most advanced radio interferometers (such as the Square Kilometre Array, SKA; see Section 9.2.1) should be able to provide two-dimensional images of the 21 cm signal. That is, they should provide significant signal-to-noise to enable both the amplitude and phase information to be used.

Direct images of the 21 cm signal contain the complicated morphology of the hot (above-average or overdense signal) and cold (below-average or underdense signal) of the 21 cm brightness temperature throughout the history of reionization and the Cosmic Dawn. The relative sizes, shapes, and clustering of these hot/cold patches can reveal numerous insights into the underlying astrophysical processes, such as the number density of sources, their contribution to the heating/ionization of the IGM, and the shape of the emitted spectrum of radiation. The study of these geometric shapes in mathematics is referred to as topology.

Topological studies of reionization and the Cosmic Dawn are complementary to the methods described previously. For example, reionization proceeds through three main stages (e.g., Gnedin 2000; Furlanetto & Oh 2016): pre-overlap, overlap, and post-overlap. In pre-overlap, the first ionized H II regions (or bubbles) grow completely in isolation roughly until $x_{HI} \geqslant 0.1$. Overlap ($0.9 \geqslant x_{HI} \geqslant 0.1$) describes the merging of these ionized bubbles into essentially a single large connected ionized region. Finally, post-overlap $x_{HI} \geqslant 0.9$ corresponds to the breaking down of the last remaining patches of neutral IGM into smaller and smaller islands. Topological studies are capable of breaking down these transitions by describing the ratios of ionized and neutral regions, how the ionized (or neutral) regions are connected together, and how they are embedded in the larger structures as they form. This provides unique insights into the reionization epoch not available from statistical methods.

Unfortunately, we cannot perform a full pixel-by-pixel analysis of a measured 21 cm image; therefore, we must still compress our images into some form of statistical measurement. There are numerous methods to attempt to characterize the topology of the 21 cm signal. Below, we summarize several of the main approaches taken in the literature. Fundamental to topological studies is the definition of how to identify regions of interest. Typically, a threshold value is required, with the quantity above/below this threshold being used to distinguish the two regions.

4.2.7.1 Genus or the Euler Characteristic
The genus, g, is a topological property that defines the number of cuts one can make to an object (i.e., the H II region) without dividing it into independent disconnected subregions. It can simply be expressed as

$$g = N_{>th} - N_{<th}, \tag{4.8}$$

where $N_{>th}$ and $N_{<th}$ are the number of connected (or fully enclosed) regions above and below the threshold value for identification. By gradually increasing the threshold value from some initial starting value, a genus curve is constructed, which is a measure of the connectedness of the quantity as a function of different threshold

values (e.g., x_{HI}, δT_b). Typically, these threshold values are expressed in units of the standard deviation from the mean.

The genus has been explored, both in two and three dimensions, either in the context of the ionized (or neutral) field (Gleser et al. 2006; Lee et al. 2008; Friedrich et al. 2011) or the 21 cm brightness temperature field (Hong et al. 2014; Wang et al. 2015). However, it has yet to be explored in the context of the heating epoch (i.e., $T_S \gg T_{CMB}$ is typically assumed). For a purely Gaussian field, the genus curve is symmetric around zero. Thus, deviations from symmetry highlight the non-Gaussianity of the 21 cm signal.

Differences in the evolution in the amplitude of the genus as a function of threshold density can distinguish different source biases and ionizing efficiencies. For example, reionization driven by larger, more biased sources exhibits a different topology than one driven by numerous fainter sources. This appears as changes in the amplitude of the genus as a function of threshold. When the ionized regions are isolated, the genus amplitude is higher than when they begin to overlap (as the total number of isolated ionized regions decreases).

4.2.7.2 Minkowski Functionals

A more generalized description of the geometry or topology of the 21 cm signal can be obtained from what are referred to as Minkowski functionals. These are well-known concepts from the branch of mathematics known as integral geometry. In n dimensions, there exist $n + 1$ independent Minkowski functionals, which means that in three-dimensional, space we have four functionals to describe the topology. Used heavily in cosmology, in particular geometrical features of the Galaxy distribution (e.g., Gott 1986; Schmalzing & Buchert 1997) and non-Gaussianity of the CMB (e.g., Komatsu et al. 2009), recently they have gained favor for describing the topology of reionization (Gleser et al. 2006; Friedrich et al. 2011; Yoshiura et al. 2017; Chen et al. 2018).

For a zero-mean scalar function, $u(x)$, (e.g., δT_b) within a volume, V, and standard deviation, u, we can define an excursion set, F_v, which contains all points that satisfy the threshold, $u(x) \geqslant v\sigma$, where $v = u_{th}/\sigma$ and u_{th} is the threshold value. Mathematically, this gives rise to the following Minkowski functionals,

$$V_0(v) = \frac{1}{V} \int_V d^3x \, \Theta[u(x) - v\sigma], \qquad (4.9)$$

$$V_1(v) = \frac{1}{6V} \int_{\partial F_v} ds, \qquad (4.10)$$

$$V_2(v) = \frac{1}{6\pi V} \int_{\partial F_v} ds \, [\kappa_1(x) + \kappa_2(x)], \qquad (4.11)$$

$$V_3(v) = \frac{1}{4\pi V} \int_{\partial F_v} ds \, \kappa_1(x)\kappa_2(x). \qquad (4.12)$$

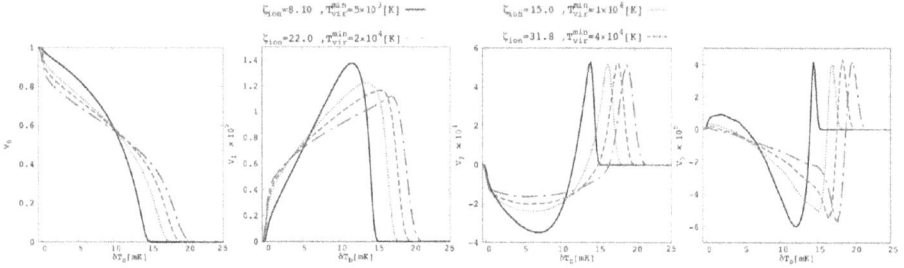

Figure 4.5. The impact of varying the astrophysical parameterization for a fixed neutral fraction ($x_{\mathrm{H_I}} \approx 0.5$) and redshift ($z = 8.6$). The colored curves highlight the impact of varying either the ionizing efficiency ζ or the minimum halo mass for star-forming galaxies, T_{vir}, on the four Minkowski functionals. Reproduced from Yoshiura et al. (2017), by permission of Oxford University press on behalf of the Royal Astronomical Society.

Here, Θ is the Heaviside step function, ∂F_ν is the surface of the excursion set, ds is the surface element, and $\kappa_1(x)$ and $\kappa_2(x)$ are the principal curvatures (inverse of the principal radii) at x. The zeroth Minkowski functional, V_0, corresponds simply to the total volume of the excursion set (i.e., volume above the threshold value), V_1 and V_2 correspond to the total surface and mean curvature of the excursion set, while V_3 is the integrated Gaussian curvature over the surface or the Euler characteristic (also χ). The Euler characteristic is related to the genus, g, via $V_3 = 2(1 - g)$, thus it effectively describes the shape of the excursion set. Thus, the full set of Minkowski functionals contains additional information beyond that of just the genus.

In Figure 4.5, we show the four Minkowski functionals for the 21 cm brightness temperature when varying the underlying astrophysical processes from Yoshiura et al. (2017) at a fixed neutral fraction ($x_{\mathrm{H_I}} \approx 0.5$) and redshift ($z = 8.6$). Here, these authors consider variations in either the ionizing efficiency, ζ, or the minimum halo mass hosting star-forming galaxies, T_{vir}. Clearly, different reionization histories are distinguishable by the Minkowski functionals.

Generally speaking, the following behavior is expected of the Minkowski functionals throughout reionization and the Cosmic Dawn. V_0 describes the volume contained above/below the threshold value. For example, if $V_0 \sim 0.5$ at $\delta T_b = 0$, this implies that the number of patches above/below the average 21 cm signal are roughly equal. The V_0 curve will move from left to right (to increasing δT_b) as heating of the IGM occurs. V_1 (reflected in V_2) exhibits a similar shift to higher δT_b; however, it is initially strongly peaked with a high-density tail containing the heated regions. This peak smooths out over a broader range of δT_b as IGM heating continues. During reionization, V_1, V_2, and V_3 will shift toward $\delta T_b = 0$ as the higher amplitude δT_b regions ionize first.

4.2.7.3 Shape Finders

An extension to Minkowski functionals, shape finders (Sahni et al. 1998) are a way to characterize the shapes of compact surfaces. Applied to reionization (Bag et al. 2018, 2019), these shape finders can provide a means to characterize how the ionized regions grow. For example, they are useful in being able to distinguish between whether the topology is planar or filamentary. Shape finders are derived directly from the Minkowski functionals via

$$\text{Thickness: } T = \frac{3V_0}{V_1}, \tag{4.13}$$

$$\text{Breadth: } B = \frac{V_1}{V_2}, \tag{4.14}$$

$$\text{Length: } L = \frac{V_3}{4\pi}. \tag{4.15}$$

These shape finders are interpreted as providing the three principal axes of a physical object. The morphology of the ionized region can then be defined by either the planarity or its filamentarity,

$$\text{Planarity: } P = \frac{B - T}{B + T}, \tag{4.16}$$

$$\text{Filamentarity: } F = \frac{L - B}{L + B}, \tag{4.17}$$

where $P \gg F$ corresponds to planar objects (i.e., sheets), while the opposite $F \gg P$ corresponds to a filament.

During the reionization epoch, percolation theory shows that a single infinitely large, multiply connected ionized region will rapidly form (e.g., Furlanetto & Oh 2016). When describing the largest singly connected ionized region, Bag et al. (2018, 2019) found that both T and B evolve slowly whereas L increases rapidly. Thus, this large ionized region grows only along its "length," implying a highly filamentary structure.

4.2.7.4 Persistent Homology Theory

Homology characterizes the topology of the ionization bubble network into its fundamental components: ionized regions, tunnels (enclosed neutral filaments), and cavities (patches of neutral hydrogen). The persistence then quantifies the significance of the feature, for example its lifetime, by computing a birth and death date for an object. Thus far, it has only been applied to the ionization field (Elbers & van de Weygaert 2019). These ionized regions (β_0), tunnels (β_1), and cavities (β_2) can be described by the so-called Betti numbers, β_n, which contain the total number of each type of structure. These can be related to the earlier Euler characteristic via $\chi = \beta_0 - \beta_1 + \beta_2$. Breaking the Euler characteristic into the constituent components and tracking their individual growth reveal additional information on the topology; thus, it is a more generalized method than either the genus of the Minkowski functionals.

4.2.7.5 Fractal Dimensions

An alternative to classifying the ionized (neutral) regions embedded in the 21 cm signal is through a fractal dimensions analysis. Applied to reionization (Bandyopadhyay et al. 2017), this provides a direct means to quantify the deviation away from a homogeneous distribution, as well as the degree of clustering and lacunarity (a measure of the size of the ionized regions). The fractal dimension, D_q, also known as the Minkowski–Bouligand dimension, is a measure of how complicated the topology of the field in question is. A homogeneous distribution in three dimensions has a $D_q = 3$. Bandyopadhyay et al. (2017) showed that the topology of reionization exhibits a significant multifractal behavior. These authors find that the fractal dimension is relatively insensitive to the minimum halo mass of the star-forming galaxies; however, it was sensitive to the mass-averaged ionization fraction. Thus, the correlation dimension can be useful for constraining the global neutral fraction. Additionally, it is a strong discriminant of models of outside-in and inside-out reionization.

4.2.7.6 Contour Minkowski Tensor

In Kapahtia et al. (2018, 2019), these authors introduced the rank-2 contour Minkowski tensor (e.g., McMullen 1997; Alesker 1999; Beisbart et al. 2002; Hug et al. 2008; Schröder-Turk et al. 2010; Schröder-Turk et al. 2013) in two dimensions, which can probe both the length and timescales of the ionized regions during reionization. The Minkowski tensors are a generalization of the scalar Minkowski functionals. The contour Minkowski tensor provides information on both the alignment of structures in two dimensions and their anisotropy. Because the ionized regions are not perfectly spherical, their shape anisotropy can be explored by the ratio of the two eigenvalues of the contour Minkowski tensor while the amplitude of the eigenvalues describes their size.

In this analysis, the number of connected regions and holes (e.g., the Betti numbers) given a specific threshold value are tracked. In addition, a characteristic radius of the structures and their shape anisotropy can be determined. For a description of the evolution of δT_b, we refer the reader to Kapahtia et al. (2019), ignoring it here owing to its complexity due to the definition of the connected regions and holes as a function of the threshold value as the 21 cm signal transitions from above-/below-average signal regions in the heating epoch to neutral/ionized regions during reionization. However, we emphasize that these authors explored varying the minimum mass hosting star-forming halos and clearly show that different astrophysical parameters can be distinguishable.

4.2.8 Bubble-size Distributions

Throughout reionization and the Cosmic Dawn, the morphology of the 21 cm signal is driven by processes that embed a morphological signature in the 21 cm signal. For example, the ionized H II regions or the hot (above-average signal) or cold (below-average signal) spots in the 21 cm brightness temperature during the heating epoch. Quite simply, if we could measure the distribution of these "bubbles" and how they evolve over cosmic time, we would have a strong discriminant of the populations of sources responsible for the heating and ionization of the IGM and also the spectrum of their emitted radiation. Effectively, this would behave as a statistical distribution function (number of bubbles given a physical scale) analogous to a halo mass function. However, the bubbles do not remain isolated, very quickly overlapping into increasingly large and topologically complex structures. Thus, there is no unique way to characterize these bubbles. Nevertheless, several methods have been explored in order to be able to construct a probabilistic distribution of the bubble sizes.

The simplest is a friends-of-friends approach (e.g., Iliev et al. 2006; Friedrich et al. 2011), which simply connects all cells above (below) a threshold value. However, very rapidly a single large ionized structure exists, filling most of the volume with only a small fraction of isolated regions remaining. The relative volume of this large ionized region and the distribution of the smaller regions can still differentiate reionization morphologies; however, it contains less statistical weight. Alternatively, in Zahn et al. (2007), a sphere is placed on every pixel, averaging the signal across increasingly larger spheres until a radius is found where the average signal is above the threshold value. While this generates a more statistically meaningful distribution of bubbles, these sizes tend to overestimate the size of the topological feature of interest, due to the assumed spherical symmetry.

Recently, more statistically robust methods have been introduced to measure the bubble-size distributions. The first of these is the mean free path method, which uses a Monte Carlo approach by considering a large number of random positions and determining the distance to the edge of the bubble from different random directions (e.g., Mesinger & Furlanetto 2007). This results in an unbiased estimator of the bubble-size distribution (e.g., Lin et al. 2016).

The Watershed method (e.g., Lin et al. 2016) is a more sophisticated approach and has been readily used in the search for cosmological voids. It is a well-known two-dimensional image segmentation algorithm creating contours of constant value (i.e., δT_b), which are treated as levels of a tomographic map. These are then "flooded" to obtain unique locations for the minima (e.g., ionized regions). Remaining in the image-processing regime, Giri et al. (2018) introduced the superpixels method. This uses a region-based method to identify regions of complex shapes (i.e., ionized regions), segmenting these regions into smaller segments called superpixels. The bubble-size distribution is then obtained by averaging the value of the 21 cm brightness temperature within each superpixel before constructing the PDF. Finally, granulometry (Kakiichi et al. 2017) has been investigated, which effectively performs a series of sieving operations to construct a distribution of the sizes of objects which pass through sieves of various sizes and shapes.

In Figure 4.6, we highlight the observed variation in the bubble-size distribution at essentially a fixed redshift/neutral fraction from Geil et al. (2016). The bubble-size distributions here show the characteristic log-normal distribution, with the width of the peak and the relative extents of the asymmetric tails providing sufficient constraining information to distinguish between the various astrophysical models. While several curves appear to produce very similar bubble-size distributions,

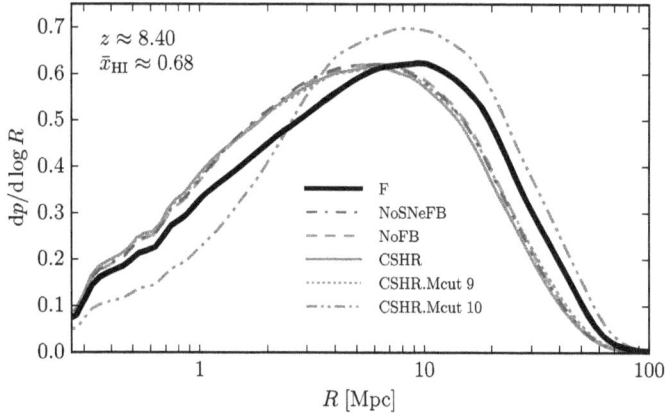

Figure 4.6. Variation in the bubble-size distribution with changes in the underlying astrophysics of the source model. Reproduced from Geil et al. (2016), by permission of Oxford University press on behalf of the Royal Astronomical Society.

folding in multiple-epoch data should be enough to discriminate among various astrophysical parameters.

4.2.9 Individual Images

Tomographic images of the 21 cm signal provide a direct tangible link to the process of reionization, revealing the exact locations of ionized regions and potentially even directly observing the sources responsible with targeted follow-up observations. For example, individual ionized regions can distinguish between ionization driven by galaxies and quasars (e.g., Datta et al. 2012; Majumdar et al. 2012) and also between other astrophysical sources such as galaxies containing either population II or III stars, mini-quasars, high-mass X-ray binaries, or mini halos (Ghara et al. 2016, 2017). This arises either directly from the size or shape of the ionized region (i.e., larger, more spherical regions in the case of active galactic nuclei, AGNs) or from the properties of the 21 cm signal in the immediate vicinity of the ionized region (i.e., sharp or gradual changes in the 21 cm signal indicative of the spectrum of emitted ionizing or X-ray radiation).

However, the signal-to-noise on a 21 cm image is considerably reduced as we cannot perform an averaging to boost the signal, and further, we observe the differential brightness temperature, which is not necessarily a zero-mean quantity, making the definition of an ionized (zero brightness temperature) region complicated. In order to counteract this, matched filters have been proposed (Datta et al. 2007, 2008, 2012; Majumdar et al. 2012; Malloy & Lidz 2013; Datta et al. 2016), which act to minimize the contributions from the noise and foregrounds while maximizing the signal by choosing a filter shape consistent with the expected feature of the signal (i.e., spherical ionized region). The 21 cm image is convolved with filters that vary in size and/or shape until the signal-to-noise of the product peaks. A peak in the signal-to-noise corresponds to a feature in the 21 cm image of the same shape as the filter. Matched filters have been explored both in the context of blind and targeted searches of ionized regions.

Alternatively, one can also extract information directly from a 21 cm image using machine-learning techniques. Rather than searching for a specific feature (i.e., ionized region), a neural network can be constructed to perform a feature classification to identify regions of interest (see Section 4.4.5 for more details). Because the 21 cm data is in the form of a 2D (or 3D) image, the preferred network of choice is a convolutional neural network (CNN). The network is constructed using a training set of either 2D or 3D images (i.e., simulated images varying the astrophysical source properties), which undergo a series of downsamplings, convolutions, and linear transformations, which determine the weights for the various network layers that are used to identify specific features. The network architecture is both user and application specific, and will output user-defined properties or parameters. Once the network is constructed, passing an image of the 21 cm signal to the network outputs the desired properties.

In recent years, the usage of CNNs have gained considerable traction. For example, Hassan et al. (2019) developed a CNN to distinguish between either AGN- or galaxy-driven reionization, La Plante & Ntampaka (2018) extracted the global history of reionization from their CNN, and both Hassan et al. (2019) and Gillet et al. (2019) developed CNNs to extract astrophysical or cosmological parameters directly from the input 21 cm image.

4.2.10 Stacked Images

Owing to the expected low signal-to-noise measurement for a 21 cm image and that the individual ionized bubbles may be too small to be directly observed (compared to the resolution of the radio interferometer), Geil et al. (2017) explored stacking redshifted 21 cm images centered on the known positions of high-redshift galaxies. Such an approach requires a precise determination of the galaxies' redshifts on which the stack is centered; otherwise, the signal will be smeared out.

The resultant stack-averaging of the 21 cm signal produces a notably higher signal-to-noise detection for the mean ionization profile by averaging out the statistical fluctuations within the IGM. If the IGM is in emission (i.e., heating has occurred), the stack-averaged profile is observed in absorption. In contrast, if the IGM is in absorption (i.e., little to no heating), then the stack-averaged profile is in emission. This stack-averaged profile then provides a rough estimate of the typical bubble size surrounding galaxies of known absolute UV magnitude, which is important for determining if reionization is driven by many small galaxies or larger, more biased galaxies. However, there remains a degeneracy between the bubble size and the ionization state of the IGM. A stack of small ionized bubbles can be mimicked by a stack of larger ionized bubbles in a more ionized IGM (owing to the dependence of the mean 21 cm signal on the ionization state of the IGM).

4.2.11 Multifield Approaches

Thus far, we have discussed statistics purely focused on the 21 cm signal. However, information can also be gleaned from combining the 21 cm signal with other independent tracers of the cosmological information. This can either be performed using a cross-correlation approach, where the 21 cm signal is cross-correlated with an alternative tracer of the Galaxy or matter distribution. The advantage of this approach is that the foregrounds between these two fields should be completely uncorrelated, meaning they do not impact the underlying astrophysics of interest. Alternatively, a multitracer approach has been proposed, whereby the ratio of two measured fields (one being the 21 cm signal) is taken, which results in the underlying matter perturbations canceling out, leaving behind the interesting astrophysical information.

The leading example of the former approach is the cross-correlation between the 21 cm signal and Lyα-emitting galaxies (LAEs; Wyithe & Loeb 2007; Wiersma et al. 2013; Sobacchi et al. 2016; Vrbanec et al. 2016; Heneka et al. 2017; Hutter et al. 2017, 2018; Kubota et al. 2018). Here, the idea is that LAEs reside within the ionized regions, where the 21 cm signal is essentially zero (i.e., very little neutral hydrogen). Outside of these regions, the resonant scattering of the Lyα photons by the neutral hydrogen in the IGM strongly attenuates the Lyα line, making these LAEs more difficult to detect; however, the IGM is visible through the 21 cm signal. Thus, on radii smaller than the typical sizes of ionized regions, the signal is anticorrelated. The anticorrelation then decreases to zero, or is slightly positive on much larger radii. The amplitude of this cross-correlation signal and the rate at which the signal transitions from anticorrelation to zero can be used to determine the neutral fraction of the IGM as well as to distinguish different reionization morphologies. Alternatives to LAEs have additionally been explored in the literature (Furlanetto & Lidz 2007; Lidz et al. 2009; Park et al. 2014; Beardsley et al. 2015).

In the multitracer approach, two or more tracers of the same underlying field (i.e., the large-scale matter density) are used to extract astrophysical information. In taking the ratio of these fields, the matter density field cancels, leaving the astrophysics and cosmological terms When using the 21 cm signal in combination with a field tracing the high-redshift galaxies, Fialkov et al. (2019) found that the anisotropy in the ratio can recover the sky-averaged 21 cm signal, distinguishing various models of the spectral energy distribution of the X-ray sources or the Galaxy bias of the high-redshift galaxies. Importantly, in the absence an overlapping high-redshift galaxy survey, any alternative probe of the high-redshift universe can be used, including, for example, planned CO or [C ɪɪ] line intensity mapping of high-redshift galaxies (e.g., Kovetz et al. 2017; Moradinezhad et al. 2019; Moradinezhad & Keating 2019).

4.3 Modeling the 21 cm Signal

The 21 cm signal contains a wealth of cosmological and astrophysical information, too complex to be able to interpret without numerical methods. Our ability to learn about the underlying physical processes driving reionization and the Cosmic Dawn hinges on being able to perform simulations as physically accurate as possible. However, such simulations require an enormous dynamic range, simultaneously resolving the small scales (subkiloparsec) in order to model the individual sources while also exploring the large-scale (~100s of megaparsecs) radiative transfer effects of the high energy (e.g., X-ray) astrophysical processes responsible for heating and ionizing the intergalactic medium. Further,to be able to produce an accurate representation of the observed 21 cm signal requires performing multiple simulations to explore the allowed parameter space.

In this section, we explore the various approaches taken within the literature to be able to simulate the 21 cm signal, with the ultimate goal of learning as much about the underlying physics as possible. These will include describing the various existing approaches to simulate the 21 cm signal, while others will describe novel methods to inform where in parameter space to concentrate our efforts or methods to bypass performing the simulations all together.

4.3.1 Numerical Simulations

Fully numerical simulations are designed to be the most physically accurate approach to investigate the underlying astrophysical processes. These generally consist of simulating the matter (baryons and dark-matter) either through N-body or hydrodynamical methods, and can additionally couple these with radiative transfer (either on-the-fly or post-processing) in order to simulate the radiation transport of the photons responsible for ionizing/heating the IGM. The sheer complexity of the dynamic range required to accurately simulate the reionization process often limits the physical volume of the simulation. However, through advances in computer design and processing power, along with the ongoing development of more sophisticated computational algorithms, we are continually able to push the boundaries with these types of simulations.

The most physically accurate approach is to perform full radiation hydrodynamical simulations capable of modeling the ionizing sources and their interplay with the IGM (e.g., Ciardi et al. 2001; Gnedin & Fan 2006; Finlator et al. 2011; Gnedin 2014; Wise et al. 2014; So et al. 2014; O'Shea et al. 2015; Norman et al. 2015; Ocvirk et al. 2016; Pawlik et al. 2017; Ocvirk et al. 2018; Rosdahl et al. 2018; Wu et al. 2019). However, depending on the mass and spatial resolution of the small scales, these are very restrictive in their physical volume (<100 Mpc). A computationally cheaper approach is to couple a dark-matter only or hydrodynamical simulation with coarser radiative transfer performed in post-processing (e.g., Iliev et al. 2006; McQuinn et al. 2007; Trac & Cen 2007; Ciardi et al. 2012; Iliev et al. 2014; Dixon et al. 2016). Such an approach enables notably larger simulation volumes to be explored (<500 Mpc), which are better suited for exploring the large-scale astrophysical processes; however, they typically require subgrid modeling of the astrophysics.

It is through these classes of simulations where we will gain the largest insights into the astrophysical processes driving reionization and the Cosmic Dawn. However, the computational cost of running these simulations is too prohibitive to perform a proper parameter exploration of the astrophysical processes. Thus, fully numerical simulations will need to be informed about interesting regions of astrophysical parameter space by analytic or seminumerical simulations (Section 4.3.2).

4.3.2 Seminumerical and Analytic Models of the 21 cm Signal

Rather than attempting to self-consistently model all the astrophysical processes, one can instead judiciously make a number of simplifying approximations in order to drastically increase the computational efficiency of the simulations. This can enable (i) huge cosmological volumes (several gigaparsecs) and (ii) large numbers of simulations to be performed for rapid exploration of the astrophysical parameter space. It's with the approaches discussed below that a lot of progress can be made through being able to perform probabilistic searchers in the full astrophysical parameter space.

Seminumerical simulations bypass radiative transfer all together, replacing it with an approximate scheme from which the ionization field can be determined. One of the main approaches to do this is through the excursion-set approach (e.g., Furlanetto et al. 2004), which spatially distributes the ionizing radiation by comparing the number of ionizations against recombinations in decreasing sizes of spherical shells (e.g., Mesinger & Furlanetto 2007; Zahn et al. 2007; Geil & Wyithe 2008; Alvarez et al. 2009; Santos et al. 2010; Mesinger et al. 2011; Visbal et al. 2012; Kim et al. 2013; Fialkov et al. 2014; Majumdar et al. 2014; Choudhury et al. 2015; Hassan et al. 2016; Kulkarni et al. 2016; Mutch et al. 2016; Hutter 2018; Park et al. 2019). The determination of the number of ionizing photons within each grid cell can either be determined from the underlying density field using excursion-set analytic halo mass functions or from identifying the discrete sources directly. Alternatively, one can calibrate a relation between the density field obtained from numerical simulations with properties of reionization. For example, Battaglia et al. (2013) used the relation between the redshift of reionization and the bias of the underlying density field, while Kim et al. (2016) used a relation between the ionization fraction and the density.

Instead of bypassing the radiative transfer altogether, one can instead replace the three-dimensional radiative transfer with a simple one-dimensional radiative transfer and assume spherical symmetry for the distribution of the ionization fronts (Thomas et al. 2009; Ghara et al. 2015) to boost the computational efficiency of the simulations.

Finally, if we are not interested in the three-dimensional structure of reionization, we can construct simplified semianalytic models which can describe the global history of reionization and the Cosmic Dawn. Realistic reionization histories can be obtained by solving the reionization equation,

$$\frac{dQ}{dt} = \frac{n_{\rm ion}}{dt} - \frac{Q}{\bar{t}_{\rm rec}}, \tag{4.18}$$

where Q is the volume average filling factor of the universe, n_{ion} is the number of ionizing photons produced per baryon, and $\bar{t}_{rec}$ is the average recombination timescale for neutral hydrogen. Using the excursion-set approach applied in one dimension (e.g., Furlanetto et al. 2004), we can determine the fraction of collapsed mass above some threshold level (barrier) given some mass threshold (e.g., halo mass). This analytic approach to solve Equation (4.18) has been extensively used in the literature as it gives a rapid and simple estimate of the number of ionizing photons required to reionize the universe (Choudhury & Ferrara 2005, 2006; Haardt & Madau 2012; Kuhlen & Faucher-Giguère 2012; Bouwens et al. 2015; Mitra et al. 2015; Robertson et al. 2015; Khaire et al. 2016; Madau 2017; Mitra et al. 2018; Finkelstein et al. 2019; Mason et al. 2019; Naidu et al. 2019).

Extending from the excursion-set approach applied in one dimension (e.g., Furlanetto et al. 2004), other works have sought semianalytic approaches to construct statistics describing the reionization epoch. For example, Paranjape & Choudhury (2014) developed a model that provides expressions for the bubble-size distribution of the ionized regions, while McQuinn et al. (2005, 2006) explored analytic expressions to describe the 21 cm power spectrum (equivalently, Barkana 2007 explored the two-dimensional correlation function). An alternative semianalytic approach to describe the global 21 cm signal was developed by Mirocha (2014) and Mirocha et al. (2017, 2018).

4.3.3 Intelligent Sampling of the Parameter Space

Understanding the astrophysics of reionization and the Cosmic Dawn will require an exploration of astrophysical parameter space in order to be able to reveal the physical insights describing the observed 21 cm signal. Increasing the complexity, i.e., increasing the number of astrophysical processes or parameters that are simulated can make even these relatively computationally inexpensive seminumerical simulations inefficient for parameter exploration. However, rather than exploring the entire astrophysical parameter space, we can instead make intelligent choices about which combinations of parameters we choose to sample within our simulations to minimize the computational costs. Such approaches can be useful for obtaining astrophysical parameter constraints directly (when combined with a metric such as a distance relation or likelihood which describes how well the model matches an observation), or for optimal designs for constructing training sets for machine-learning approaches (see, e.g., Section 4.4.5).

The most naïve approach is to construct a fixed grid of simulations, sampling evenly along each dimension of the astrophysical parameter space. However, as the number of dimensions increases, even this approach can become computationally intractable. An alternative approach is to consider sampling the parameter grid using a Latin-Hypercube approach (McKay et al. 1979). Here, the idea is to place points in the parameter grid to ensure no astrophysical parameter is sampled twice (see e.g., Kern et al. 2017; Schmit & Pritchard 2018). This approach minimizes the overlap among the astrophysical parameters in the parameter set. Depending on our purpose, we can improve further on the Latin-Hypercube approach. If we have a reasonable idea with regard to the region of parameter space we expect the signal to occur, we can apply a spherical prior on the parameter space (e.g., Schneider et al. 2011). This sphericity drastically reduces the amount of volume in the hypersurface that needs to be filled with samples (i.e., we ignore the edges of the parameter space). As we increase the dimensionality, the gains in reduction in volume become considerable (see, e.g., the discussion in Kern et al. 2017).

Alternatively, rather than directly sampling the astrophysical parameter space by drawing from the astrophysical parameters, one can instead adopt a Jeffreys prior (Jeffreys 1946). Such an approach searches for regions of the parameter space where the observable (e.g., statistic of the 21 cm signal) varies maximally and increases the sampling within such a region, producing coarser sampling elsewhere. Eames et al. (2018) explored use of this Jeffreys prior in sampling the astrophysical parameter space for reionization simulations. In addition to the Latin-Hypercube approach described above, they also explored sampling the parameter space using average-eigenvector sampling and adaptive grid-free sampling. The latter two use a hyper-surface distance-based metric to inform the placement of points of interest in the parameter space. Such approaches can drastically improve the performance of neural-network-based approaches by ensuring optimal designs for the training sets (see Section 4.4.5).

4.3.4 Emulators

If we are only interested in a statistical description of the 21 cm signal (e.g., 21 cm power spectrum), we can bypass performing the entire numerical or seminumerical simulations in favor of constructing an emulator. Emulators are a machine-learning technique which aim to replicate the desired output of a model using either a series of functional curves (for example, polynomials) or a neural network. Once constructed, the emulator provides the desired output statistic describing the signal almost instantaneously, given a set of input astrophysical parameters, which can drastically improve parameter exploration. Emulators have been used within astrophysics for a while (see, e.g., Heitmann et al. 2009; Agarwal et al. 2012; Heitmann et al. 2014, 2016); however, only recently have they been explored in the context of reionization. The construction of an emulator benefits from the intelligent sampling of the astrophysical parameter space (e.g., Section 4.3.3) to minimize the size of the required training set.

Kern et al. (2017) constructed a Gaussian process (GP) based emulator of the 21 cm power spectrum for the seminumerical simulation code 21cmFAST (Mesinger & Furlanetto 2007; Mesinger et al. 2011). This emulator takes as input 11 parameters, 5 cosmological and 6 astrophysical, and outputs the 21 cm power spectrum at any redshift during the reionization and Cosmic Dawn epochs. In order to accelerate the training and construction of the emulator, rather than using the raw 21 cm power spectrum outputs ($\Delta_{21}^2(k,z)$), a data compression step can be performed to minimize the number of features that the emulator needs to learn. In this work, a principal component analysis (PCA) approach was adopted, which minimizes the number of independent pieces of information required to describe the 21 cm power spectrum (i.e., replace the full correlated k-bin range with the sum of a few PCA components). The emulator is then constructed using GP regression to minimize a GP generator function, which is completely defined by its mean and covariance within the astrophysical parameter space. For further details, refer to Kern et al. (2017).

Alternatively, Schmit & Pritchard (2018) constructed an emulator of the 21 cm power spectrum from 21cmFAST using an artificial neural network (see Section 4.4.5 for further details and applications). This neural network takes as input the astrophysical parameters describing the model and directly returns an estimate of the 21 cm power spectrum. Evaluating the network to obtain a new 21 cm power spectrum is effectively instantaneous.

Further, Jennings et al. (2019) explored several different possible techniques to construct an emulator of seminumerical simulations. In addition to two simplistic interpolation techniques (i.e., interpolate the result between points in the training set of data), they also explored neural networks, GPs, and a support vector machine (SVM). They find that a neural network approach performs best (e.g., Schmit & Pritchard 2018); however, note that the more sophisticated GP and SVM approaches could be optimized to outperform a neural network emulator.

Instead of simply emulating a function describing the 21 cm signal statistics, Chardin et al. (2019) recently developed an emulator for the radiative transfer process within reionization simulations. This approach uses deep learning (another machine-learning technique) to output three-dimensional maps of the reionization time in each cell given an input two-dimensional map of the number density of stars and gas. Specifically, it uses a trained auto encoder convolutional neural network, which uses layers of two-dimensional convolution kernels to describe the system that is being emulated.

4.3.5 Characterizing Our Ignorance

The trade-off for increased computational efficiency with the semianalytic and seminumerical approaches described in Section 4.3.2 is the reduction in numerical accuracy. When it comes to extracting information about the astrophysics of reionization and the Cosmic Dawn from the 21 cm signal, we must therefore be fully aware of the shortcomings of our simulations in order to be able to interpret the results. Further, we must understand in what regimes we can confidently trust these approximate simulations.

For example, in Santos et al. (2008), some of the analytic models described in Section 4.3.2 were compared against a hybrid N-body and radiative transfer simulation of cosmic reionization. Under certain regimes, these analytic models are shown to perform relatively well at matching the statistics of the 21 cm power spectrum. Following on from this, Zahn et al. (2011) explored the comparison between radiative transfer simulations and seminumerical simulations. In terms of the 21 cm power spectrum, differences between these approaches were found to be of order of 10% in the power spectrum amplitude. With these sorts of comparisons, we can gain confidence that parameter explorations using approximate techniques can reveal useful astrophysical insights. However, for these algorithm comparisons, only single astrophysical models were considered. To truly characterize our ignorance, a larger, more detailed suite of simulations would be required to fully ascertain how good an approximation they are.

This tens of percent level uncertainty can be added as an additional modeling uncertainty in attempts to recover astrophysical parameters from the 21 cm signal. This effectively acts as an uncertainty floor, with parameter constraints only available where the impact of the astrophysical parameters on the 21 cm signal is larger than this modeling uncertainty. Greig & Mesinger (2015) and Greig et al. (2020) adopted a 25% modeling uncertainty error to the 21 cm power spectrum, finding no biases in the recovery of the astrophysical parameters.

Seminumerical simulations based on the excursion-set formalism are explicitly photon nonconserving. That is, not all ionizing photons are exhausted (i.e., they are lost to the ether) within the simulation. The basis of this is that the analytic solutions from excursion-set theory are photon conserving in one dimension; however, the three-dimensional application in seminumerical simulations is not. When bubble overlap occurs, ionizing photons are not redistributed from the overlap region; they are just unused. This photon nonconservation can result in notable biases in the amplitude of the 21 cm power spectrum (Choudhury & Paranjape 2018) when comparing simulation outputs. However, this photon nonconservation can be trivially accounted for by rescaling the production rates of ionizing photons to match the expected global reionization histories (e.g., Zahn et al. 2011; Majumdar et al. 2014). Alternatively, more robust corrections can be considered (e.g., Paranjape et al. 2016; Choudhury & Paranjape 2018; Molaro et al. 2019).

Provided we are aware of the issues and account for the biases or limitations of the approximate schemes, in most cases we should be able to confidently use these approaches for detailed astrophysical parameter exploration. Of course, in practice it is difficult to verify this for each possible astrophysical model or summary statistic.

4.4 Inference Methods for the 21 cm Signal

In the previous chapters, we have discussed the astrophysics and cosmology encoded within the 21 cm signal. In Section 4.2, we discussed numerous ways to characterize the 21 cm signal to tease out the interesting astrophysics, while in Section 4.3 we discussed the various approaches to modeling the 21 cm signal. The final piece to unlocking the astrophysical information from a 21 cm observation is through performing a robust probabilistic exploration of our simulated astrophysical parameter space. This requires comparing the observed 21 cm signal (or a statistic characterizing it) against the synthetic output from our simulations, taking into account all forms of possible uncertainties (both observational and theoretical).

Ultimately, we are interested in obtaining the PDF of the entire astrophysical parameter space from our simulated model (or the posterior distribution, $P(\theta|\mathbf{d})$; the probability of the model astrophysical parameter set, θ, given the observational data, $\mathbf{d}$). This is what is referred to as the posterior distribution which is obtained from Bayesian statistics through Bayes' theorem,

$$P(\theta|\mathbf{d}) = \frac{P(\mathbf{d}|\theta)P(\theta)}{P(\mathbf{d})}, \tag{4.19}$$

where $P(\mathbf{d}|\theta)$ is the likelihood that describes how likely the astrophysical model described by the parameter set θ describes the data, $P(\theta)$ contains all the prior information we have about the specific astrophysical parameters within our model, and $P(\mathbf{d})$ is the evidence which measures how likely the data is given the model. Throughout this section, we will discuss the various approaches considered in the literature for obtaining the posterior PDF.

4.4.1 Fisher Matrices

One of the simplest and easiest to implement approaches to obtain astrophysical constraints is from the Fisher information matrix (Fisher 1935; see e.g., Tegmark et al. 1997; Coe 2009 for examples how to implement it). This provides a method to quantify the amount of information that an observation contains about any of the unknown parameters in the model parameter set, θ. The Fisher information matrix, $\mathbf{F}$, is calculated via

$$\mathbf{F}_{ij} = \left\langle \frac{\partial^2 \ln \mathcal{L}}{\partial \theta_i \partial \theta_j} \right\rangle = \sum_{\mathbf{x}} \frac{1}{\varepsilon^2(\mathbf{x})} \frac{\partial f(\mathbf{x})}{\partial \theta_i} \frac{\partial f(\mathbf{x})}{\partial \theta_j}, \tag{4.20}$$

where $\mathcal{L}$ is the likelihood function (probability distribution of the observed data given the astrophysical parameter set) and ε characterizes the error on the measurement of the function, $f(\mathbf{x})$, where $\mathbf{x}$ is the data vector describing the function (i.e., for the 21 cm power spectrum, this would be (k,z), the Fourier wavenumber, k, and the redshift, z). Here, θ is the astrophysical parameter set, and we sum the contribution of the partial derivatives of the measured function with each parameter. Parameters that result in large variations in the partial derivatives contain considerable weight and thus highlight which model parameters are sensitive to the function describing the observational data.

Evaluating the Fisher matrix first requires the determination of the maximum likelihood model. We can either assume a fiducial parameter set that maximizes the model, or we can find the model parameter set which is maximal given the observational uncertainties. In the latter case, this can be somewhat computationally expensive, as it requires determining the maximum of our likelihood function.

Once the Fisher information matrix has been calculated, the resultant errors on the model parameters, θ, given the observation can be obtained by inverting $\mathbf{F}_{ij}$. That is,

$$\mathbf{C}_{ij} = \frac{1}{\mathbf{F}_{ij}},$$ (4.21)

where $\mathbf{C}$ is the covariance matrix, with the diagonal entries, $\mathbf{C}_{ii}$, containing the errors on the model parameters (i.e., $\mathbf{C}_{ii} = \sigma_{ii}^2$, where σ is the standard deviation), and the off-diagonal entries describing the two-dimensional joint probabilities which highlights the degeneracies between those two specific model parameters (i.e., how much similar information each parameter holds).

Fundamentally, the Fisher matrix approach assumes that the observation has been performed optimally, where the uncertainty, ε, contains a full description of all sources of error. Further, the inversion of the Fisher matrix to obtain parameter uncertainties assumes that the model parameter set is fully described by a Gaussian likelihood (which is rarely the case in reality). Despite these shortcomings, the Fisher matrix provides an excellent and computationally efficient means to provide astrophysical parameter constraints given an observation of the 21 cm signal.

Forecasting of astrophysical or cosmological parameters from during reionization and the Cosmic Dawn using Fisher matrices has been extensively used in the literature. For example, with the 21 cm power spectrum, Pober et al. (2014) explored the forecasts for parameters responsible for reionization, Liu et al. (2016) explored similar parameters but coupled with cosmological parameters, while Ewall-Wice et al. (2016) considered the astrophysical parameters responsible for X-ray heating. Alternatively, Kubota et al. (2016) explored astrophysics from the variance and skewness of the 1D PDF of 21 cm fluctuations while Shimabukuro et al. (2017) instead investigated the 21 cm bispectrum and Pritchard & Loeb (2010) explored the global 21 cm signal. Pure cosmology or joint cosmology and astrophysics were additionally investigated using analytic expressions of the reionization epoch by McQuinn et al. (2006), Mao et al. (2008), Barger et al. (2009), Visbal et al. (2009), ang Liu et al. (2016).

4.4.2 Fixed Grid Sampling

The simplest approach to recover the true PDF of our astrophysical parameters (i.e., not under the Gaussian approximation applied in the case of the Fisher matrix) is to construct a grid of astrophysical models which are sampled along the dimensions of the allowed astrophysical parameters. This can either be in a fixed, evenly sampled grid along each dimension or a more informed grid sampling as discussed in Section 4.3.3, which reduces the number of models required. Once the grid has been constructed, at each grid point we then compare the observed 21 cm signal against the simulated output given that set of astrophysical parameters to assign it a probability (e.g., the likelihood of it being the correct description of the observed data). With this grid of probabilities, we can then interpolate it to generate a full (continuous) description of the underlying PDF (e.g., $P(\theta|\mathbf{d})$) for this specific astrophysical setup. With this PDF, we can then obtain constraints on any specific astrophysical parameter within our model by marginalizing (integrating) over the uncertainties in all other parameters,

$$P(\theta_1|\mathbf{d}) = N^{-1} \int P(\theta|\mathbf{d})d\theta_2, \, d\theta_3, \, \dots \, , \, d\theta_n, \qquad (4.22)$$

where $\theta = (\theta_1, \theta_2, \dots, \theta_n)$ is the astrophysical parameter set, and N is the normalization constant which ensures $\int P(\theta|\mathbf{d})d\theta = 1$.

Grid-based sampling of astrophysical models has been used throughout the literature both for parameter forecasting, as well as inference from observed upper limits on the 21 cm signal. For example, limits on astrophysical parameters during reionization and the Cosmic Dawn using the 21 cm global signal have been explored with the Experiment to Detect the Global EoR Signature (EDGES; Monsalve et al. 2017, 2018, 2019) as well as the Long Wavelength Array (LWA; Fialkov & Barkana 2019). Grids of seminumerical simulations of reionization have also been used to interpret existing constraints on reionization (e.g., such as the optical depth, τ_e, Planck Collaboration 2018; or limits on the IGM neutral fraction, McGreer et al. 2015) in the context of PDFs of astrophysical model parameters (e.g., Mesinger et al. 2012, 2013; Greig & Mesinger 2017b). The equivalent has also been considered for analytic methods (Choudhury & Ferrara 2005; Barkana 2009; Zahn et al. 2012; Mirocha et al. 2018).

While the fixed grid approach recovers the true underlying PDF and thus is more accurate than the Fisher matrix, it is considerably more computationally expensive due to the increase in the number of simulations required. Further, this assumes that the likelihood space is well behaved and varies smoothly. If the likelihood varies sharply, then finer resolution sampling would be required around those regions of parameter space. It is tractable for a low number of astrophysical parameters, but once this goes beyond just a few free parameters, it can become infeasible. In the next few sections, we discuss techniques to circumvent this.

4.4.3 Bayesian MCMC

Once the dimensionality of our astrophysical parameter space becomes too large to directly sample, we must shift to more approximate methods to recover the true PDF of our astrophysical model. In statistics, this is achieved through Markov Chain Monte Carlo (MCMC) methods, where we can obtain an estimate of our posterior distribution, $P(\theta|\mathbf{d})$, through random sampling. To demonstrate the basic idea, we outline one of the simplest MCMC approaches, the Metropolis–Hastings algorithm (Metropolis et al. 1953; Hastings 1970). We start with a set of initial positions within our astrophysical parameter space, compute the product of the prior and the likelihood (e.g., the numerator of Equation (4.19)), and then take a random jump to a new position in the probability space. In this new position, we compute the product of the prior and likelihood corresponding to the new position, and compare against the previous position. If the new quantity is higher, we keep the new parameter set; if lower, we keep it some fraction of the time according to a probability check (e.g., generate a random number between zero and one, and if its higher than the ratio (which is less than one), we keep it). Following this procedure through a large number of iterations, eventually the chain will converge to the peak of the posterior distribution as it must move to regions where the likelihood is higher (i.e., higher probability). To ensure robustness of the sampling (i.e., avoid local minima in our probability space), we perform many Markov Chains. Once this is complete, simply constructing a histogram of all the sampled points returns an estimate of the posterior distribution (as the most frequent data points in the parameter space are those in regions of higher likelihood).

Returning to Bayes' theorem (Equation (4.19)), all we require for performing an MCMC is the likelihood ($P(\mathbf{d}|\theta)$) and the prior information ($P(\theta)$). Because at all points within the MCMC we take the ratio of the likelihood multiplied by the prior, we never require the evidence ($P(\mathbf{d})$). This is the advantage of the MCMC approach, as it is the evidence that is the most computationally expensive component of Bayes' theorem (to calculate the evidence we need to perform a multidimensional integral over our entire astrophysical parameter space). The last remaining component is the MCMC sampler itself, which is how to determine the new position within the MCMC chain. Within the field of statistics, there are many different types of MCMC samplers, all with their own pros and cons, with plenty of literature to assist in deciding which approach is most suitable given the problem.

Over the past decade, the use of MCMC techniques for the reionization and the Cosmic Dawn have gained considerable attention. For analytic models that generate reionization histories that can be coupled with CMB data and other observational constraints, MCMC approaches have been well established (e.g., Pritchard et al. 2010; Clesse et al. 2012; Morandi & Barkana 2012; Mitra et al. 2015; Gorce et al. 2018; Finkelstein et al. 2019; Mason et al. 2019; Naidu et al. 2019). Analytic models of the global 21 cm signal have also been explored with MCMC, both for interpreting observational limits and also for parameter forecasting for future 21 cm experiments (e.g., Pritchard & Loeb 2010; Harker et al. 2012; Mirocha et al. 2015; Bernardi et al. 2016; Harker et al. 2016).

Only relatively recently have computational resources become efficient enough to be directly applied to seminumerical simulations of the 21 cm signal (e.g., Greig & Mesinger 2015, 2017a, 2018; Greig et al. 2020; Park et al. 2019), that is, to be able to perform a three-dimensional simulation of the 21 cm signal at each set within the MCMC. Alternatively, one can also interpolate over a fixed grid of simulations within an MCMC framework (Hassan et al. 2017). Emulators can additionally be coupled to MCMC techniques, whereby the seminumerical simulation is bypassed with an emulated function describing the 21 cm signal (e.g., Kern et al. 2017; Schmit & Pritchard 2018), drastically increasing the computational efficiency. An alternative hybrid approach is to instead train an emulator during the MCMC, to decide whether a new parameter position is close enough to previously sampled positions from which we can emulate the expected result or whether we need to perform the actual likelihood (simulation) call (e.g., van der Velden et al. 2019).

4.4.4 Model Selection and Nested Sampling

Let's return to Bayes' theorem (Equation (4.19)), and instead write it explicitly as a function of our chosen astrophysical model, **M**,

$$P(\theta|\mathbf{M},\mathbf{d}) = \frac{P(\mathbf{d}|\mathbf{M},\theta)P(\theta|\mathbf{M})}{P(\mathbf{d}|\mathbf{M})}. \tag{4.23}$$

All terms remain as in Equation (4.19); however, let's focus explicitly on the Bayesian evidence, $P(\mathbf{d}|\mathbf{M})$ (which is often expressed as $\mathcal{Z}$). This evidence quantifies how likely the observation data was, given our astrophysical model. In traditional MCMC techniques (see previous section), the evidence is ignored as it can be computationally expensive to evaluate and also it is a redundant calculation as we consistently take the ratio of the likelihood to estimate our positions in our astrophysical parameter space. However, if we instead evaluate the evidence term, $\mathcal{Z}$, we can use this to perform model selection among a variety of potentially plausible astrophysical models (i.e., determine which model provides a better representation of the observational data). Herein lies the value of estimating the Bayesian evidence.

Model selection can be performed by taking the ratio of the Bayesian evidence for each model (known as the Bayes factor),

$$\mathcal{B}_{12} = \frac{P(\mathbf{d}|\mathbf{M}_1)}{P(\mathbf{d}|\mathbf{M}_2)}. \tag{4.24}$$

The Bayes factor informs us of the strength of the evidence for one model against another. In other words, if $\mathcal{B}_{12}$ is greater than unity, then the evidence suggests that model 1 is better than model 2 at modeling our observational data. In Jeffreys (1961), the Jeffreys scale was introduced to provide a means to classify how strongly the evidence for one model is relative to another. Since then, this scale has been modified several times, meaning there is no unique criterion. Here, we adopt the scaling provided by Lee & Wagenmakers (2014), which is broken down as follows: (i) if $\mathcal{B}_{12} > 100$, there is extremely strong evidence for model 1 compared to model 2, (ii) if $30 < \mathcal{B}_{12} < 100$, there is very strong evidence, (iii) if $10 < \mathcal{B}_{12} < 30$, there is strong evidence, (iv) if $3 < \mathcal{B}_{12} < 10$, there is moderate evidence, (v) if $1 < \mathcal{B}_{12} < 3$, there is anecdotal evidence, and (vi) $\mathcal{B}_{12} = 1$, there is no evidence.

Model selection in the context of reionization simulations has only relatively recently been explored in Binnie & Pritchard (2019). Here, various seminumerical simulations of reionization were explored within the context of a mock 21 cm observation. The models differ in how reionization proceeded (i.e., inside out compared to outside in) along with simpler prescriptions for simulating reionization (i.e., excursion-set compared to a simpler pixel-by-pixel definition; Miralda-Escudé et al. 2000). Using model selection, certain models could be ruled out with mock observations from next generation radio interferometers.

In order to be able to perform model selection, we must be able to compute the Bayesian evidence. This can be achieved using the nested sampling algorithm (e.g., Skilling 2004), which performs transformations of the astrophysical parameter space to collapse the multidimensional integral for the evidence into a series of more computationally feasible one-dimensional integrals. A convenient byproduct of the nested sampling algorithm is that in order to compute the evidence, one generates samples from the posterior distribution (e.g., $P(\theta|\mathbf{M},\mathbf{d})$), thus it can perform the same task as a traditional MCMC algorithm. In fact, the particular approach to sampling the astrophysical parameter space within nested sampling can be notably more efficient with regard to the number of required model calls to estimate the posterior distribution. As such, nested sampling is often preferred over traditional MCMC algorithms.

4.4.5 Neural Networks

We have already briefly touched upon neural networks (see Section 4.3.4); however, that was in the context of constructing an emulator of the 21 cm power spectrum given input astrophysical parameters. Instead, in this section we flip the problem around and focus on the usage of neural networks to recover estimates of the underlying astrophysical parameters given some input observational data set, that is, bypass MCMC techniques all together and infer astrophysics directly from a neural network. There have been several works in the literature exploring the validity of using neural networks to perform this task, and we will touch upon the similarities and differences of each of these different approaches.

The fundamental idea of a neural network is to construct a computing system which mimics the behavior of the brain, containing multiple layers of neurons (see Figure 4.7 for an example). A neural network must contain an input layer (which processes the input data), any number of hidden layers, and a final output layer which produces the desired user-defined output (i.e., astrophysical parameters).

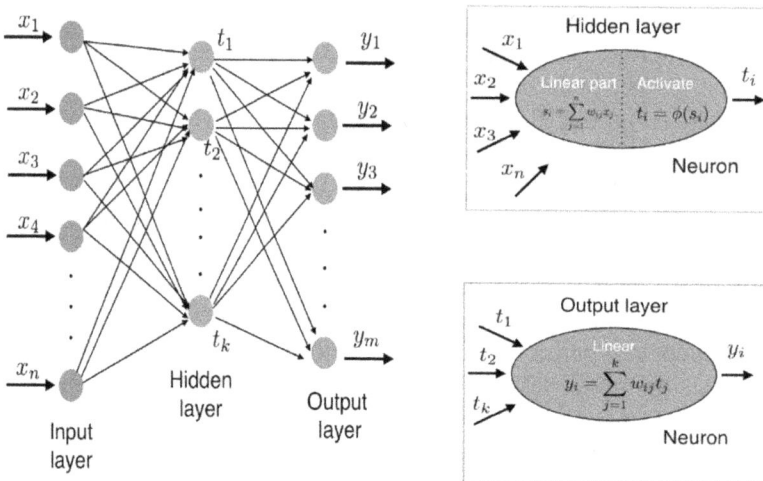

Figure 4.7. An example architecture of an artificial neural network. Reproduced from Shimabukuro & Semelin (2017), by permission of Oxford University press on behalf of the Royal Astronomical Society.

Each neuron in a layer is connected to all other neurons in adjoining layers (it is not connected within its own layer), and the strength of the connection is driven by what is referred to as the activation function (which takes as input the sum of all results from all other neurons multiplied by the weight of each). We must then train the neural network (the weights returned by each neuron) given the inputs and outcomes of the training set we seek to learn. In order to be able to learn the weights, what's referred to as back propagation is often applied. Here, the aim is to estimate the weights for the network by minimizing a cost function. This is an iterative procedure requiring many epochs, and we must be careful not to overfit our network. Thus, the number of epochs is not predetermined but instead is typically taken to be the value when the cost function first begins to plateau. We then validate the accuracy of the neural network by comparing the expected outputs from a new data set (validation data set, which must differ from the training set) against the returned output from the neural network. Once constructed and validated, the network then almost instantaneously returns the desired user-defined outputs given the preferred input format.

Shimabukuro & Semelin (2017) explored the use of artificial neural networks (ANN) in the context of astrophysical parameter recovery from the 21 cm power spectrum. The network was constructed to take as input a training set of seventy 21 cm power spectra varying three astrophysical parameters, and to return the expected value given an input 21 cm power spectrum. Doussot et al. (2019) significantly improved upon this initial ANN approach, considering both a larger training set for the same astrophysical model (2400 models) and supervised learning techniques (techniques to improve the accuracy and optimization of the constructed neural network; see Doussot et al. 2019 for more details).

Rather than only using a statistical descriptor of the 21 cm signal (i.e., power spectrum), we could instead use the expected full two- or three-dimensional 21 cm signal. To do this, we use a CNN, whose network architecture is designed to work with images. The main differences between an ANN and CNN is that the CNN requires feature extraction in order to break down the volume of data into a more manageable set. Feature extraction is performed by a series of convolutional and pooling layers on the input image, with each convolutional layer convolving the result with a number of filters to break down the image (see Figure 4.8 for an example of a CNN) into a simpler set of values to be passed to the neurons of the network. Following feature extraction, the network of neurons is constructed in a similar fashion to that in the case of an ANN.

CNNs have been used in a few different ways for reionization and the Cosmic Dawn. For example, Gillet et al. (2019) used two-dimensional light cones of the simulated 21 cm signal in order to extract the eight underlying astrophysical parameters (from reionization and X-ray heating) from a mock observation. Similarly, Hassan et al. (2019) jointly constrained three cosmological parameters along with three astrophysical parameters from reionization using two-dimensional images of the 21 cm signal at several redshift snapshots. Alternatively, La Plante & Ntampaka (2018) explored recovering the reionization history from

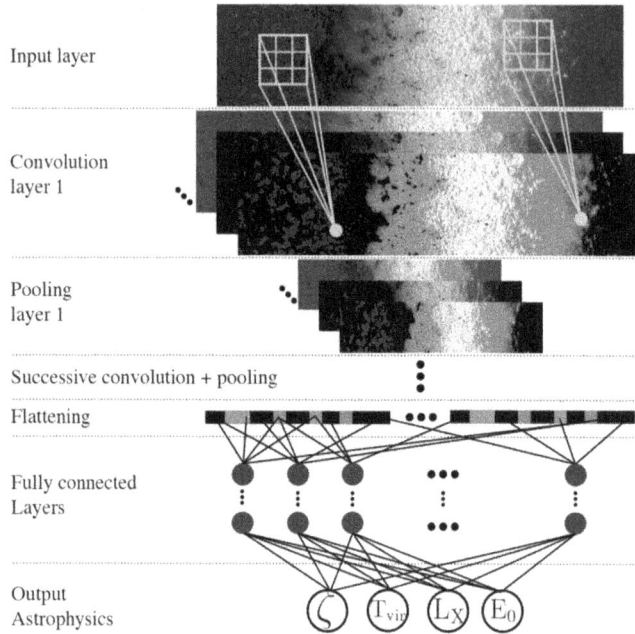

Figure 4.8. An example architecture of a convolutional neural network. Showing the architecture from an input two-dimensional light cone of the 21 cm signal down to the output astrophysical parameters. Reproduced from Gillet et al. (2019), by permission of Oxford University press on behalf of the Royal Astronomical Society.

foreground-dominated two-dimensional 21 cm images at several redshifts while Hassan et al. (2019) used a CNN to perform image classification to determine whether features in the 21 cm image could be distinguished as being driven by galaxies or AGNs.

Neural networks have been shown to perform extremely well at recovering the expected astrophysics from mock observations, and the evaluation of the neural network for parameter estimation is more computationally efficient than MCMC techniques.[1] However, the fundamental issue with neural networks is their relative inability to provide meaningful uncertainties on the recovered parameters, that is, to characterize how inherent uncertainties in the network construction propagate through into the recovered astrophysical constraints. Contrast this to the recovered posterior distributions from MCMC techniques. Nevertheless, neural networks are an extremely useful and valuable tool for inferring astrophysics about the reionization and Cosmic Dawn.

---

[1] Although the construction of the training set can be as slow or slower than an MCMC, the training set may only need to be constructed once whereas an MCMC must be performed each time.

References

Agarwal, S., Abdalla, F. B., Feldman, H. A., Lahav, O., & Thomas, S. A. 2012, MNRAS, 424, 1409

Alesker, S. 1999, Geometriae dedicata, 74, 241

Alvarez, M. A., Busha, M., Abel, T., & Wechsler, R. H. 2009, ApJ, 703, L167

Baek, S., Semelin, B., Di Matteo, P., Revaz, Y., & Combes, F. 2010, A&A, 523, A4

Bag, S., Mondal, R., Sarkar, P., et al. 2019, MNRAS, 485, 2235

Bag, S., Mondal, R., Sarkar, P., Bharadwaj, S., & Sahni, V. 2018, MNRAS, 477, 1984

Bandyopadhyay, B., Choudhury, T. R., & Seshadri, T. R. 2017, MNRAS, 466, 2302

Barger, V., Gao, Y., Mao, Y., & Marfatia, D. 2009, PhLB, 673, 173

Barkana, R. 2007, MNRAS, 376, 1784

Barkana, R. 2009, MNRAS, 397, 1454

Barkana, R., & Loeb, A. 2005, ApJ, 624, L65

Barkana, R., & Loeb, A. 2008, MNRAS, 384, 1069

Battaglia, N., Trac, H., Cen, R., & Loeb, A. 2013, ApJ, 776, 81

Beardsley, A. P., Morales, M. F., Lidz, A., Malloy, M., & Sutter, P. M. 2015, ApJ, 800, 128

Beisbart, C., Dahlke, R., Mecke, K., & Wagner, H. 2002, in Springer Lecture Notes in Physics, Vol. 600, Morphology of Condensed Matter (Berlin: Springer), 238

Bernardi, G., Zwart, J. T. L., Price, D., et al. 2016, MNRAS, 461, 2847

Bharadwaj, S., & Pandey, S. K. 2005, MNRAS, 358, 968

Binnie, T., & Pritchard, J. R. 2019, MNRAS, 487, 1160

Bouwens, R. J., Illingworth, G. D., Oesch, P. A., et al. 2015, ApJ, 811, 140

Bowman, J. D., Morales, M. F., & Hewitt, J. N. 2006, ApJ, 638, 20

Chardin, J., Uhlrich, G., Aubert, D., et al. 2019, arXiv:1905.06958

Chen, Z., Xu, Y., Wang, Y., & Chen, X. 2018, arXiv:1812.10333

Choudhury, T. R., & Ferrara, A. 2005, MNRAS, 361, 577

Choudhury, T. R., & Ferrara, A. 2006, MNRAS, 371, L55

Choudhury, T. R., & Paranjape, A. 2018, MNRAS, 481, 3821

Choudhury, T. R., Puchwein, E., Haehnelt, M. G., & Bolton, J. S. 2015, MNRAS, 452, 261

Ciardi, B., Bolton, J. S., Maselli, A., & Graziani, L. 2012, MNRAS, 423, 558

Ciardi, B., Ferrara, A., Marri, S., & Raimondo, G. 2001, MNRAS, 324, 381

Clesse, S., Lopez-Honorez, L., Ringeval, C., Tashiro, H., & Tytgat, M. H. G. 2012, PhRvD, 86, 123506

Coe, D. 2009, arXiv:0906.4123

Cohen, A., Fialkov, A., & Barkana, R. 2018, MNRAS, 478, 2193

Cohen, A., Fialkov, A., Barkana, R., & Lotem, M. 2017, MNRAS, 472, 1915

Cooray, A., Li, C., & Melchiorri, A. 2008, PhRvD, 77, 103506

Datta, K. K., Bharadwaj, S., & Choudhury, T. R. 2007, MNRAS, 382, 809

Datta, K. K., Friedrich, M. M., Mellema, G., Iliev, I. T., & Shapiro, P. R. 2012, MNRAS, 424, 762

Datta, K. K., Ghara, R., Majumdar, S., et al. 2016, JApA, 37, 27

Datta, K. K., Majumdar, S., Bharadwaj, S., & Choudhury, T. R. 2008, MNRAS, 391, 1900

Dixon, K. L., Iliev, I. T., Mellema, G., Ahn, K., & Shapiro, P. R. 2016, MNRAS, 456, 3011

Doussot, A., Eames, E., & Semelin, B. 2019, arXiv:1904.04106

Eames, E., Doussot, A., & Semelin, B. 2018, arXiv:1811.12198

Elbers, W., & van de Weygaert, R. 2019, MNRAS, 486, 1523

Ewall-Wice, A., Hewitt, J., Mesinger, A., et al. 2016, MNRAS, 458, 2710

Fialkov, A., & Barkana, R. 2014, MNRAS, 445, 213

Fialkov, A., & Barkana, R. 2019, MNRAS, 486, 1763

Fialkov, A., Barkana, R., & Cohen, A. 2018, PhRvL, 121, 011101

Fialkov, A., Barkana, R., & Jarvis, M. 2019, arXiv:1904.10857

Fialkov, A., Barkana, R., & Visbal, E. 2014, Natur, 506, 197

Finkelstein, S. L., D'Aloisio, A., Paardekooper, J.-P., et al. 2019, ApJ, 879, 36

Finlator, K., Davé, R., & Özel, F. 2011, ApJ, 743, 169

Fisher, R. A. 1935, JRSS, 98, 39

Friedrich, M. M., Mellema, G., Alvarez, M. A., Shapiro, P. R., & Iliev, I. T. 2011, MNRAS, 413, 1353

Furlanetto, S. R. 2006, MNRAS, 371, 867

Furlanetto, S. R., & Lidz, A. 2007, ApJ, 660, 1030

Furlanetto, S. R., McQuinn, M., & Hernquist, L. 2006, MNRAS, 365, 115

Furlanetto, S. R., & Oh, S. P. 2016, MNRAS, 457, 1813

Furlanetto, S. R., Oh, S. P., & Briggs, F. H. 2006, PhR, 433, 181

Furlanetto, S. R., Zaldarriaga, M., & Hernquist, L. 2004, ApJ, 613, 1

Geil, P. M., Mutch, S. J., Poole, G. B., et al. 2016, MNRAS, 462, 804

Geil, P. M., Mutch, S. J., Poole, G. B., et al. 2017, MNRAS, 472, 1324

Geil, P. M., & Wyithe, J. S. B. 2008, MNRAS, 386, 1683

Ghara, R., Choudhury, T. R., & Datta, K. K. 2015, MNRAS, 447, 1806

Ghara, R., Choudhury, T. R., & Datta, K. K. 2016, MNRAS, 460, 827

Ghara, R., Choudhury, T. R., Datta, K. K., & Choudhuri, S. 2017, MNRAS, 464, 2234

Gillet, N., Mesinger, A., Greig, B., Liu, A., & Ucci, G. 2019, MNRAS, 484, 282

Giri, S. K., Mellema, G., & Ghara, R. 2018, MNRAS, 479, 5596

Gleser, L., Nusser, A., Ciardi, B., & Desjacques, V. 2006, MNRAS, 370, 1329

Gluscevic, V., & Barkana, R. 2010, MNRAS, 408, 2373

Gnedin, N. Y. 2000, ApJ, 535, 530

Gnedin, N. Y. 2014, ApJ, 793, 29

Gnedin, N. Y., & Fan, X. 2006, ApJ, 648, 1

Gorce, A., Douspis, M., Aghanim, N., & Langer, M. 2018, A&A, 616, A113

Gorce, A., & Pritchard, J. R. 2019, arXiv:1903.11402

Gott III, J. R., Melott, A. L., & Dickinson, M. 1986, ApJ, 306, 341

Greig, B., & Mesinger, A. 2015, MNRAS, 449, 4246

Greig, B., & Mesinger, A. 2017a, MNRAS, 472, 2651

Greig, B., & Mesinger, A. 2017b, MNRAS, 465, 4838

Greig, B., & Mesinger, A. 2018, MNRAS, 477, 3217

Greig, B., Mesinger, A., & Koopmans, L. V. E. 2020, MNRAS, 491, 1398

Haardt, F., & Madau, P. 2012, ApJ, 746, 125

Harker, G., Zaroubi, S., Bernardi, G., et al. 2010, MNRAS, 405, 2492

Harker, G. J. A., Mirocha, J., Burns, J. O., & Pritchard, J. R. 2016, MNRAS, 455, 3829

Harker, G. J. A., Pritchard, J. R., Burns, J. O., & Bowman, J. D. 2012, MNRAS, 419, 1070

Harker, G. J. A., Zaroubi, S., Thomas, R. M., et al. 2009, MNRAS, 393, 1449

Hassan, S., Andrianomena, S., & Doughty, C. 2019, arXiv:1907.07787

Hassan, S., Davé, R., Finlator, K., & Santos, M. G. 2016, MNRAS, 457, 1550

Hassan, S., Davé, R., Finlator, K., & Santos, M. G. 2017, MNRAS, 468, 122

Hassan, S., Liu, A., Kohn, S., & La Plante, P. 2019, MNRAS, 483, 2524

Hastings, W. K. 1970, Biometrika, 57, 97

Heitmann, K., Bingham, D., Lawrence, E., et al. 2016, ApJ, 820, 108

Heitmann, K., Higdon, D., White, M., et al. 2009, ApJ, 705, 156

Heitmann, K., Lawrence, E., Kwan, J., Habib, S., & Higdon, D. 2014, ApJ, 780, 111

Heneka, C., Cooray, A., & Feng, C. 2017, ApJ, 848, 52

Hong, S. E., Ahn, K., Park, C., et al. 2014, JKAS, 47, 49

Hug, D., Schneider, R., & Schuster, R. 2008, St. Petersburg Mathematical Journal, 19, 137

Hutter, A. 2018, MNRAS, 477, 1549

Hutter, A., Dayal, P., Müller, V., & Trott, C. M. 2017, ApJ, 836, 176

Hutter, A., Trott, C. M., & Dayal, P. 2018, MNRAS, 479, L129

Hutter, A., Watkinson, C. A., Seiler, J., et al. 2019, arXiv:1907.04342

Ichikawa, K., Barkana, R., Iliev, I. T., Mellema, G., & Shapiro, P. R. 2010, MNRAS, 406, 2521

Iliev, I. T., Mellema, G., Ahn, K., et al. 2014, MNRAS, 439, 725

Iliev, I. T., Mellema, G., Pen, U. L., et al. 2006, MNRAS, 369, 1625

Jeffreys, H. 1961, Theory of Probability (3rd ed.; Oxford: Clarendon)

Jeffreys, H. 1946, RSPSA, 186, 453

Jennings, W. D., Watkinson, C. A., Abdalla, F. B., & McEwen, J. D. 2019, MNRAS, 483, 2907

Kakiichi, K., Majumdar, S., Mellema, G., et al. 2017, MNRAS, 471, 1936

Kapahtia, A., Chingangbam, P., & Appleby, S. 2019, arXiv:1904.06840

Kapahtia, A., Chingangbam, P., Appleby, S., & Park, C. 2018, JCAP, 2018, 011

Kern, N. S., Liu, A., Parsons, A. R., Mesinger, A., & Greig, B. 2017, ApJ, 848, 23

Khaire, V., Srianand, R., Choudhury, T. R., & Gaikwad, P. 2016, MNRAS, 457, 4051

Kim, H.-S., Wyithe, J. S. B., Park, J., et al. 2016, MNRAS, 455, 4498

Kim, H.-S., Wyithe, J. S. B., Raskutti, S., Lacey, C. G., & Helly, J. C. 2013, MNRAS, 428, 2467

Kittiwisit, P., Bowman, J. D., Jacobs, D. C., Thyagarajan, N., & Beardsley, A. P. 2016, arXiv:1610.06100

Komatsu, E., Dunkley, J., Nolta, M. R., et al. 2009, ApJS, 180, 330

Kovetz, E. D., Viero, M. P., Lidz, A., et al. 2017, arXiv:1709.09066

Kubota, K., Yoshiura, S., Shimabukuro, H., & Takahashi, K. 2016, PASJ, 68, 61

Kubota, K., Yoshiura, S., Takahashi, K., et al. 2018, MNRAS, 479, 2754

Kuhlen, M., & Faucher-Giguère, C.-A. 2012, MNRAS, 423, 862

Kulkarni, G., Choudhury, T. R., Puchwein, E., & Haehnelt, M. G. 2016, MNRAS, 463, 2583

La Plante, P., & Ntampaka, M. 2018, arXiv:1810.08211

Lee, K.-G., Cen, R., Gott, J. R. III, & Trac, H. 2008, ApJ, 675, 8

Lee, M. D., & Wagenmakers, E.-J. 2014, Bayesian Cognitive Modeling: A Practical Course (Cambridge: Cambridge Univ. Press)

Lewis, A. 2011, JCAP, 2011, 026

Lidz, A., Zahn, O., Furlanetto, S. R., et al. 2009, ApJ, 690, 252

Lidz, A., Zahn, O., McQuinn, M., Zaldarriaga, M., & Hernquist, L. 2008, ApJ, 680, 962

Lin, Y., Oh, S. P., Furlanetto, S. R., & Sutter, P. M. 2016, MNRAS, 461, 3361

Liu, A., Pritchard, J. R., Allison, R., et al. 2016, PhRvD, 93, 043013

Madau, P. 2017, ApJ, 851, 50

Majumdar, S., Bharadwaj, S., & Choudhury, T. R. 2012, MNRAS, 426, 3178

Majumdar, S., Mellema, G., Datta, K. K., et al. 2014, MNRAS, 443, 2843

Majumdar, S., Pritchard, J. R., Mondal, R., et al. 2018, MNRAS, 476, 4007

Malloy, M., & Lidz, A. 2013, ApJ, 767, 68

Mao, Y., Tegmark, M., McQuinn, M., Zaldarriaga, M., & Zahn, O. 2008, PhRvD, 78, 023529

Mason, C. A., Naidu, R. P., Tacchella, S., & Leja, J. 2019, arXiv:1907.11332

McGreer, I. D., Mesinger, A., & D'Odorico, V. 2015, MNRAS, 447, 499

McKay, M. D., Beckman, R. J., & Conover, W. J. 1979, Technometrics, 21, 239

McMullen, P. 1997, Supplemento Ai Rendiconti Circ Mat Palermo, 50, 259

McQuinn, M., Furlanetto, S. R., Hernquist, L., Zahn, O., & Zaldarriaga, M. 2005, ApJ, 630, 643

McQuinn, M., Lidz, A., Zahn, O., et al. 2007, MNRAS, 377, 1043

McQuinn, M., Zahn, O., Zaldarriaga, M., Hernquist, L., & Furlanetto, S. R. 2006, ApJ, 653, 815

Mesinger, A., Ferrara, A., & Spiegel, D. S. 2013, MNRAS, 431, 621

Mesinger, A., & Furlanetto, S. 2007, ApJ, 669, 663

Mesinger, A., Furlanetto, S., & Cen, R. 2011, MNRAS, 411, 955

Mesinger, A., Greig, B., & Sobacchi, E. 2016, MNRAS, 459, 2342

Mesinger, A., McQuinn, M., & Spergel, D. N. 2012, MNRAS, 422, 1403

Metropolis, N., Rosenbluth, A. W., Rosenbluth, M. N., Teller, A. H., & Teller, E. 1953, JChPh, 21, 1087

Miralda-Escudé, J., Haehnelt, M., & Rees, M. J. 2000, ApJ, 530, 1

Mirocha, J. 2014, MNRAS, 443, 1211

Mirocha, J., Furlanetto, S. R., & Sun, G. 2017, MNRAS, 464, 1365

Mirocha, J., Harker, G. J. A., & Burns, J. O. 2013, ApJ, 777, 118

Mirocha, J., Harker, G. J. A., & Burns, J. O. 2015, ApJ, 813, 11

Mirocha, J., Mebane, R. H., Furlanetto, S. R., Singal, K., & Trinh, D. 2018, MNRAS, 478, 5591

Mitra, S., Choudhury, T. R., & Ferrara, A. 2015, MNRAS, 454, L76

Mitra, S., Choudhury, T. R., & Ratra, B. 2018, MNRAS, 479, 4566

Molaro, M., Davé, R., Hassan, S., Santos, M. G., & Finlator, K. 2019, arXiv:1901.03340

Monsalve, R. A., Fialkov, A., Bowman, J. D., et al. 2019, ApJ, 875, 67

Monsalve, R. A., Greig, B., Bowman, J. D., et al. 2018, ApJ, 863, 11

Monsalve, R. A., Rogers, A. E. E., Bowman, J. D., & Mozdzen, T. J. 2017, ApJ, 847, 64

Moradinezhad, D. A., & Keating, G. K. 2019, ApJ, 872, 126

Moradinezhad, D. A., Keating, G. K., & Fialkov, A. 2019, ApJ, 870, L4

Morandi, A., & Barkana, R. 2012, MNRAS, 424, 2551

Mutch, S. J., Geil, P. M., Poole, G. B., et al. 2016, MNRAS, 462, 250

Naidu, R. P., Tacchella, S., Mason, C. A., et al. 2019, arXiv:1907.13130

Norman, M. L., Reynolds, D. R., So, G. C., Harkness, R. P., & Wise, J. H. 2015, ApJS, 216, 16

Obreschkow, D., Power, C., Bruderer, M., & Bonvin, C. 2013, ApJ, 762, 115

Ocvirk, P., Aubert, D., Sorce, J. G., et al. 2018, arXiv:1811.11192

Ocvirk, P., Gillet, N., Shapiro, P. R., et al. 2016, MNRAS, 463, 1462

O'Shea, B. W., Wise, J. H., Xu, H., & Norman, M. L. 2015, ApJ, 807, L12

Pan, T., & Barkana, R. 2012, arXiv:1209.5751

Paranjape, A., & Choudhury, T. R. 2014, MNRAS, 442, 1470

Paranjape, A., Choudhury, T. R., & Padmanabhan, H. 2016, MNRAS, 460, 1801

Park, J., Kim, H.-S., Wyithe, J. S. B., & Lacey, C. G. 2014, MNRAS, 438, 2474

Park, J., Mesinger, A., Greig, B., & Gillet, N. 2019, MNRAS, 484, 933

Patil, A. H., Zaroubi, S., Chapman, E., et al. 2014, MNRAS, 443, 1113

Pawlik, A. H., Rahmati, A., Schaye, J., Jeon, M., & Dalla Vecchia, C. 2017, MNRAS, 466, 960

Pillepich, A., Porciani, C., & Matarrese, S. 2007, ApJ, 662, 1

Planck Collaboration 2018, arXiv:1807.06209

Pober, J. C., Liu, A., Dillon, J. S., et al. 2014, ApJ, 782, 66

Pritchard, J. R., & Furlanetto, S. R. 2007, MNRAS, 376, 1680

Pritchard, J. R., & Loeb, A. 2010, PhRvD, 82, 023006

Pritchard, J. R., & Loeb, A. 2012, RPPh, 75, 086901

Pritchard, J. R., Loeb, A., & Wyithe, J. S. B. 2010, MNRAS, 408, 57

Robertson, B. E., Ellis, R. S., Furlanetto, S. R., & Dunlop, J. S. 2015, ApJ, 802, L19

Rosdahl, J., Katz, H., Blaizot, J., et al. 2018, MNRAS, 479, 994

Ross, H. E., Dixon, K. L., Iliev, I. T., & Mellema, G. 2017, MNRAS, 468, 3785

Sahni, V., Sathyaprakash, B. S., & Shandarin, S. F. 1998, ApJ, 495, L5

Santos, M. G., Amblard, A., Pritchard, J., et al. 2008, ApJ, 689, 1

Santos, M. G., Cooray, A., & Knox, L. 2005, ApJ, 625, 575

Santos, M. G., Ferramacho, L., Silva, M. B., Amblard, A., & Cooray, A. 2010, MNRAS, 406, 2421

Schmalzing, J., & Buchert, T. 1997, ApJ, 482, L1

Schmit, C. J., & Pritchard, J. R. 2018, MNRAS, 475, 1213

Schneider, M. D., Holm, Ó., & Knox, L. 2011, ApJ, 728, 137

Schröder-Turk, G. E., Mickel, W., Kapfer, S. C., et al. 2013, NJPh, 15, 083028

Schröder-Turk, G. E., Kapfer, S., Breidenbach, B., Beisbart, C., & Mecke, K. 2010, JMic, 238, 57

Seiler, J., Hutter, A., Sinha, M., & Croton, D. 2019, MNRAS, 487, 5739

Shimabukuro, H., & Semelin, B. 2017, MNRAS, 468, 3869

Shimabukuro, H., Yoshiura, S., Takahashi, K., Yokoyama, S., & Ichiki, K. 2015, MNRAS, 451, 467

Shimabukuro, H., Yoshiura, S., Takahashi, K., Yokoyama, S., & Ichiki, K. 2016, MNRAS, 458, 3003

Shimabukuro, H., Yoshiura, S., Takahashi, K., Yokoyama, S., & Ichiki, K. 2017, MNRAS, 468, 1542

Skilling, J. 2004, in AIP Conf. Proc. 735, Bayesian Inference and Maximum Entropy Methods in Science and Engineering: 24th International Workshop on Bayesian Inference and Maximum Entropy Methods in Science, ed. R. Fischer, R. Preuss, & U. Von Toussain (Melville, NY: AIP), 395

So, G. C., Norman, M. L., Reynolds, D. R., & Wise, J. H. 2014, ApJ, 789, 149

Sobacchi, E., Mesinger, A., & Greig, B. 2016, MNRAS, 459, 2741

Tegmark, M., Taylor, A. N., & Heavens, A. F. 1997, ApJ, 480, 22

Thomas, R. M., Zaroubi, S., Ciardi, B., et al. 2009, MNRAS, 393, 32

Trac, H., & Cen, R. 2007, ApJ, 671, 1

Trott, C. M. 2016, MNRAS, 461, 126

Trott, C. M., Watkinson, C. A., Jordan, C. H., et al. 2019, PASA, 36, e023

van der Velden, E., Duffy, A. R., Croton, D., Mutch, S. J., & Sinha, M. 2019, ApJS, 242, 22

Visbal, E., Barkana, R., Fialkov, A., Tseliakhovich, D., & Hirata, C. M. 2012, Natur, 487, 70

Visbal, E., Loeb, A., & Wyithe, S. 2009, JCAP, 2009, 030

Vrbanec, D., Ciardi, B., Jelić, V., et al. 2016, MNRAS, 457, 666

Wang, Y., Park, C., Xu, Y., Chen, X., & Kim, J. 2015, ApJ, 814, 6

Watkinson, C. A., Mesinger, A., Pritchard, J. R., & Sobacchi, E. 2015, MNRAS, 449, 3202

Watkinson, C. A., & Pritchard, J. R. 2014, MNRAS, 443, 3090

Watkinson, C. A., & Pritchard, J. R. 2015, MNRAS, 454, 1416

Watkinson, C. A., Giri, S. K., Ross, H. E., et al. 2019, MNRAS, 482, 2653

Watkinson, C. A., Majumdar, S., Pritchard, J. R., & Mondal, R. 2017, MNRAS, 472, 2436

Wiersma, R. P. C., Ciardi, B., Thomas, R. M., et al. 2013, MNRAS, 432, 2615

Wise, J. H., Demchenko, V. G., Halicek, M. T., et al. 2014, MNRAS, 442, 2560

Wu, X., Kannan, R., Marinacci, F., Vogelsberger, M., & Hernquist, L. 2019, MNRAS, 488, 419

Wyithe, J. S. B., & Loeb, A. 2007, MNRAS, 375, 1034

Wyithe, J. S. B., & Morales, M. F. 2007, MNRAS, 379, 1647

Yoshiura, S., Shimabukuro, H., Takahashi, K., & Matsubara, T. 2017, MNRAS, 465, 394

Yoshiura, S., Shimabukuro, H., Takahashi, K., et al. 2015, MNRAS, 451, 266

Zahn, O., Lidz, A., McQuinn, M., et al. 2007, ApJ, 654, 12

Zahn, O., Mesinger, A., McQuinn, M., et al. 2011, MNRAS, 414, 727

Zahn, O., Reichardt, C. L., Shaw, L., et al. 2012, ApJ, 756, 65

The Cosmic 21-cm Revolution
Charting the first billion years of our universe
Andrei Mesinger

Chapter 5

21 cm Observations: Calibration, Strategies, Observables

Gianni Bernardi

This chapter aims to provide a review of the basics of 21 cm interferometric observations and its methodologies. A summary of the main concepts of radio interferometry and their connection with the 21 cm observables—power spectra and images—is presented. I then provide a review of interferometric calibration and its interplay with foreground separation, including the current open challenges in the calibration of 21 cm observations. Finally, a review of 21 cm instrument designs in light of calibration choices and observing strategies follows.

5.1 Interferometry Overview

The Van Cittert–Zernike theorem expresses the fundamental relationship between the sky spatial brightness (or brightness distribution) I and the quantity measured by an interferometer, i.e., the visibility V (e.g., Thompson et al. 2017):

$$V_{ij}(\mathbf{b},\nu) = \int_\Omega \bar{I}(\hat{\sigma},\nu)\, e^{-2\pi i \nu \frac{\mathbf{b}\cdot\hat{\sigma}}{c}} d\hat{\sigma}, \tag{5.1}$$

where $\mathbf{b}$ is the baseline vector that separates the antenna i and antenna j, ν is the observing frequency, $\hat{\sigma}$ is the observing direction (see Figure 5.1), c is the speed of light, and the integral is taken over the source size Ω. It can be seen in Figure 5.1 that the celestial signal travels an extra path between the two antennas, and that the length corresponds to a geometrical time delay $\tau = (\mathbf{b}\cdot\hat{\sigma})/c$, where the word "geometrical" refers to the fact that the delay depends upon the source position in the sky and the relative separation between the two antennas. Equation (5.1) can be derived as the output of the correlator, the digital equipment responsible for combining signals from antenna pairs. The voltage induced by a celestial source at any antenna can be written in a generic form as $(v \cos 2\pi\nu t)$, and the correlator

doi:10.1088/2514-3433/ab4a73ch5

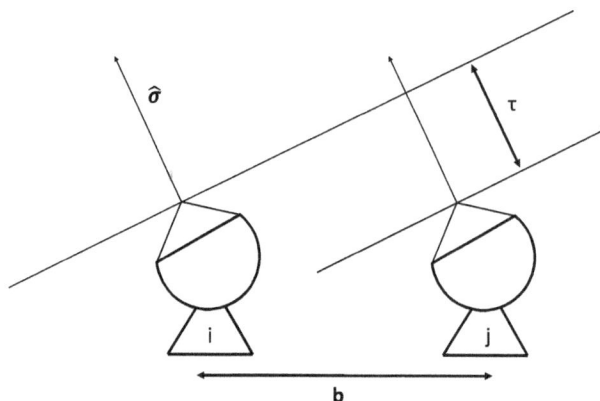

Figure 5.1. A standard schematic of the two-element interferometer.

will output the time average product of the voltages r measured by antennas i and j, respectively,

$$r = \langle v_i(t)v_j(t)\rangle = \langle v^2 \cos(2\pi\nu t)\cos[2\pi\nu(t-\tau)]\rangle$$
$$= \left\langle v^2 \frac{\cos(2\pi\nu\tau) + \cos(4\pi\nu t - 2\pi\nu\tau)}{2}\right\rangle. \tag{5.2}$$

While the first term of the right-hand side of Equation (5.2) varies slowly with Earth's rotation, the second term oscillates rapidly for any typical radio observation ($\nu > 10$ MHz) and averages to zero, leading to

$$r(\tau) \approx \frac{v^2}{2}\cos 2\pi\nu\tau, \tag{5.3}$$

which is a sinusoidal pattern termed "fringe." Equation (5.3) actually represents the contribution to the fringe from the pointing direction. We can obtain the contribution from the whole source by integrating over the source size and adding the odd (sine) to the even (cosine) fringe component to form a general, complex-valued fringe R,

$$R(\mathbf{b},\nu) = \int_\Omega r(\tau)d\tau = \int_\Omega \frac{v^2}{2}(\cos 2\pi\nu\tau - i\sin 2\pi\nu\tau)$$
$$= \int_\Omega \frac{v^2}{2}e^{-2\pi i\nu\frac{\mathbf{b}\cdot\hat{\sigma}}{c}}d\hat{\sigma}, \tag{5.4}$$

where I have substituted the definition of geometrical delay in the last step. If we note that the $v^2/2$ voltage square term depends upon the direction in the sky as it is proportional to the source brightness, Equation (5.4) is essentially equivalent to Equation (5.1) and shows how the correlator outputs directly the spatial coherence function of the sky emission, i.e., the visibility.

The sky brightness distribution I does not appear directly in the Van Cittert–Zernike theorem, but filtered by the antenna primary beam response A that depends upon the direction in the sky and the wavelength, i.e., $\bar{I}(\hat{\sigma},\nu) = A(\hat{\sigma},\nu)\,I(\hat{\sigma},\nu)$. The response of the primary beam attenuates the sky emission away from the pointing direction, effectively reducing the field of view Ω_F of the instrument. Generally speaking, the size of the field of view is essentially given by the antenna diameter D,

$$\Omega_F \approx \frac{\lambda}{D}, \tag{5.5}$$

where λ is the observing wavelength.

The Van Cittert–Zernike theorem that defines the visibility function is often rewritten in a different coordinate system, i.e., using the components of the baseline vector (u,v,w), where (u,v) are the components of the baseline vector in the plane of the array and w is the component along the pointing direction σ_0 (Figure 5.2). The sky position in the $\hat{\sigma}$ direction can be decomposed into the (l,m) components parallel to the plane of the sky and the n component along the w axis. In this system, coordinates can be rewritten as in Thompson et al. (2017):

$$\frac{\nu\,\mathbf{b}\cdot\hat{\sigma}}{c} = ul + vm + wn,$$

$$\frac{\nu\,\mathbf{b}\cdot\hat{\sigma}_0}{c} = w, \tag{5.6}$$

$$d\Omega = \frac{dl\,dm}{n} = \frac{dl\,dm}{\sqrt{1 - l^2 - m^2}},$$

and Equation (5.1) then becomes

$$V_{ij}(u,v,w,\nu) = \int_\Omega \bar{I}(l,m,\nu)\,e^{-2\pi i\left(ul+vm+w\left(\sqrt{1-l^2-m^2}-1\right)\right)}$$

$$\times \frac{dl\,dm\,dn}{\sqrt{1 - l^2 - m^2}}. \tag{5.7}$$

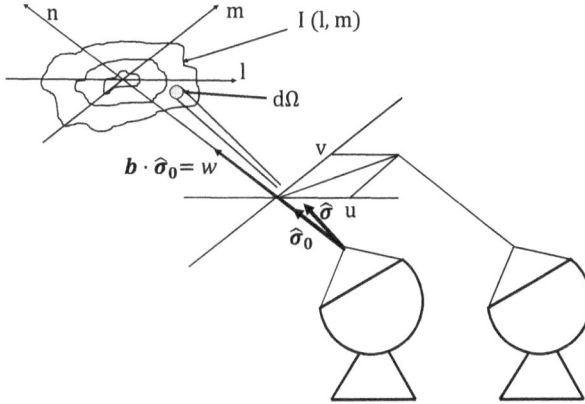

Figure 5.2. Cartoon representation of the coordinate system used for interferometric imaging. The (l,m) plane is tangent to the sky.

Although low-frequency radio observations are intrinsically wide field, for the purpose of studying the 21 cm observables, we can reduce Equation (5.7) to a two-dimensional Fourier transform,

$$V_{ij}(u,v,\nu) = \int_\Omega \bar{I}(l,m,\nu)\, e^{-2\pi i(ul+vm)} dl\, dm. \tag{5.8}$$

Equation (5.8) indicates that an interferometer measures the two-dimensional Fourier transform of the spatial sky brightness distribution. If our goal is to reconstruct the sky brightness distribution, Equation (5.8) can be inverted into its corresponding Fourier pair,

$$\bar{I}(l,m,\nu) = \int V_{ij}(u,v,\nu)\, e^{2\pi i(ul+vm)} du\, dv. \tag{5.9}$$

Equation (5.9) is, however, a poor reconstruction of the sky brightness distribution as only one Fourier mode is sampled at a single time instance. Strictly speaking, indeed, all of the quantities in Equations (5.8) and (5.9) are time variable. In most cases, the time dependence of the primary beam and the sky brightness distribution can be ignored; however, this is not the case for the visibility V as the projection of the baseline vector with respect to the source direction changes significantly throughout a long (e.g., a few hour) track. In this way, many measurements of the visibility coherence function V can be made as (u,v) changes with time, allowing for a better reconstruction of the $\bar{I}(l,m,\nu)$ function. This method is commonly referred to as filling the uv plane via the Earth rotation synthesis and was invented by Ryle & Hewish (1960). The other (complementary) way to fill the uv plane is to deploy more antennas on the ground in order to increase the number of instantaneous measurements of independent Fourier modes. If N antennas are connected in an interferometric array, $N(N-1)/2$ instantaneous measurements are made.

The combination of a large number of antennas and the Earth rotation synthesis defines the sampling function $S(u,v)$ in the uv plane. In any real case, Equation (5.9) can therefore be rewritten as

$$\bar{I}_D(l,m,\nu) = \int S(u,v,\nu)V(u,v,\nu)\, e^{2\pi i(ul+vm)} du\, dv, \tag{5.10}$$

where $\bar{I}_D$ indicates the sky brightness distribution sampled at a finite number of (u,v) points (often termed dirty image) and where the explicit dependence on the antenna pair was dropped for simplicity. Using the convolution theorem, Equation (5.10) can be rewritten as

$$\bar{I}_D(l,m,\nu) = \tilde{S}V = \tilde{S}*\tilde{V} = \text{PSF}(l,m,\nu)*\bar{I}(l,m,\nu), \tag{5.11}$$

where the tilde indicates the Fourier transform, * the convolution operation, and PSF is the point-spread function, i.e., the response of the interferometric array to a point source, which, in our case, is also the Fourier transform of the uv coverage.

The sampling function always effectively reduces the integral over a finite (often not contiguous) area of the uv plane. In particular, the sampled uv plane is restricted between a minimum uv distance that cannot be shorter than the antenna[1] size and the largest separation between antennas, i.e., the maximum baseline $\mathbf{b}_{max}$. The maximum baseline also sets the maximum angular resolution θ_b,

$$\theta_b \approx \frac{\lambda}{|\mathbf{b}_{max}|}. \tag{5.12}$$

The incomplete sampling of the uv space leads to a PSF that has "side lobes," i.e., nulls and secondary lobes that can often contaminate fainter true sky emission. The best reconstruction of the sky brightness distribution $\bar{I}$ requires deconvolution of the dirty image from the PSF.

Figures 5.3 and 5.4 provide an example of the sampling $S(u,v)$ and the corresponding point-spread function. A single baseline essentially imprints a (sinusoidal) fringe pattern on the sky, whose period and phase depend upon the baseline length and orientation, respectively (Equation (5.1)). The combination of more baselines of different lengths and orientations improves the sampling function until a good-quality point-spread function is obtained.

5.2 21 cm Observables: Power Spectra and Images

The ultimate goal of 21 cm observations is to image the spatial distribution of the 21 cm signal as a function of redshift, also known as 21 cm tomography. Given the current theoretical predictions, such observations need to achieve millikelvin sensitivity on a few arcminute angular scales (see Chapters 1–3 in this book). Most of the current arrays, however, only have the sensitivity to perform a statistical detection of the 21 cm signal, i.e., to measure its power spectrum. Given an intensity field T, a function of the three-dimensional spatial coordinate $\mathbf{x}$, its power spectrum $P(k)$ is defined as

$$\langle \tilde{T}^*(\mathbf{k})\tilde{T}(\mathbf{k}')\rangle = (2\pi)^3 P(k)\delta^3(\mathbf{k} - \mathbf{k}'), \tag{5.13}$$

where $\langle\rangle$ indicates the ensemble average, $\mathbf{k}$ is the Fourier conjugate of $\mathbf{x}$, tilde the Fourier transform, $*$ the conjugate operator, k the magnitude of the $\mathbf{k}$ vector, and δ the Dirac delta function. In 21 cm observations, power spectra can be computed from interferometric image cubes after deconvolution of the dirty image $\bar{I}_D(l,m,\nu)$ from the point-spread function (e.g., Pen et al. 2009; Harker et al. 2010; Beardsley et al. 2016; Patil et al. 2017). Alternatively, the 21 cm power spectrum can be estimated directly from the interferometric visibilities. Equation (5.8) already shows that the interferometer is a "natural" spatial power-spectrum instrument (e.g., White et al. 1999 and Figures 5.3, 5.4). Visibilities can be further Fourier transformed along the frequency axis (the so-called delay transform; Parsons et al. 2012):

$$\tilde{V}_{ij}(u,v,\tau) = \int_B V(u,v,\nu)\, e^{-2\pi i\nu\tau} d\nu, \tag{5.14}$$

---

[1] In this chapter, I use the words "antenna" and "station" interchangeably to indicate the correlated elements even if, in the literature, they are normally used to indicate a dish and a cluster of dipoles, respectively.

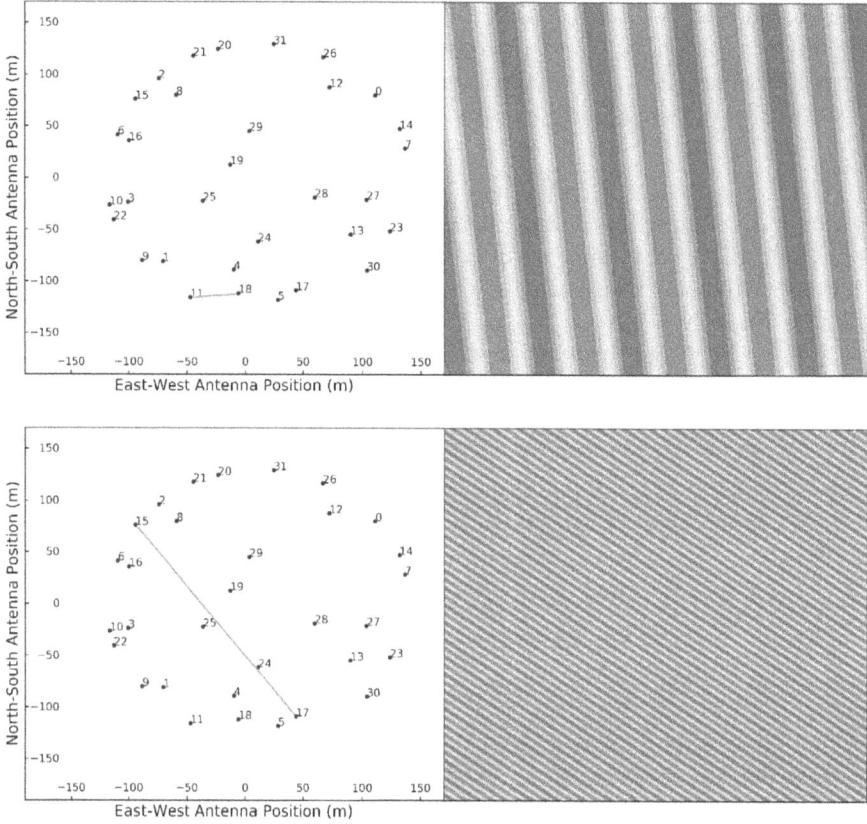

Figure 5.3. Left panel: one baseline selected (indicated with a red line) from 32 antennas distributed within a 350 m circle (taken from Jacobs et al. 2011). Right panel: corresponding point-spread function. For a single baseline case, the point-spread function is essentially a sinusoidal fringe pattern whose period is inversely proportional to the baseline length, i.e., the pattern corresponding to ~50 m baseline (top) oscillates approximately seven times slower than a ~350 m baseline (bottom). The fringe phase (i.e., the pattern orientation) is given by the baseline orientation.

where B is the observing bandwidth and the delay τ is the Fourier conjugate of ν.[2] The delay transform is therefore proportional to the three-dimensional power spectrum (Parsons et al. 2012),

$$P(k) \propto \tilde{V}_{ij}(|\mathbf{b}|, \tau), \qquad (5.15)$$

where the proportionality constant transforms the visibility units into power units (Parsons et al. 2012). The observer units $(\mathbf{b}, \tau)$ map directly to k modes perpendicular and parallel to the line of sight (e.g., Morales & Hewitt 2004),

---

[2] The delay variable here is almost equivalent to the geometrical delay and that is why I used the same symbol (see Parsons et al. 2012 for details).

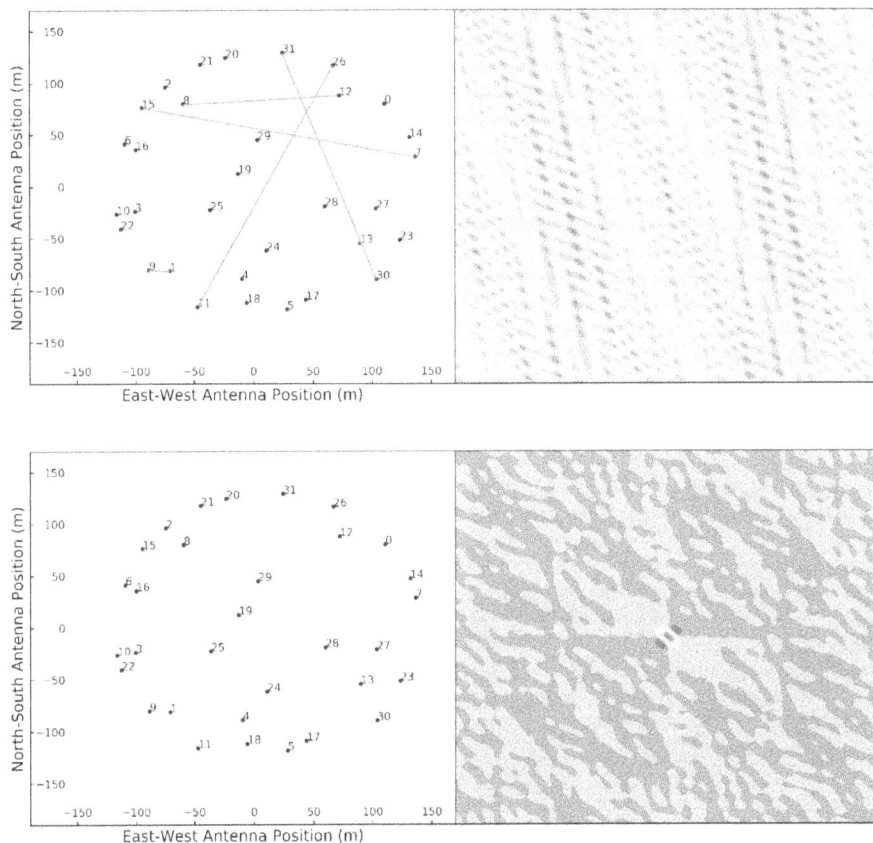

Figure 5.4. Same as Figure 5.3, but including five baselines with different lengths and orientations (red lines, top panel) and all the baselines (for $N = 32$ there are 496 baselines; bottom panel) simultaneously. The fringe pattern is already noticeably different when five baselines are included with respect to the single baseline, although a clean point-spread function only appears when all the baselines are used simultaneously (bottom panel).

$$k_\perp = \frac{2\pi|\mathbf{b}|}{D_c} = \frac{2\pi\sqrt{u^2 + v^2}}{D_c}, \quad k_\parallel = \frac{2\pi f_{21} H_0 E(z)}{c(1 + z)^2} \tau, \tag{5.16}$$

where D_c is the transverse comoving distance, $f_{21} = 1421$ MHz, H_0 is the Hubble constant, and $E(z) = \sqrt{\Omega_m(1 + z)^3 + \Omega_k(1 + z)^2 + \Omega_\Lambda}$. Due to the dependence of the baseline length on frequency, Equation (5.15) is only valid for short baselines, typically shorter than a few hundred meters, for which the baseline length can be considered constant across the bandwidth, and lines of constant $k_\parallel$ are essentially orthogonal to the $k_\perp$ axis (Parsons et al. 2012).

Equation (5.14) does not only provide a link between visibilities and three-dimensional power spectra, but also introduces the concept of "horizon limit," i.e., the maximum physical delay allowed $\tau_{max} = \frac{|\mathbf{b}|}{c}$, where c is the speed of light. The most relevant implication of the existence of a horizon limit is the definition of a region in the two-dimensional $(k_\parallel, k_\perp)$ power-spectrum space, where smooth-spectrum

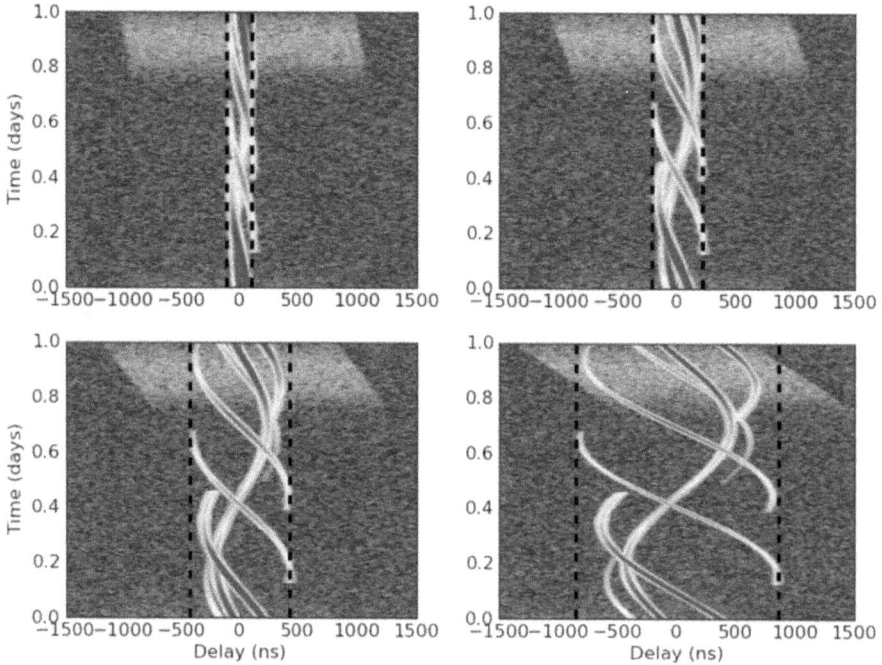

Figure 5.5. Amplitude of delay transformed visibilities as a function of time and delay for a 32 (top let), 64 (top right), 128 (bottom left), and 256 m (bottom right) baseline, respectively (reproduced from Parsons et al. 2012 © 2012. The American Astronomical Society. All rights reserved). A number of smooth-spectrum point sources are simulated as foregrounds, and their tracks are clearly bound within the horizon limit (black dashed line). The cyan emission is a fiducial 21 cm model that has power up to high delays regardless of the baseline length. The 21 cm signal is, in principle, directly detectable outside the horizon limit (EoR window) without foreground contamination.

foregrounds are confined, leaving the remaining area uncontaminated in order to measure the 21 cm signal (the so-called "Epoch of Reionization (EoR) window"; Figure 5.5). Foregrounds can therefore be "avoided" with no requirements for subtraction (e.g., Morales et al. 2012; Vedantham et al. 2012; Pober et al. 2013; Thyagarajan et al. 2013; see also Chapter 6 in this book). The choice of a foreground avoidance strategy versus subtraction plays an important role in planning an experiment, its related observing strategy, and the array calibration strategy.

The requirements for image tomography are the same as for high brightness sensitivity observations of diffuse emission like the cosmic microwave background (e.g., Halverson et al. 2002; Dickinson et al. 2004; Readhead et al. 2004). The 21 cm spatial distribution throughout cosmic reionization has structures on 5′–10′ up to degree scales (e.g., Majumdar et al. 2012; Datta et al. 2012; Mellema et al. 2013; Kakiichi et al. 2017). In order to image 21 cm fluctuations, a maximum baseline of the order of a few kilometers is required to obtain a resolution of a few arcminutes in the 100–200 MHz range, together with a filled *uv* plane in order to accurately reconstruct their complex spatial structure. A filled *uv* plane also leads to a point-spread function with very low side lobes, making the deconvolution process easier (see the bottom-right panel of Figure 5.4 for an example of a densely sampled uv plane that leads to a good-quality point-spread function). The most stringent requirements for image tomography remain the accurate foreground separation and, as I will review in the next section, the related instrumental calibration.

5.3 Interferometric Calibration and 21 cm Observations

Celestial radio signals always experience a corruption when observed with an interferometric array, due to the nonideal instrumental response that is corrected in post-processing in a process that is known as interferometric calibration. Calibration relies on the definition of a data model where the corruptions are described by antenna-based quantities known as Jones matrices. Such data model is known as the interferometric measurement equation (Hamaker et al. 1996; Smirnov 2011a, 2011b, 2011c).

If antenna 1 and antenna 2 measure two orthogonal, linear polarizations x and y, the cross-polarization visibility products can be grouped in a 2×2 complex matrix **V**,

$$\mathbf{V}_{12}(u,v,\nu) \equiv \begin{bmatrix} V_{12,xx}(u,v,\nu) & V_{12,xy}(u,v,\nu) \\ V_{12,yx}(u,v,\nu) & V_{12,yy}(u,v,\nu) \end{bmatrix}. \tag{5.17}$$

The sky brightness distribution I can also be written as a 2×2 matrix **B** using the Stokes parameters as a polarization basis,

$$\mathbf{B}_I(l,m,\nu) \equiv \frac{1}{2} \begin{bmatrix} I(l,m,\nu) + Q(l,m,\nu) & U(l,m,\nu) + iV(l,m,\nu) \\ U(l,m,\nu) - iV(l,m,\nu) & I(l,m,\nu) - Q(l,m,\nu) \end{bmatrix}. \tag{5.18}$$

At this point, Equation (5.7) can be written by including the corruptions represented by the complex Jones matrices **J** (Hamaker et al. 1996; Smirnov 2011a),

$$\mathbf{V}_{12}(u,v,\nu) = \mathbf{J}_1 \left(\int_\Omega \mathbf{B}_I(l,m,\nu) \, e^{-2\pi i(ul+vm)} dl \; dm \right) \mathbf{J}_2^H, \tag{5.19}$$

where H is the Hermitian operator.

Equation (5.19) is known as the measurement equation and is the core of interferometric calibration. For an array with N antennas, Equation (5.19) can be written for each of the $N(N-1)/2$ visibilities forming an overdetermined system of equations. The development of algorithms to solve the calibration system of equations is a very active line of research (Mitchell et al. 2008; Kazemi et al. 2011; Tasse 2014; Yatawatta 2015; Smirnov & Tasse 2015); although beyond the scope of this chapter, we mention it here for completeness.

The solution of the measurement equation requires some knowledge of the sky brightness distribution $\mathbf{B}_I$, in other words, a sky model. Traditionally, this is achieved by observing a calibration source, i.e., a bright, unresolved point source with known spectral and polarization properties. Calibration solutions are then applied to the observed field that is then used to improve the sky model $\mathbf{B}_I$ which, in turn, leads to more accurate calibration solutions **J**. This loop is traditionally called self-calibration (Cornwell & Wilkinson 1981; Pearson & Readhead 1984) and can lead to a highly accurate calibration (e.g., Bernardi et al. 2010; Smirnov 2011b).

The advantage of the measurement equation formalism is that it can factorize different physical terms into different matrices. For example, the frequency response of the telescope electronics and its time variations essentially affects only the two polarization responses and are modeled with a diagonal Jones matrix $\mathbf{B}$,

$$\mathbf{B}(t,\nu) \equiv \begin{bmatrix} b_x(t,\nu) & 0 \\ 0 & b_y(t,\nu) \end{bmatrix}, \qquad (5.20)$$

where we made it explicit that $\mathbf{B}$ can vary with time and frequency. The undesired instrumental leakage between the two orthogonal polarizations can be written as a $\mathbf{D}$ Jones matrix of the form,

$$\mathbf{D}(t,\nu) \equiv \begin{bmatrix} 1 & d_x(t,\nu) \\ -d_y(t,\nu) & 1 \end{bmatrix}, \qquad (5.21)$$

and the measurement equation can be written as

$$\mathbf{V}_{12}(u,v,\nu) = \mathbf{B}_1 \, \mathbf{D}_1 \left(\int_{\Omega} \mathbf{B}_I(l,m,\nu) \, e^{-2\pi i(ul+vm)} dl \; dm \right) \mathbf{D}_2^H \, \mathbf{B}_2^H . \qquad (5.22)$$

We note that, in principle, the primary beam response should appear as an additional 2×2 Jones matrix before the $\mathbf{D}$ matrix. I have ignored it for now, although I will discuss it later in this section.

Retaining only the first-order terms, Equation (5.22) can be written as (Sault et al. 1996)

$$V_{12,xx}(u,v,\nu) = b_{1,x} \, b_{2,x}^*[V_I(u,v,\nu) - V_Q(u,v,\nu)], \qquad (5.23)$$

$$V_{12,xy}(u,v,\nu) = b_{1,x} \, b_{2,y}^*\left[\left(d_{1,x} - d_{2,y}^*\right)V_I(u,v,\nu) + V_U(u,v,\nu) + iV_V(u,v,\nu)\right], \qquad (5.24)$$

$$V_{12,yx}(u,v,\nu) = b_{1,y} \, b_{2,x}^*\left[\left(d_{2,x} - d_{1,y}^*\right)V_I(u,v,\nu) + V_U(u,v,\nu) - iV_V(u,v,\nu)\right], \qquad (5.25)$$

$$V_{12,yy}(u,v,\nu) = b_{1,y} \, b_{2,y}^*[V_I(u,v,\nu) - V_Q(u,v,\nu)], \qquad (5.26)$$

where I dropped the explicit dependence on time and wavelength from the gain terms for notation clarity, and where $V_{i=I,Q,U,V}$ are the Fourier transforms of the elements of the sky brightness matrix $\mathbf{B}_I$.

This form of the measurement equation offers an intuitive understanding as to why calibration is of paramount importance in 21 cm observations. The observed visibilities are essentially a measurement of foreground emission and, in the ideal case, their amplitudes would vary smoothly with frequency, and foregrounds could either be avoided or subtracted. However, the instrumental response inevitably corrupts this smoothness in several ways: because the telescope primary beam is not sufficiently smooth in frequency, because of the electronic response or because of reflections along the signal path. Although calibration will correct for these effects and restore the intrinsic foreground frequency smoothness, calibration errors (i.e., deviations from the true **B** and **D** solutions) will still corrupt the foreground spectra. In practice, calibration errors result in foreground power leaking out of the horizon limit and jeopardizing (part of) the EoR window. The corruption of foreground spectra will limit the accuracy of any subtraction method (see discussion in Chapter 6 in this book). *The effectiveness of foreground separation, proven in ideal cases, depends significantly on the accuracy of interferometric calibration.*

The form of the measurement equation written in Equations (5.22) and (5.26) is often referred to as a direction-independent calibration as it implicitly assumes that a single Jones matrix is sufficient to describe corruptions across the whole sky area of interest. This assumption is often invalid at low frequencies, mostly because of the changing primary beam response over a wide field of view, frequency, and over the course of the observation, and the position and time-dependent corruptions introduced by Earth's ionosphere. In this case, the measurement equation becomes direction dependent, i.e., a different Jones matrix is written and solved for a certain number of directions in the sky:

$$\mathbf{V}_{12}(u,v,\nu) = \sum_s \left[\mathbf{B}_{1,s} \left(\int_\Omega \mathbf{B}_{I,s}(l,m,\nu) \, e^{-2\pi i(ul+vm)} dl \, dm \right) \mathbf{B}_{2,s}^H \right], \qquad (5.27)$$

where the sum is over the number of directions s. We note that we have used the **B** matrix for pedagogical purposes here, regardless of the physical origin of the direction-dependent effect. Direction-dependent effects also impact foreground separation, in a similar way to the direction-independent effects.

Accurate direction-independent and -dependent calibration of 21 cm observations is at the forefront of current research and can be grouped into a few main topics:

- *Sky models.* Ideally, the sky brightness model matrix $\mathbf{B}_I$ (Equations (5.22) and (5.26)) would include the whole sky emission. This is practically impossible as part of the sky signal is the unknown of interest (the 21 cm signal), and the detailed properties of the foreground sky are not known sufficiently well. Sky models are normally constituted of a catalog of compact sources of known (or measured) properties, often covering an area significantly larger than the telescope field of view (e.g., Yatawatta et al. 2013; Pober et al. 2016). Nevertheless, sky models remain essentially always incomplete at some level, as source catalogs are limited in depth, source characterization, and—often—sky coverage. Grobler et al. (2014), Wijnholds et al. (2016), and Grobler et al. (2016) showed that incomplete catalogs used as sky models bias the calibration and eventually lead to artifacts in the form of ghost-like sources in interferometric images, most of the times fainter than the image noise level. The ghost pattern is stronger for regularly spaced arrays and if the sky model is less complete. In terms of power spectrum, Ewall-Wice et al. (2017) and Barry et al. (2016) showed that the calibration bias introduced by incomplete sky models leads to an overall leakage of foreground power in the EoR window (Figure 5.6). A similar foreground leakage may occur because of the

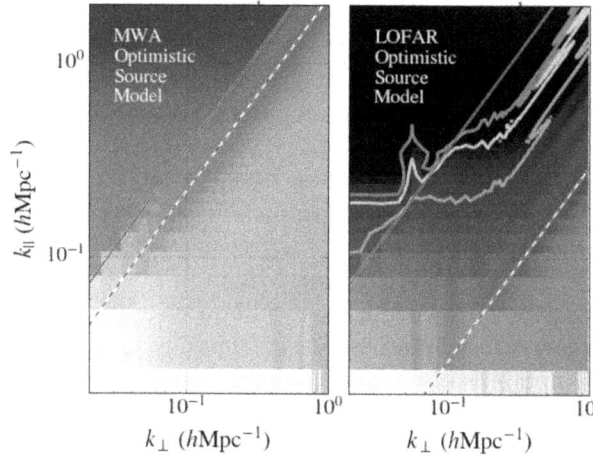

Figure 5.6. Example of power-spectrum bias introduced by calibration errors due to an incomplete sky model for the Murchison Widefield Array (MWA, left) and the Low Frequency Array (LOFAR, right) cases, respectively (adapted from Ewall-Wice et al. 2017, by permission of Oxford University Press on behalf of the Royal Astronomical Society). Power spectra are shown in their two-dimensional ($k_\perp$,$k_\parallel$) form in order to display the foreground-dominated region below the horizon limit (gray solid line). Cyan, orange, and red lines are the locii where a fiducial 21 cm model power spectrum is 1, 5, and 10 times higher than the bias level. In an ideal case with perfectly smooth foregrounds and no calibration errors, the 21 cm power spectrum should be detectable just outside the horizon limit. The errors introduced by an incomplete sky model leak foreground power in the EoR window at a level that may completely prevent a detection in the MWA case.

finite angular resolution of interferometric observations: for example, two sources whose sizes are, respectively, 1/3 and 1/10 of the instrument angular resolution will both be modeled as point like even if the first source is only barely unresolved. This biased catalog would again lead to a leakage of foreground power in the EoR window (Procopio et al. 2017). In this case, the bias can be mitigated by obtaining a sky model with an angular resolution that is much higher than the scales at which the 21 cm signal is expected (Procopio et al. 2017).

Sky models that include only compact sources are not adequate for baselines shorter than a few tens of meters as they are sensitive to Galactic diffuse emission, which contributes to most of the power on angular scales $\theta > 10'-20'$ (e.g., Bernardi et al. 2009; Choudhuri et al. 2017). Excluding short baselines from the calibration solutions prevents the problem of modeling diffuse emission, but can bias the solutions (Patil et al. 2016) if the system of calibration equations is not properly constrained, e.g., via regularization (Mouri Sardarabadi & Koopmans 2019).

In summary, different analysis approaches provide evidence that imperfect sky models (either because of missing catalog sources, misestimating source properties, or missing diffuse emission) are a source of calibration bias that has the general effect of corrupting the foreground properties, leaking their power well beyond the ideal horizon limit and requiring additional modeling and subtraction. For this reason, significant efforts are currently ongoing in order to improve sky models via wider and deeper low-frequency surveys (e.g., Hurley-Walker et al. 2017; Intema et al. 2017; Shimwell et al. 2019), more accurate low-frequency catalogs (Carroll et al. 2016), and even better observations of Galactic diffuse emission (Zheng et al. 2017; Dowell et al. 2017);

- *Instrument/primary beam models.* A complete knowledge of a sky model may not be, by itself, sufficient for an accurate calibration of 21 cm observations as the brightness matrix $\mathbf{B}_l$ is multiplied by the antenna primary beam (Equations (5.8) and (5.19)) and the measurement of an intrinsic sky model requires the separation from the primary beam effect.

Unlike steerable dishes, most 21 cm interferometers are constituted of dipoles fixed on the ground, in some cases clustered together to form larger stations whose beams can be digitally pointed to a sky direction by introducing different delays to the dipoles (e.g., like the MWA and LOFAR arrays). As station beams are formed in order to track a source on the sky, the station projected area changes with time and the shape of the primary beam changes noticeably (Figure 5.7). This is a typical direction-dependent effect that can be casted in the measurement equation as

$$\mathbf{V}_{12}(u,v,\nu) = \int_\Omega \mathbf{E}_1(t,l,m,\nu)\mathbf{B}_{l,s}(l,m,\nu)\, e^{-2\pi i(ul+vm)} \\ \times \mathbf{E}_2^H(t,l,m,\nu)dl\; dm, \tag{5.28}$$

were $\mathbf{E}(t,l,m,\nu)$ is the Jones matrix describing the primary beam which, in the simplest cases, is a diagonal matrix,

$$\mathbf{E}(t,l,m,\nu) \equiv \begin{bmatrix} e_x(t,l,m,\nu) & 0 \\ 0 & e_y(t,l,m,\nu) \end{bmatrix}. \tag{5.29}$$

We note that we have written the explicit dependence on the time due to the change in projected area for dipole stations and that the direction dependence of $\mathbf{E}$ is encoded in its (l,m) dependence.

Time- and frequency-variable primary beams lead to apparent time-variable sky models with variations that are larger away from the pointing direction due to the greater changes in the side-lobe pattern. For example, sky sources that are well within the main lobe of the primary beam in Figure 5.7 will experience relatively negligible variations throughout an observation; the opposite will occur to sources located well outside the main lobe as they run through primary beam side lobes.

Primary beams are also frequency variable and, to first order, their size scales with the observing wavelength (Equation (5.5)), i.e., rather smoothly. However, in the side-lobe region, variations become rather abrupt as the source can be located on a side-lobe peak at a certain frequency and in the side-lobe null at another frequency. As a final remark, stations that include several dipoles are not perfectly equal to each other, due to manufacturing reasons or mutual coupling between their elements (e.g., Sokolowski et al. 2017), leading to $\mathbf{E}_1 \neq \mathbf{E}_2$. As primary beams are different, even visibilities for baselines that have the same length and orientation will be different, rather

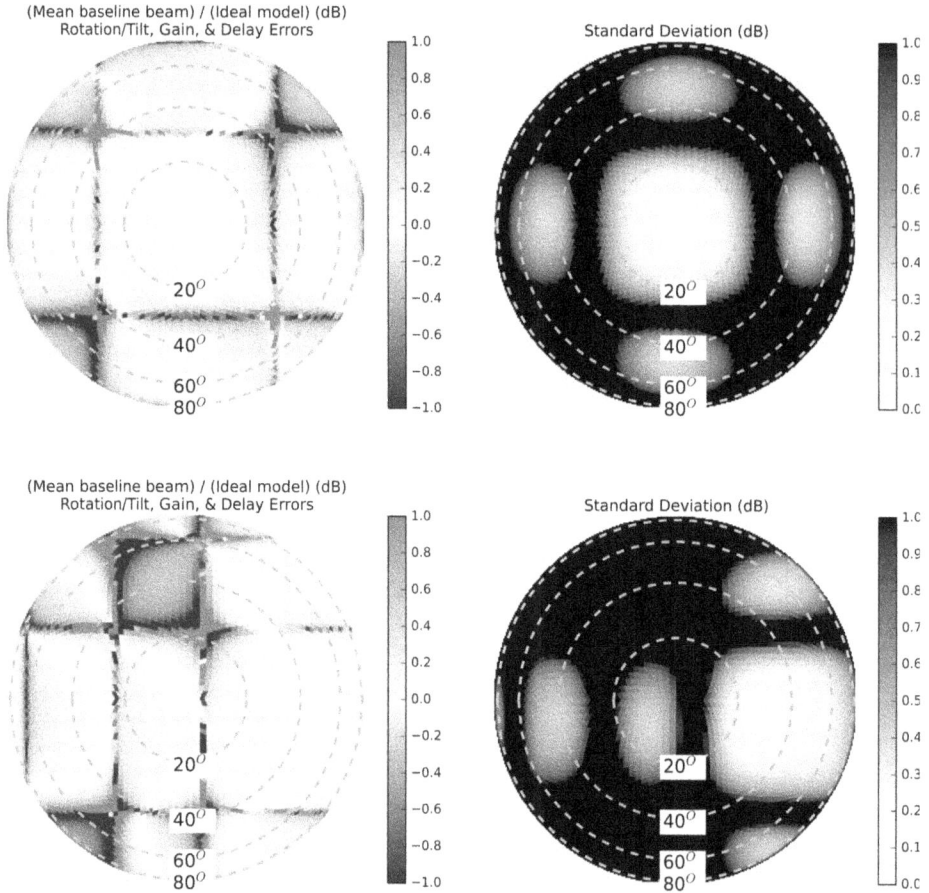

Figure 5.7. Example of primary beam variations as an MWA station points at zenith (top right) and ~30°
away from zenith (bottom right) at 150 MHz. The left column shows the fractional variation of individual
station beam models, with respect to the nominal primary beam (right column, from Neben et al. 2016 © 2016.
The American Astronomical Society. All rights reserved). It is visible how different the side-lobe pattern is
when pointing toward two different directions. The ~10% magnitude of the first lobe and the large null regions
around the side lobes should also be noticed. The specific pattern is due to the regular shape of the MWA
station, where 16 dipoles are arranged in a square 4 × 4 grid.

than identical, as expected. The left panel of Figure 5.7 shows an example of
how much primary beams vary for different stations due to mutual coupling
interactions: variations in the side-lobe region can be as large as ~30%.

If not accurately modeled and taken into account, primary beam effects
can bias the calibration solution and, again, corrupt the foreground frequency
smoothness. Bhatnagar et al. (2008), Bernardi et al. (2011), Sullivan et al.
(2012), and Tasse et al. (2013) have developed methods to incorporate time-
and frequency-variable primary beams in interferometric images; however,
the accuracy of the correction is limited by the accuracy of the primary beam
model. Increasing effort is therefore being placed in precise modeling and

measurements of primary beams (e.g., Pupillo et al. 2015; Trott & Wayth 2016; de Lera Acedo et al. 2017; Trott et al. 2017; Jacobs et al. 2017; de Lera Acedo et al. 2018);

- *Polarization leakage calibration.* Equations (5.19) and (5.26) show that, even if the 21 cm signal is unpolarized, care needs to be taken against the contamination from polarized foreground emission. Most point sources are unpolarized below 200 MHz (Bernardi et al. 2013; Lenc et al. 2016; Van Eck et al. 2018), therefore the assumption of an unpolarized sky model is well justified. However, calibration errors (in the **B** matrix) would lead to a relative miscalibration of the *xx* and *yy* polarizations and, in turn, to leakage of polarized emission into total intensity. This effect may be particularly strong on short baselines (e.g., shorter than a ~1 km), where polarized foregrounds are brighter (Bernardi et al. 2009; Iacobelli et al. 2013; Jelić et al. 2015; Lenc et al. 2016). Polarized foregrounds that are Faraday rotated by the interstellar medium and leak to total intensity are a severe contamination to the 21 cm signal: they have a characteristic frequency dependence similar to the 21 cm signal therefore have power across the whole EoR window and cannot be subtracted using standard methods (e.g., Jelić et al. 2010; Moore et al. 2013; Nunhokee et al. 2017).

Even if calibration errors are negligible, low-frequency antennas have a nonnegligible polarized response across their wide field of view, i.e., the primary beam Jones matrix **E** is no longer diagonal. The measurement equation with a full polarized primary beam response can be written as (Nunhokee et al. 2017)

$$
\begin{bmatrix} V_{12,I}(u,v,\nu) \\ V_{12,Q}(u,v,\nu) \\ V_{12,U}(u,v,\nu) \\ V_{12,V}(u,v,\nu) \end{bmatrix} = \int_\Omega \mathbf{S}^{-1}[\mathbf{E}_1 \otimes \mathbf{E}_2^H]\mathbf{S} \begin{bmatrix} I(l,m,\nu) \\ Q(l,m,\nu) \\ U(l,m,\nu) \\ V(l,m,\nu) \end{bmatrix} e^{-2\pi i(ul+vm)} dl\ dm
$$

$$
= \int_\Omega \mathbf{A}(l,m,\nu) \begin{bmatrix} I(l,m,\nu) \\ Q(l,m,\nu) \\ U(l,m,\nu) \\ V(l,m,\nu) \end{bmatrix} e^{-2\pi i(ul+vm)} dl\ dm,
$$

(5.30)

where **S** is the matrix that relates the intrinsic Stokes parameters to the observer *x–y* frame (Hamaker et al. 1996) and $\otimes$ is the outer product. Visibilities are written as a four-element vector as this form shows that the **A** matrix maps the intrinsic (unprimed) Stokes parameters into the observed (primed) ones:

$$
\begin{pmatrix} I' \leftarrow I & I' \leftarrow Q & I' \leftarrow U & I' \leftarrow V \\ Q' \leftarrow I & Q' \leftarrow Q & Q' \leftarrow U & Q' \leftarrow V \\ U' \leftarrow I & U' \leftarrow Q & U' \leftarrow U & U' \leftarrow V \\ V' \leftarrow I & V' \leftarrow Q & V' \leftarrow U & V' \leftarrow V \end{pmatrix}.
$$

(5.31)

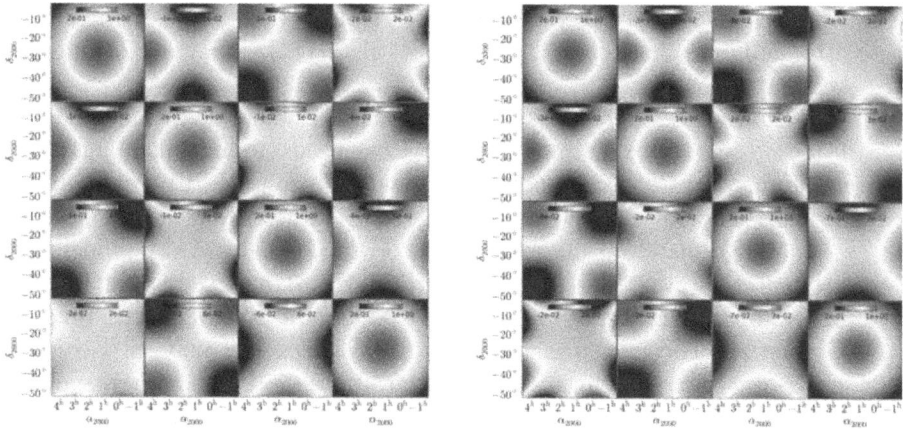

Figure 5.8. Examples of **A** matrices that model the dipole of the Precision Array to Probe the Epoch of Reionization (PAPER; Parsons et al. 2010) at 130 (left) and 150 MHz (right), respectively. They map the intrinsic Stokes parameters into observed ones: the diagonal terms represent the standard primary beam patterns, whereas the off-diagonal terms are the leakage terms The second, the third, and the fourth elements of the first row show how Stokes parameters Q, U, and V, respectively, contaminate the total intensity signal (reproduced from Nunhokee et al. 2017 © 2017. The American Astronomical Society. All rights reserved).

An example of the **A** matrix is shown in Figure 5.8. The first row of the matrix shows how the four intrinsic Stokes parameters contribute to the observed total intensity and, therefore, how polarized foregrounds leak into the 21 cm signal even in the absence of any calibration errors: the magnitude of the contaminating Stokes Q and U foregrounds increases away from the pointing direction. Wide-field polarization is another textbook example of a direction-dependent calibration problem.

Calibration of polarization leakage remains a challenging task. Instruments with narrow fields of view are less prone to polarization leakage (Asad et al. 2015, 2016, 2018). Another way of mitigating polarization leakage is to extend the sky model to include polarization (e.g., Geil et al. 2011), although modeling the diffuse Galactic foreground—the brightest component—is not straightforward and requires accurate imaging. Nunhokee et al. (2017) showed, however, that the magnitude of the Galactic polarization leakage may be below the 21 cm signal at high $k_{\parallel}$ values ($k_{\parallel} > 0.3$ Mpc^{-1}) and, potentially, an avoidance strategy is not completely excluded. A more extensive characterization of the polarized foreground properties is needed in order to generalize their results.

- *Ionospheric distortions.* The ionosphere is the partially ionized layer situated between ~50 and 1000 km above the surface of Earth, where the density of free electrons changes with time and position. At low frequencies, the ionosphere is no longer transparent to radio waves and, to first order, it delays the wave propagation by an amount that is proportional to the integral of the electron density along the line of sight (e.g., Thompson et al. 2017; Intema et al. 2009):

$$\phi(t,\nu) \propto \frac{1}{\nu} \int n_e(t)dl, \qquad (5.32)$$

where ϕ is the extra delay, n_e the electron density, and the integral is the total electron content (TEC) along the line of sight. When the delay is different for two different antennas, visibilities measure an additional, time-variable delay. In the measurement equation formalism, ionospheric delays can be modeled by a scalar term $Z \propto e^{i\phi(t,\nu)}$; however, ionospheric effects are another textbook example of direction-dependent calibration as the Z is different for different directions. Given the size S of a characteristic ionospheric patch where the TEC is constant, direction-dependent effects occur when either the field of view is much larger than S or the baseline separation is much larger than S, i.e., different antennas "see through" different TEC values (see Intema et al. 2009 for an extensive discussion on the different ionospheric regimes). In this case, the measurement equation takes a form similar to Equation (5.27),

$$V_{12}(u,v,t,\nu) = \sum_s Z_{1,s}(t,\nu)\left(\int_\Omega B_{I,s}(l,m,\nu)\, e^{-2\pi i(ul+vm)}\, dl\, dm \right) \\ \times Z_{2,s}^H(t,\nu), \qquad (5.33)$$

leading to images where sources are convolved with a position- and time-dependent point-spread function. An example of this effect is shown in Figure 5.9: the column on the left shows sources after the standard self-calibration, still surrounded by artifacts due to the ionosphere; moving toward the right, iterative direction-dependent corrections lead to virtually artifact-free images on the right column (see van Weeren et al. 2016 for further details).

Trott et al. (2018) analyzed the effects of ionospheric perturbations on MWA observations, whose maximum baseline is a factor of ∼30 shorter than the LOFAR example displayed in Figure 5.9, but with a field of view ∼4 times larger. They found that direction-dependent ionospheric distortions can affect the sky coherence up to degree scales (i.e., scales relevant for 21 cm observations); however, due to the relatively short baselines, these effects occur only in 8% of the observations and it is relatively straightforward to monitor the ionospheric activity and exclude the most affected observations.

An extensive modeling of the impact of ionospheric errors on the two-dimensional $(k_\perp,k_\parallel)$ power spectrum has been carried out by Vedantham & Koopmans (2016). They found that most of the residual effects due to the ionosphere on baselines shorter than a few kilometers are confined within the horizon limit, therefore not impacting foreground avoidance. Moreover, the frequency coherence of the ionospheric residual errors is such that they will likely be removed by foreground subtraction algorithms.

Current investigations seem therefore to suggest that ionospheric effects are not going to be a showstopper for both 21 cm power-spectrum observations and, likely, 21 cm tomography.

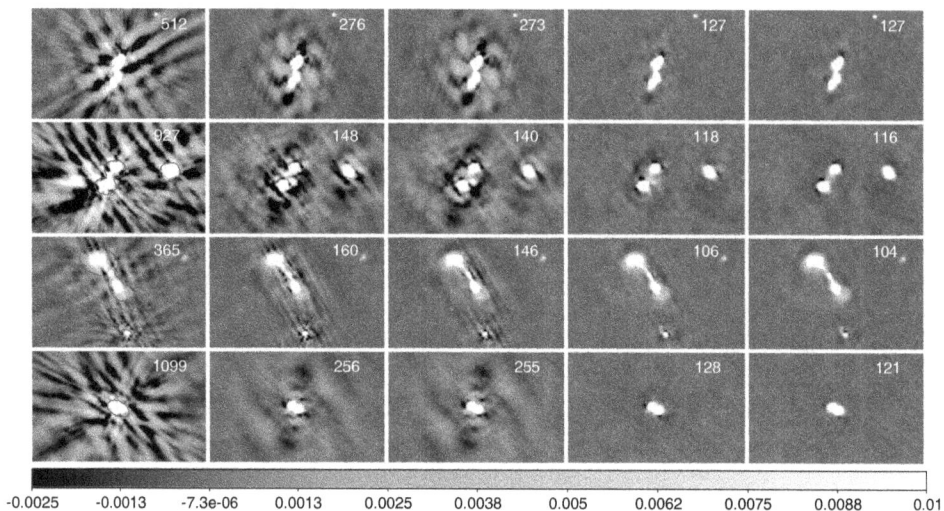

-0.0025   -0.0013   -7.3e-06   0.0013   0.0025   0.0038   0.005   0.0062   0.0075   0.0088   0.01

Figure 5.9. Calibration of ionospheric effects in LOFAR observations using a faceting algorithm (reproduced from van Weeren et al. 2016 © 2016. The American Astronomical Society. All rights reserved). The image resolution is $8'' \times 6''.5$ averaged over the 120–180 MHz bandwidth. The left column shows zoom-in images around sources without direction-dependent calibration, which is, in turn, applied incrementally toward the right panels. For each source, a sky model and a direction-dependent Jones scalar Z is improved at each iteration until an artifact-free image is obtained (right column). Solutions were computed every 10 s. An additional amplitude calibration to account for primary beam variations was determined on scales of 10 minutes. The color scale is in units of Jy beam^{-1}.

5.3.1 Redundant Calibration

An interferometric array where most of the baselines have the same length and orientation is called redundant, as these baselines measure the same Fourier mode of the sky brightness distribution. Redundant array configurations are often not appealing as they have poor imaging performances because they do not measure sufficient Fourier modes to reconstruct accurate sky images. However, a maximally redundant array where the antennas are laid out in a regularly spaced square grid offers the maximum power-spectrum sensitivity on the $k_\perp$ modes corresponding to the most numerous baselines. This criterion has inspired the highly redundant layouts of the MIT Epoch of Reionization experiment (Zheng et al. 2014) and PAPER (Parsons et al. 2012), and partly driven the updated MWA (Wayth et al. 2018).

One of the advantages of a redundant array is that it enables a different calibration strategy, i.e., redundant calibration. In redundant calibration, the form of the measurement equation does not change and can be written, for a single polarization, like Equation (5.26),

$$V_{12,xx}(u,v,\nu) = b_{1,x}\, b_{2,x}^{*}\, Y_{12,xx}(u,v,\nu), \qquad (5.34)$$

with the difference now that the model visibility Y is not tied to a sky model, but it is solved for, simply assuming that it is the same for each group of redundant baselines

(Wieringa 1992; Liu et al. 2010). In other words, redundant calibration is independent of the sky model and, therefore, bypasses entirely the biases related to sky model incompleteness described in Section 5.3. However, as redundant calibration is not tied to any physical (i.e., sky-based) spatial or spectral model, its solutions have degeneracies that need to be solved for by using a sky model (e.g., Zheng et al. 2014; Byrne et al. 2019). In particular, spectral calibration, which is critical for foreground separation, cannot currently be obtained using redundant calibration and requires a sky-based calibration. Byrne et al. (2019) suggested that sky model incompleteness can bias this calibration step, in a way similar to what happens with a traditional calibration scheme. Moreover, as redundant calibration is agnostic of the polarization state of the sky brightness distribution, mitigation of polarization leakage remains an open question in the framework of redundant calibration (Dillon et al. 2018).

Finally, redundant calibration is prone to effects that break the assumption of redundancy, the most common being errors in the antenna positions and different antenna primary beams. Antenna position errors can be reduced to have a negligible impact on redundant calibration (Joseph et al. 2018). The effect of primary beam variations among the different antennas on redundant calibration is likely more severe, although new calibration schemes are being developed to mitigate it (Orosz et al. 2019).

5.4 Array Design and Observing Strategies

I will conclude this chapter by discussing how the various interferometric effects discussed so far impact the choice of array designs and the consequent observing strategies. Morales (2005) and Parsons et al. (2012), for example, investigated how instrumental choices like the array layout, the antenna size, and the bandwidth (do not) affect measurements of the 21 cm power spectrum. Here I would rather emphasize the interdependence between instrumental choices, calibration, and foreground separation strategies. If the total collecting area is kept fixed, there are two main elements that impact calibration and foreground separation strategies:

- *Station size.* The choice of the station size determines the minimum $k_\perp$ value accessible and the footprint of each *uv* measurement. Each visibility is not a single point in the *uv* plane but has a footprint corresponding to the two-dimensional Fourier transform of the primary beam. This can be seen using the convolution theorem to rewrite Equation (5.1),

$$V_{ij}(\mathbf{b},\nu) = \tilde{A}(\mathbf{b},\nu) * \tilde{I}(\mathbf{b},\nu). \tag{5.35}$$

Smaller stations have smaller footprints in the *uv* plane (see Figure 5.10) and can, therefore, sample the *uv* plane more accurately than larger stations. They also allow smaller $k_\perp$ values to be probed (as the minimum possible *uv* length is essentially the station size) for which the avoidance strategy is more effective (see Figure 5.5). If smaller stations are preferred for power-spectrum measurements, they are generally more challenging in terms of calibration: they have wider fields of view that require a more accurate sky model for

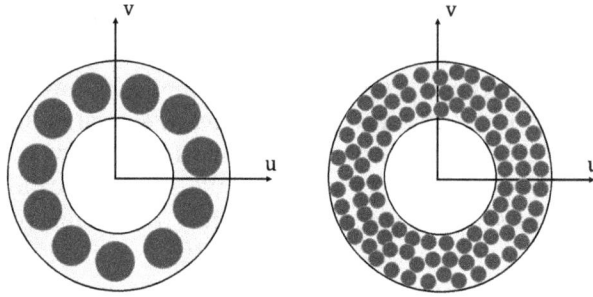

Figure 5.10. Cartoon illustration of the *uv* footprint due to the primary beam. The purple circles are the *uv* footprints for a large (left panel) and small (right panel) station, respectively. The minimum and maximum baselines are the same for both cases, in order to sample the same *uv* annulus (yellow area).

calibration and that suffer from more severe ionospheric distortions and polarization leakage contamination. Given the smaller size, their visibilities have a lower signal-to-noise ratio compared to larger stations, possibly limiting the calibration of high time-variable effects. On the other hand, they do not necessarily require sources with a high time cadence to be tracked but can use drift scan strategies (where they are pointed to a fixed direction and the sky drifts overhead) or a mix of drift scan and pointed observations to maximize sensitivity (Trott 2014). The advantage of drift scan over pointed observations is that the primary beams remain constant in time, avoiding some of the effects described in the Section 5.3.

- *Array layout.* Beyond the obvious sensitivity requirement that prefers compact arrays due to their better brightness sensitivity, layout choices are also intrinsically related to calibration and foreground separation strategies. A pseudo-random station distribution that leads to a filled *uv* coverage (between the minimum and the maximum station separation) is highly desirable for imaging, modeling, and subtracting foregrounds. It is not a stringent requirement for power-spectrum measurement and for the avoidance strategy. It is probably necessary for 21 cm tomography, in order to provide reconstruction of the low-brightness neutral hydrogen regions.

  On the opposite side of the spectrum of choices, redundant arrays are the most sensitive power-spectrum machines. They obviously leverage on redundant calibration, which is precluded in imaging arrays. Their drawbacks are the poor imaging performances that prevent accurate foreground modeling and essentially only allow foreground avoidance. For the same reason, if redundant calibration is not sufficient, redundant arrays have limited options to improve calibration by reconstructing the sky brightness sensitivity.

I will use three existing low-frequency arrays as examples of the range of cases of interest:

- *Low Frequency Array* (LOFAR; van Haarlem et al. 2013). LOFAR is an array of 40 stations located in The Netherlands and several remote stations across Europe. Twenty-four stations are located in a 2 km core from the array center, and the remaining stations are distributed in a logarithmic spiral layout up to ~100 km, providing a very dense uv coverage in a few hours of tracked observation (see Chapter 8 in this book for an image of the LOFAR array layout and the other arrays discussed here).

Stations are formed by two types of receptors sensitive to the 30–90 and 110–200 MHz range. The 110–200 MHz stations are the most sensitive to 21 cm observations, and we will only consider them in this discussion. They constitute 48 clusters of dipoles (each of them being a 4×4 square grid) arranged in a regular ~30 m diameter grid, leading to a ~4° field of view at 150 MHz.

LOFAR is an example of a traditional interferometric array, with excellent point-source sensitivity that favors sky-based calibration and a very dense uv coverage for high fidelity imaging. Its large station size has a large uv footprint and a relatively narrow field of view that essentially requires tracking a sky patch. The narrow field of view allows sky patches with low foreground (including polarization) contamination to be selected and wide-field foreground emission to be rejected. Unwanted sky emission far from the pointing direction is further suppressed by rotating each station grid with respect to another, while rotating the dipoles back to a common polarization frame: this operation makes the station primary beams all different, and their side-lobe patterns, which would otherwise be reinforced by the regular station grid, tend to average out.

The calibration of LOFAR 21 cm observations relies on an accurate sky model where compact sources are modeled using the longest baselines available (i.e., ~100 km). Direction-dependent calibration corrects for ionospheric effects that corrupt visibilities on baselines longer than a few kilometers, and for the effect of variable primary beams on compact sources (Yatawatta et al. 2013). The sky model is then subtracted from the visibilities and residual foregrounds are subtracted in the image domain (see details in Chapter 6 in this book).

The LOFAR design is suited for 21 cm tomography on large angular scales, providing foregrounds are adequately subtracted (Zaroubi et al. 2012).

- *Murchison Widefield Array* (MWA; Tingay et al. 2013; Wayth et al. 2018). The MWA is an array located in Western Australia, operating between 80 and ~200 MHz. It employs the same LOFAR dipoles, although they are assembled in stations of 4×4 elements arranged in a regular grid. The station size is therefore ~6 times smaller compared to LOFAR, with an equivalent increase of the field of view. The MWA underwent a recent upgrade to phase II (to distinguish it from the initial deployment, named phase I) and is now constituted of 256 stations (out of which only 128 can be simultaneously correlated) in a hybrid configuration: 128 stations are deployed in a pseudo-random configuration out to a ~3 km baseline (the phase I telescope), 72 stations in two highly redundant hexagons next to the core of the array and 56 stations to extend the maximum baseline up to ~5 km.

MWA phase II is a fairly versatile instrument: in its compact, redundant configuration, it is optimized for power-spectrum observations and can leverage redundant calibration (Li et al. 2018); its small stations give a good sampling in the *uv* plane (right panel case in Figure 5.10). In its extended configuration, it has an exceptionally good instantaneous *uv* coverage (due to the high number of stations instantaneously correlated) with low side-lobe levels, which is good for imaging and foreground modeling, and a large field of view, which allows the sky to be surveyed very quickly. The wide field of view does not allow low foreground patches to be isolated, but it allows opting for drift scan observations or a mix of drift scan and pointed observations (Trott 2014), which have the advantage of more time-stable primary beams. Wide field ionospheric effects are somewhat mitigated by the array compactness (Jordan et al. 2017). The MWA can therefore leverage on the strength of both redundant and traditional calibrations, and can adopt a mixture of foreground subtraction and avoidance strategies.

The MWA approach has, however, limitations, too: the regular station grid (without any rotation, unlike LOFAR) generates strong side lobes (see Figure 5.7), which make calibration and foreground separation more challenging; the large field of view requires more comprehensive sky models for calibration and is more susceptible to polarization leakage; and the relatively short maximum baseline may be insufficient to derive accurate, high-angular-resolution sky models (Procopio et al. 2017).

- *Precision Array to Probe the Epoch of Reionization* (PAPER; Parsons et al. 2010). PAPER was an array located in the South Africa, operating in the 100–200 MHz range and now decommissioned in favor of its successor (the Hydrogen Epoch of Reionization Array; see Chapter 9 in this book). It employed custom-designed ~2 m dipoles that were deployed and rearranged in several configurations up to a 128 element array. Dipoles were always individually correlated with no clustering into larger stations, implying a nearly all-sky field of view. In order to maximize power-spectrum sensitivity, dipoles were always deployed in maximally redundant configurations with very short baselines (up to a maximum of 350 m), enabling the advantages of redundant calibration (Parsons et al. 2014; Ali et al. 2015; Jacobs et al. 2015). In the final 128 element deployment, ~20 dipoles were placed as outriggers outside the regular grid in order to partially improve the *uv* coverage for foreground characterization and calibration.

In some sense, PAPER represents the choice opposite to the LOFAR case: an almost fully redundant array that works using essentially only foreground avoidance and without any spatial characterization of foregrounds for either calibration or subtraction. PAPER is a full drift scan array with primary beams that are fairly stable with time, but also with an all-sky field of view where no selection of low foreground regions is possible, for which polarization leakage and ionospheric effects are the most severe, although the latter are mitigated by the very compact configuration.

As pointed out earlier in this chapter, a redundant array like PAPER is not suited for 21 cm tomography.

5.5 Conclusions

This chapter presented a summary of interferometry and calibration in light of 21 cm observations. I started from the basics of interferometry to show how they are related to observations of the 21 cm power spectrum and its tomographic images. I reviewed calibration of 21 cm observations, highlighting how foreground separation—the biggest challenge of 21 cm observations—critically depends on various calibration effects (sky models, primary beam modeling and calibration, polarization leakage, the ionosphere). I also attempted to show how the various array designs adopted by current experiments enable different calibration and observational strategies—neither of which is clearly winning, at the present point. The field is rapidly developing, and both current and upcoming instruments (see Chapter 9 in this book) will address some of the open questions presented in this chapter.

5.6 Acknowledgments

It is my pleasure to thank C.D. Nunhokee for useful discussions and help with Figures 5.3 and 5.4.

References

Ali, Z. S., Parsons, A. R., Zheng, H., et al. 2015, ApJ, 809, 61
Asad, K. M. B., Koopmans, L. V. E., Jelić, V., et al. 2015, MNRAS, 451, 3709
Asad, K. M. B., Koopmans, L. V. E., Jelić, V., et al. 2016, MNRAS, 462, 4482
Asad, K. M. B., Koopmans, L. V. E., Jelić, V., et al. 2018, MNRAS, 476, 3051
Barry, N., Hazelton, B., Sullivan, I., Morales, M. F., & Pober, J. C. 2016, MNRAS, 461, 3135
Beardsley, A. P., Hazelton, B. J., Sullivan, I. S., et al. 2016, ApJ, 833, 102
Bernardi, G., de Bruyn, A. G., Brentjens, M. A., et al. 2009, A&A, 500, 965
Bernardi, G., de Bruyn, A. G., Harker, G., et al. 2010, A&A, 522, A67
Bernardi, G., Greenhill, L. J., Mitchell, D. A., et al. 2013, ApJ, 771, 105
Bernardi, G., Mitchell, D. A., Ord, S. M., et al. 2011, MNRAS, 413, 411
Bhatnagar, S., Cornwell, T. J., Golap, K., & Uson, J. M. 2008, A&A, 487, 419
Byrne, R., Morales, M. F., Hazelton, B., et al. 2019, ApJ, 875, 70
Carroll, P. A., Line, J., Morales, M. F., et al. 2016, MNRAS, 461, 4151
Choudhuri, S., Bharadwaj, S., Ali, S. S., et al. 2017, MNRAS, 470, L11
Cornwell, T. J., & Wilkinson, P. N. 1981, MNRAS, 196, 1067
Datta, K. K., Friedrich, M. M., Mellema, G., Iliev, I. T., & Shapiro, P. R. 2012, MNRAS, 424, 762
de Lera Acedo, E., Bolli, P., Paonessa, F., et al. 2018, ExA, 45, 1
de Lera Acedo, E., Trott, C. M., Wayth, R. B., et al. 2017, MNRAS, 469, 2662
Dickinson, C., Battye, R. A., Carreira, P., et al. 2004, MNRAS, 353, 732
Dillon, J. S., Kohn, S. A., Parsons, A. R., et al. 2018, MNRAS, 477, 5670
Dowell, J., Taylor, G. B., Schinzel, F. K., Kassim, N. E., & Stovall, K. 2017, MNRAS, 469, 4537
Ewall-Wice, A., Dillon, J. S., Liu, A., & Hewitt, J. 2017, MNRAS, 470, 1849
Geil, P. M., Gaensler, B. M., & Wyithe, J. S. B. 2011, MNRAS, 418, 516
Grobler, T. L., Nunhokee, C. D., Smirnov, O. M., van Zyl, A. J., & de Bruyn, A. G. 2014, MNRAS, 439, 4030

Grobler, T. L., Stewart, A. J., Wijnholds, S. J., Kenyon, J. S., & Smirnov, O. M. 2016, MNRAS, 461, 2975

Halverson, N. W., Leitch, E. M., Pryke, C., et al. 2002, ApJ, 568, 38

Hamaker, J. P., Bregman, J. D., & Sault, R. J. 1996, A&AS, 117, 137

Harker, G., Zaroubi, S., Bernardi, G., et al. 2010, MNRAS, 405, 2492

Hurley-Walker, N., Callingham, J. R., Hancock, P. J., et al. 2017, MNRAS, 464, 1146

Iacobelli, M., Haverkorn, M., Orrú, E., et al. 2013, A&A, 558, A72

Intema, H. T., Jagannathan, P., Mooley, K. P., & Frail, D. A. 2017, A&A, 598, A78

Intema, H. T., van der Tol, S., Cotton, W. D., et al. 2009, A&A, 501, 1185

Jacobs, D. C., Aguirre, J. E., Parsons, A. R., et al. 2011, ApJ, 734, L34

Jacobs, D. C., Burba, J., Bowman, J. D., et al. 2017, PASP, 129, 035002

Jacobs, D. C., Pober, J. C., Parsons, A. R., et al. 2015, ApJ, 801, 51

Jelić, V., de Bruyn, A. G., Pandey, V. N., et al. 2015, A&A, 583, A137

Jelić, V., Zaroubi, S., Labropoulos, P., et al. 2010, MNRAS, 409, 1647

Jordan, C. H., Murray, S., Trott, C. M., et al. 2017, MNRAS, 471, 3974

Joseph, R. C., Trott, C. M., & Wayth, R. B. 2018, AJ, 156, 285

Kakiichi, K., Majumdar, S., Mellema, G., et al. 2017, MNRAS, 471, 1936

Kazemi, S., Yatawatta, S., Zaroubi, S., et al. 2011, MNRAS, 414, 1656

Lenc, E., Gaensler, B. M., Sun, X. H., et al. 2016, ApJ, 830, 38

Li, W., Pober, J. C., Hazelton, B. J., et al. 2018, ApJ, 863, 170

Liu, A., Tegmark, M., Morrison, S., Lutomirski, A., & Zaldarriaga, M. 2010, MNRAS, 408, 1029

Majumdar, S., Bharadwaj, S., & Choudhury, T. R. 2012, MNRAS, 426, 3178

Mellema, G., Koopmans, L. V. E., Abdalla, F. A., et al. 2013, ExA, 36, 235

Mitchell, D. A., Greenhill, L. J., Wayth, R. B., et al. 2008, ISTSP, 2, 707

Moore, D. F., Aguirre, J. E., Parsons, A. R., Jacobs, D. C., & Pober, J. C. 2013, ApJ, 769, 154

Morales, M. F. 2005, ApJ, 619, 678

Morales, M. F., Hazelton, B., Sullivan, I., & Beardsley, A. 2012, ApJ, 752, 137

Morales, M. F., & Hewitt, J. 2004, ApJ, 615, 7

Mouri Sardarabadi, A., & Koopmans, L. V. E. 2019, MNRAS, 483, 5480

Neben, A. R., Hewitt, J. N., Bradley, R. F., et al. 2016, ApJ, 820, 44

Nunhokee, C. D., Bernardi, G., Kohn, S. A., et al. 2017, ApJ, 848, 47

Orosz, N., Dillon, J. S., Ewall-Wice, A., Parsons, A. R., & Thyagarajan, N. 2019, MNRAS, 487, 537

Parsons, A., Pober, J., McQuinn, M., Jacobs, D., & Aguirre, J. 2012, ApJ, 753, 81

Parsons, A. R., Backer, D. C., Foster, G. S., et al. 2010, AJ, 139, 1468

Parsons, A. R., Liu, A., Aguirre, J. E., et al. 2014, ApJ, 788, 106

Parsons, A. R., Pober, J. C., Aguirre, J. E., et al. 2012, ApJ, 756, 165

Patil, A. H., Yatawatta, S., Koopmans, L. V. E., et al. 2017, ApJ, 838, 65

Patil, A. H., Yatawatta, S., Zaroubi, S., et al. 2016, MNRAS, 463, 4317

Pearson, T. J., & Readhead, A. C. S. 1984, ARA&A, 22, 97

Pen, U.-L., Chang, T.-C., Hirata, C. M., et al. 2009, MNRAS, 399, 181

Pober, J. C., Hazelton, B. J., Beardsley, A. P., et al. 2016, ApJ, 819, 8

Pober, J. C., Parsons, A. R., Aguirre, J. E., et al. 2013, ApJ, 768, L36

Procopio, P., Wayth, R. B., Line, J., et al. 2017, PASA, 34, e033

Pupillo, G., Naldi, G., Bianchi, G., et al. 2015, ExA, 39, 405

Readhead, A. C. S., Mason, B. S., Contaldi, C. R., et al. 2004, ApJ, 609, 498

Ryle, M., & Hewish, A. 1960, MNRAS, 120, 220

Sault, R. J., Hamaker, J. P., & Bregman, J. D. 1996, A&AS, 117, 149

Shimwell, T. W., Tasse, C., Hardcastle, M. J., et al. 2019, A&A, 622, A1

Smirnov, O. M. 2011a, A&A, 527, A106

Smirnov, O. M. 2011b, A&A, 527, A107

Smirnov, O. M. 2011c, A&A, 527, A108

Smirnov, O. M., & Tasse, C. 2015, MNRAS, 449, 2668

Sokolowski, M., Colegate, T., Sutinjo, A. T., et al. 2017, PASA, 34, e062

Sullivan, I. S., Morales, M. F., Hazelton, B. J., et al. 2012, ApJ, 759, 17

Tasse, C. 2014, A&A, 566, A127

Tasse, C., van der Tol, S., van Zwieten, J., van Diepen, G., & Bhatnagar, S. 2013, A&A, 553, A105

Thompson, A. R., Moran, J. M., & Swenson, G. W. Jr 2017, Interferometry and Synthesis in Radio Astronomy (3rd ed.; Berlin: Springer)

Thyagarajan, N., Udaya Shankar, N., Subrahmanyan, R., et al. 2013, ApJ, 776, 6

Tingay, S. J., Goeke, R., Bowman, J. D., et al. 2013, PASA, 30, e007

Trott, C. M. 2014, PASA, 31, e026

Trott, C. M., de Lera Acedo, E., Wayth, R. B., et al. 2017, MNRAS, 470, 455

Trott, C. M., Jordan, C. H., Murray, S. G., et al. 2018, ApJ, 867, 15

Trott, C. M., & Wayth, R. B. 2016, PASA, 33, e019

Van Eck, C. L., Haverkorn, M., Alves, M. I. R., et al. 2018, A&A, 613, A58

van Haarlem, M. P., Wise, M. W., Gunst, A. W., et al. 2013, A&A, 556, A2

van Weeren, R. J., Williams, W. L., Hardcastle, M. J., et al. 2016, ApJS, 223, 2

Vedantham, H., Udaya Shankar, N., & Subrahmanyan, R. 2012, ApJ, 745, 176

Vedantham, H. K., & Koopmans, L. V. E. 2016, MNRAS, 458, 3099

Wayth, R. B., Tingay, S. J., Trott, C. M., et al. 2018, PASA, 35, e033

White, M., Carlstrom, J. E., Dragovan, M., & Holzapfel, W. L. 1999, ApJ, 514, 12

Wieringa, M. H. 1992, ExA, 2, 203

Wijnholds, S. J., Grobler, T. L., & Smirnov, O. M. 2016, MNRAS, 457, 2331

Yatawatta, S. 2015, MNRAS, 449, 4506

Yatawatta, S., de Bruyn, A. G., Brentjens, M. A., et al. 2013, A&A, 550, A136

Zaroubi, S., de Bruyn, A. G., Harker, G., et al. 2012, MNRAS, 425, 2964

Zheng, H., Tegmark, M., Buza, V., et al. 2014, MNRAS, 445, 1084

Zheng, H., Tegmark, M., Dillon, J. S., et al. 2017, MNRAS, 465, 2901

Chapter 6

Foregrounds and Their Mitigation

Emma Chapman and Vibor Jelić

The low-frequency radio sky is dominated by the diffuse synchrotron emission of our Galaxy and extragalactic radio sources related to Active Galactic Nuclei and star-forming galaxies. This foreground emission is much brighter than the cosmological 21 cm emission from the Cosmic Dawn and Epoch of Reionization. Studying the physical properties of the foregrounds is therefore of fundamental importance for their mitigation in the cosmological 21 cm experiments. This chapter gives a comprehensive overview of the foregrounds and our current state-of-the-art knowledge about their mitigation.

6.1 What Are the Foregrounds?

A detection of the redshifted 21 cm emission from the Cosmic Dawn (CD) and Epoch of Reionization (EoR) is a daunting task due to a number of challenges, which are different in nature and complexity. One of them is the extremely prominent foreground emission, which dominates the sky at low radio frequencies. This emission intervenes like fog on an autumn morning and obscures our view toward the neutral hydrogen regions from the times of the first "stars" in the Universe. To clear the view and to make the detection possible, we need to study the foreground emission in great detail and acquire knowledge about its properties.

The foreground emission can be dived in two main categories: (i) galactic foregrounds, mostly associated with the diffuse synchrotron and to some extent free–free emission from the Milky Way; and (ii) extragalactic foregrounds, associated with the radio emission from star-forming galaxies and Active Galactic Nuclei, and less relevant radio halos and relics. For an illustration of different foreground components see Figure 6.1. The former component dominates at angular scales larger than a degree and its contribution to the total foreground power is estimated to about 70% at 150 MHz. The later component dominates at small angular scales and its contribution is estimated to about 30%. Both components are

doi:10.1088/2514-3433/ab4a73ch6

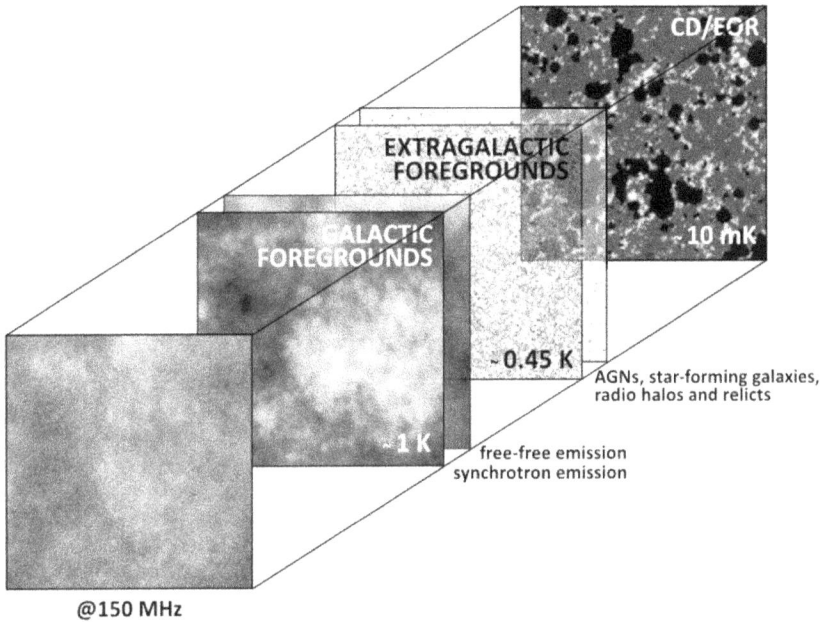

Figure 6.1. An illustration of different foreground components in the redshifted 21 cm experiments. The images are based on Jelić simulations of the foregrounds (Jelić et al. 2010, 2008) and 21cmFAST simulations (Mesinger et al. 2011).

expected to be spectrally smooth due to the dominant synchrotron nature of their emission.

In comparison to the cosmological 21 cm signal, the foreground emission is three to four orders of magnitudes brighter in total power. This amounts to two to three orders of magnitudes in fluctuations. Thus, the global redshifted 21 cm experiments, which use a single antenna for the measurement (e.g., EDGES), need to deal with an order of magnitude brighter foreground emission than the ones using interferometers (e.g., LOFAR, MWA and SKA).

The first overview of the foregrounds was outlined by Shaver et al. (1999). Since then various authors have studied the foregrounds in the context of the cosmological 21 cm measurements (Bowman et al. 2009; Cooray & Furlanetto 2004; de Oliveira-Costa et al. 2008; Di Matteo et al. 2004, 2002; Jelić et al. 2008, 2010; Liu & Tegmark 2012; Oh & Mack 2003; Petrovic & Peng 2011; Spinelli et al. 2018; Wang et al. 2006) (see also references in Section 6.2). At the beginning these studies were mainly based on simulations shaped by extrapolated statistical properties of the foregrounds from the higher radio frequencies. The most comprehensive simulation of the foregrounds was carried by Jelić et al. (2008). This simulation has been used extensively in development of the robust foreground mitigation techniques for the LOFAR-EoR project (Chapman et al. 2013, 2012; Ghosh et al. 2018; Harker et al. 2010, 2009a, 2009b; Jelić et al. 2008; Mertens et al. 2018) and more recently for the SKA CD/EoR project (Chapman et al. 2015, 2016). In addition to the dedicated foreground simulations, there are also more complex simulations of both Galactic and extragalactic emission, tailored for studies of the interstellar medium and magnetic fields in the Milky Way (Haverkorn et al. 2019; Sun & Reich 2009; Waelkens et al. 2009) or of different populations of the radio sources at low-radio frequencies (Bonaldi et al. 2019; Wilman et al. 2010, 2008), that can be used as the foreground template in the cosmological 21 cm studies as well.

In parallel to the studies based on simulations, there were also a few dedicated observations taken with the WSRT (Bernardi et al. 2009, 2010) and the GMRT (Pen et al. 2009) radio telescopes to constrain the foregrounds at low-radio frequency. However, only once the new low-frequency instruments came online (e.g., EDGES, LOFAR, MWA and PAPER) our knowledge of the foregrounds started to grow extensively. In the following sections a more comprehensive overview of the foregrounds is given both in total intensity and polarization.

6.1.1 Galactic Foregrounds in Total Intensity

Galactic diffuse synchrotron emission is a dominant foreground component from a few tens of MHz to a few tens of GHz. It is non-thermal in its nature, produced mostly by the relativistic cosmic-ray electrons and to some extent positrons that spiral around the interstellar magnetic field lines and emit radiation. Above a few tens of GHz free–free emission from diffuse ionized gas and thermal dust emission start to dominate over the synchrotron emission (see Figure 6.2).

For a fairly complete theory of the synchrotron emission please refer to e.g., Pacholczyk (1970); Rybicki & Lightman (1986), while here we outline the basics. The radiated synchrotron power emitted by a single electron is proportional to the square of the electron's relativistic kinetic energy, the magnetic energy density, and the pitch angle between the electron velocity and the magnetic field. The angular distribution of the radiation is given by the Larmor dipole pattern in the electron's frame, but in the observer's frame is beamed sharply in the direction of motion.

As the electron spirals around the magnetic field, it is in effect accelerating and emitting radiation over a range of frequencies. Its synchrotron spectrum has a logarithmic slope of 1/3 at low-frequencies, a broad peak near the critical frequency ν_c, and sharp fall off at higher frequencies. The critical frequency is directly proportional to the square of the electron energy and the strength of the perpendicular component of the magnetic field. The longer the electron travels, the more energy it loses, the narrower spiral it makes, and the critical frequency is smaller.

In the case of the Milky Way we need to take into consideration an ensemble of the cosmic-ray electrons, mainly originating from supernovae located close to the Galactic plane and then diffusing outwards. Given a typical magnetic field strength of a few μG, the cosmic-ray electrons with energies between 0.5 to 20 GeV account for the observed synchrotron radiation from tens of MHz to hundreds of GHz. Their energy distribution can be approximated with a power law with slope δ:

$$n_{CR}(E)dE \propto E^{-\delta}dE, \tag{6.1}$$

where $n_{CR}(E)dE$ is the number of cosmic-ray electrons per unit volume with energies between E and $E + dE$. A distribution of their pitch angles is assumed further to be

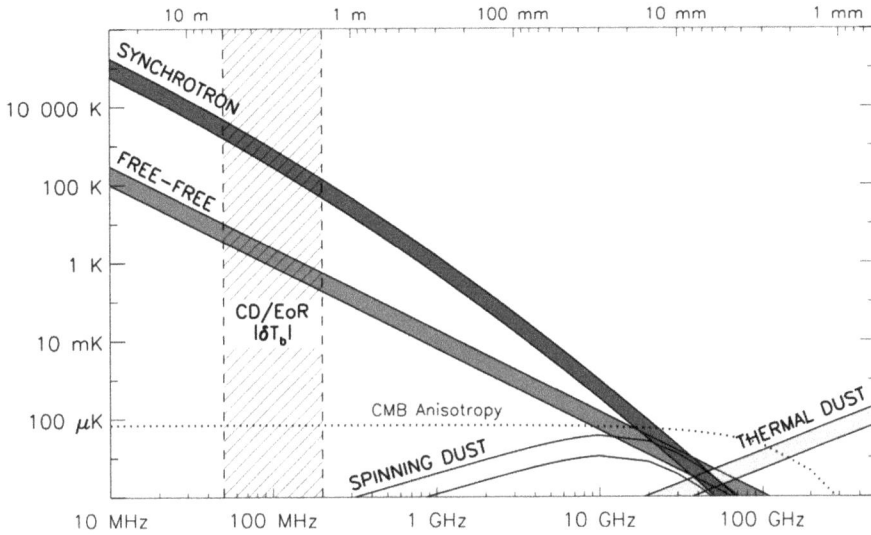

Figure 6.2. The main Galactic diffuse foreground components given as a function of frequency in the total intensity: (i) synchrotron emission from cosmic-ray electrons; (ii) free–free emission from diffuse ionized gas; and (iii) thermal dust emission. There is also a forth component associated with small rapidly spinning dust grains. Synchrotron emission dominates at frequencies below ~10 GHz, while thermal dust emission dominates at frequencies above ~100 GHz. Over the whole frequency range of the CD/EoR experiments, Galactic synchrotron emission is 3–4 orders of magnitude stronger in total power (illustrated by the dark grey area) and 2–3 orders of magnitude stronger in fluctuations than the cosmological 21 cm signal ($|\delta T_b|$). In the CMB experiments, on the contrary, there is a sweetspot around 70 GHz where the CMB anisotropies are relatively bright compared to the Galactic foreground emission.

almost random and isotropic due to relatively long timescales (up to several millions of years) over which they lose their relativistic energies and due to repeatedly scattering that occurs in their environments.

The observed synchrotron spectrum is then given by summing the emission spectra of individual electrons, which are smeared out in the observed spectrum by broad power law energy distribution of the comic-ray electrons. Thus, the synchrotron intensity at frequency ν depends only on n_{CR} and δ from Equation (6.1) and on the strength of the magnetic field component perpendicular to the line-of-sight $B_\perp$:

$$I_\nu \propto n_{CR} B_\perp^{(\delta+1)/2} \nu^{(1-\delta)/2}. \tag{6.2}$$

The observed I_ν can be also described as a featureless power law in regards to the observed intensity I_0 at a reference frequency ν_0:

$$I_\nu = I_0 \left(\frac{\nu}{\nu_0}\right)^{-\alpha}, \tag{6.3}$$

where observed spectral index α is directly connected to the cosmic-ray index δ as $\alpha = (\delta - 1)/2$. Moreover, the observed intensity is commonly expressed in terms of the brightness temperature $T_b(\nu) \sim \nu^{-\beta}$, using the Rayleigh–Jeans law which holds at radio frequencies. In this case the observed spectral index is $\beta = 2 + \alpha = 2 + (\delta - 1)/2$.

The comic-ray energy slope is estimated to $-3.0 < -\delta < -2.5$ at GeV energies (Lawson et al. 1987; Orlando & Strong 2013; Strong et al. 2011). This corresponds to the synchrotron spectral index of $-1 < -\alpha < -0.8$ or $-3 < -\beta < -2.8$ observed at GHz frequencies (Platania et al. 1998; Reich & Reich 1988). At MHz frequencies the synchrotron spectrum is flatter (Guzmán et al. 2011; Rogers & Bowman 2008). Typical values at mid and high Galactic latitudes are $-2.59 < -\beta < -2.54$ between 50 and 100 MHz (Mozdzen et al. 2019) and $-2.62 < -\beta < -2.60$ between 90 and 190 MHz (Mozdzen et al. 2017), as measured by the EDGES instrument.

A difference in the spectral index at MHz and GHz frequencies is due to ageing of the cosmic-ray energy spectrum. As the cosmic-ray electrons propagate trough the interstellar medium, they lose their energies by a number of energy loss mechanisms (Longair 2011) that involve interactions with matter, with magnetic fields and with radiation. This then depletes the population of relativistic electrons and changes their original energy (injection) spectra. For example, the energy loss trough synchrotron radiation is larger for cosmic-ray electrons with higher energies ($\sim E_{CR}^2$). The critical frequency is also proportional to $\sim E_{CR}^2$, so over time, the cosmic-ray spectra becomes steeper together with the synchrotron spectra at higher frequencies. In a similar way, as the cosmic-ray electrons diffuse away from the Galactic plane, the ageing effect also makes a steepening of the synchrotron spectrum at higher Galactic latitudes (Strong et al. 2007).

Besides the spectral index variations across the sky, brightness temperature variations of the Galactic diffuse synchrotron emission reflect spatial fluctuations of the comic-ray electron density and magnetic field strength in the interstellar medium. Synchrotron emission is hence the brightest along the Galactic plane, which has the largest concentration of supernovae, a major source of the cosmic-ray particles, while the darkest parts are within the halo. This can be seen in Landecker all-sky map obtained at 150 MHz (see Figure 6.3, Landecker & Wielebinski 1970), where typical high latitude brightness is between 150 K and 250 K. Given the low resolution of this map ($\sim 5°$), Haslam map at 408 MHz (see Figure 6.3, Haslam et al.

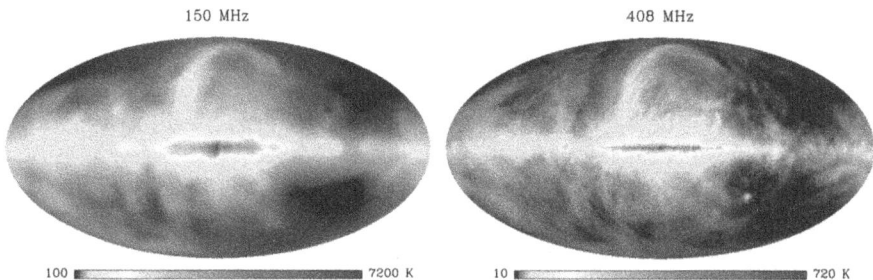

Figure 6.3. All sky maps of Galactic radio emission at 150 MHz (Landecker & Wielebinski 1970) and 408 MHz (Haslam et al. 1982, 1981; Remazeilles et al. 2015). This data is available on the Legacy Archive for Microwave Background Data Analysis (LAMBDA, https://lambda.gsfc.nasa.gov), a service of the Astrophysics Science Division at the NASA Goddard Space Flight Center.

1982, 1981; Remazeilles et al. 2015) is more commonly used as a template for emission at low radio frequencies.

A number of recent dedicated observations additionally constrained Galactic synchrotron emission in selected areas at high Galactic latitudes. The WSRT observations at 150 MHz show an excess of power attributed to the diffuse synchrotron with an rms of 3–5 K on scales greater than 30 arcmin (observations of the fields around 3C 196 and the North Celestial Pole, Bernardi et al. 2010). The LOFAR observation of the North Celestial Pole (Patil et al. 2017) also clearly shows diffuse emission on scales larger than a degree, while slightly higher levels are found on scales greater than 54 arcmin in the MWA observations at 154 MHz of the fields near the South Galactic Pole (Lenc et al. 2016).

6.1.2 Extragalactic Foregrounds in Total Intensity

Extragalactic radio sources are of composite nature. They consist mainly of the active galactic nuclei (AGNs) or the star-forming galaxies (SFGs).

Radio (synchrotron) emission in the AGNs, so called radio-loud AGNs, is related to the accretion of matter by a supermassive black hole at the centre of its host galaxy, typically an elliptical galaxy. This produces narrow jets in a direction perpendicular to the plane of the accretion. The jets can be as large as a few to ten times the size of the host galaxy and many of them have diffuse endings, so called radio lobes. Observed morphology of radio loud AGNs varies and can be classified in different ways. For example, we can classified them based on their radio luminosity and brightness of their components (nucleus, jets and lobes) Fanaroff & Riley (1974). In this case, the FR-I type galaxies have lower radio powers with an edge darkened morphology, while the FR-II type galaxies have higher radio powers with an edge brightened morphology.

Radio emission in the SFGs is produced like in the Milky Way by synchrotron radiation from supernovae related relativistic electrons and by free–free emission from Hɪɪ regions. Observed radio emission of these galaxies is usually also tightly connected, although still not well understood why, to the observed infrared luminosity measuring the star formation rate (e.g., Condon 1992; Helou et al. 1985; Jarvis et al. 2010), hence the name SFGs.

At low-radio frequencies different populations of radio galaxies are still poorly constrained, especially at the faint end of their distribution. There is a low-frequency extragalactic catalogue obtained with the MWA radio telescope in the south (GLEAM, Hurley-Walker et al. 2017) and the ongoing LOFAR Two-metre Sky Survey (LoTTs, Shimwell et al. 2017, 2019) in the north. Until we get deeper with these surveys we need to rely on the data obtained at higher radio frequencies.

Normalized differential source counts for different populations of radio sources at 1.4 GHz is given in Figure 6.4. Thanks to the recent very deep surveys (e.g., COSMOS, Bondi et al. 2008; Novak et al. 2018; Smolčić et al. 2017a, 2017b, 2017c) the extragalactic radio sources are constrained well up to the flux densities of 500 μJy. The population of the SFGs dominate at μJy levels, while the population of the radio-loud AGNs dominates at flux densities ⩾1 mJy (for a review see Prandoni

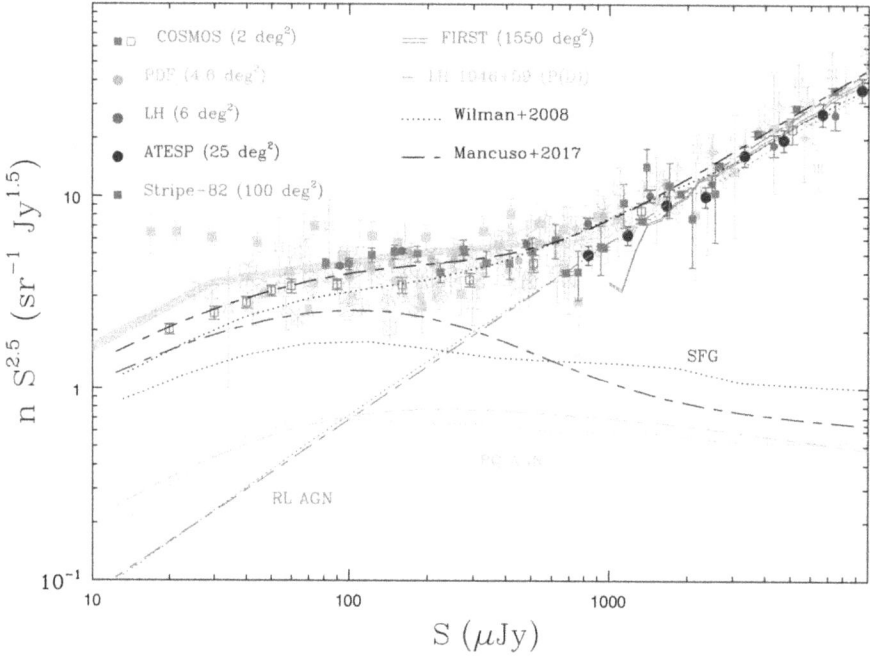

Figure 6.4. Normalized 1.4 GHz differential source counts. for different radio source populations: radio-quite (RQ) AGN, radio-loud (RL) AGN and star-forming galaxies (SFG). The dotted and dashed lines represent predicted counts from different model (Mancuso et al. 2017; Wilman et al. 2008, 2010). Different colors indicate different populations: radio-quiet (RQ) AGN, radio-loud (RL) AGN and star-forming galaxies (SFG), while their sum is given in black. Colored symbols show the counts from a number of large-scale surveys: COSMOS field (Bondi et al. 2008; Smolčić et al. 2017a); Phoenix Deep Field (PDF, Hopkins et al. 2003); the Lockman Hole (LH, Prandoni et al. 2018); the ATESP survey (Prandoni et al. 2001), the Stripe-82 region (Heywood et al. 2016); and the FIRST survey (White et al. 1997). Reproduced with permission from Prandoni (2018).

2018 and references therein). There is also a third population of the sources detected below ∼100 μJy, commonly referred to as radio-quiet AGNs. These sources do not have large scale radio jets and lobes like radio-loud AGNs. They are probably SFGs hosting also an active nucleus that contributes to the radio emission (Ceraj et al. 2018; Delvecchio et al. 2017).

In addition to the radio source counts we also need to have a good knowledge of their distribution in the sky (clustering properties) and of their radio spectra. Neglecting source clustering may result in underestimating the angular foreground power which can potentially lead to a false detection of the cosmological 21 cm signal (Murray et al. 2017, 2018), while if the radio spectra is not smooth the foreground removal will be much more demanding.

The radio spectra of the radio galaxies can be described with the power-law function with a spectral index of $\alpha \sim -0.7/-0.8$, due to the synchrotron nature of the emission. Nevertheless, there are process that can change the shape of the spectra (free–free absorption, synchrotron self-absorption, spectral ageing, etc) and make it complicated. Recent LOFAR observations of the Boötes field (Calistro et al. 2017) showed significant differences in the spectral curvature between SFG and AGN populations. The radio spectra of SFGs show a weak but statistically significant flattening, while the radio spectra of the AGNs is becoming steeper toward the lower frequencies. Therefore, different power-law slopes should be assumed for AGNs and SFGs, when modeling the radio sky at frequencies relevant for the cosmological 21 experiments.

6.1.3 Polarized Foregrounds

Galactic synchrotron emission is partially linearly polarized. Its polarized intensity PI_ν depends on a cosmic-ray electron density n_{CR}, a slope of the cosmic-ray energy spectrum δ, and a strength of the magnetic field component perpendicular to the line-of-sight $B_\perp$, in the same way as defined by Equation (6.2) in total intensity. The only difference is the amount of emission, defined by the degree of polarization (Rybicki & Lightman 1986):

$$\Pi = \frac{\delta + 1}{\delta + 7/3}. \tag{6.4}$$

For $\delta = 2.2$, which is consistent with the observed synchrotron spectral index of $-\beta = -2.6$ at 150 MHz (Mozdzen et al. 2017), we get $\Pi = 0.7$. At low radio frequencies (100–200 MHz) about 70% of Galactic synchrotron emission is intrinsically polarized, while in fact we observe only a few percent (Bernardi et al. 2013; Jelić et al. 2014, 2015; Lenc et al. 2017, 2016; Van Eck et al. 2019, 2017). To understand why we observe such a small percentage of polarized emission, we need to take a closer look at Faraday rotation and associated depolarization that occurs.

As a linearly-polarized wave, with a wavelength λ, propagates through a magnetized plasma its polarization angle θ is Faraday rotated by:

$$\frac{\Delta\theta}{[\text{rad}]} = \frac{\lambda^2}{[\text{m}^2]} \frac{\Phi}{[\text{rad m}^{-2}]} = \frac{\lambda^2}{[\text{m}^2]} \left(0.81 \int \frac{n_e}{[\text{cm}^{-3}]} \frac{B_\parallel}{[\mu\text{G}]} \frac{dl}{[\text{pc}]} \right), \tag{6.5}$$

where Φ is Faraday depth, n_e is a density of the thermal electrons, $B_\parallel$ is a strength of the magnetic field component parallel to the line-of-sight. The integral is taken over the entire path-length l, from the source to the observer. The Faraday depth is positive when $B_\parallel$ points toward the observer, while it is negative when $B_\parallel$ points away.

In the Milky Way, where distributions of thermal and comic-ray electrons are perplexed throughout the entire volume, differential Faraday rotation will occur and will depolarize the observed synchrotron emission (Sokoloff et al. 1998). As Faraday rotation is proportional to λ^2, depolarization at low radio frequencies will be significant. Nevertheless, small amounts of polarized emission that can still be observed carry valuable information about the physical properties of the intervening magnetized plasma.

First attempts to constrain diffuse polarized emission at 150 MHz were done using the GMRT (Pen et al. 2009) and WSRT observations (Bernardi et al. 2009,

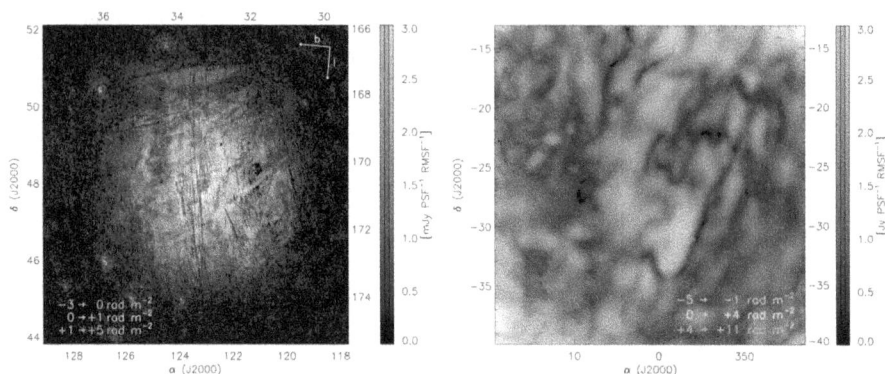

Figure 6.5. Polarized structures discovered at different Faraday depths with LOFAR (an image on the left—created using the data presented and discussed in Jelić et al. 2015 with permission of the authors) and MWA (an image on the right—created using the data presented and discussed in Lenc et al. 2016, with permission of the authors) in two fields at high Galactic latitudes.

2010). However, the full richness and complexity of polarized emission at low-radio frequencies was not revealed until LOFAR and MWA came online. Observations with these instruments discovered astonishing morphology of polarized Galactic synchrotron emission of a few Kelvin in brightness (see Figure 6.5 and Bernardi et al. 2013; Iacobelli et al. 2013; Jelić et al. 2014, 2015; Lenc et al. 2016, 2017; Van Eck et al. 2017, 2019). The discovered structures were unraveled by Rotation Measure (RM) synthesis Brentjens and de Bruyn (2005). This is a technique in radio polarimetry that disentangles the observed wavelength-dependent polarization into a Faraday spectrum, i.e., the distribution of polarized emission as a function of Faraday depth. This allow us then to preform, so called, Faraday tomography, a study of the intervening magnetized plasma as a function of Faraday depth.

Given a wide frequency coverage and a high spectral resolution available in the low-frequency instruments Faraday tomography is performed at an exquisite sensitivity and resolution in Faraday depth of ~ 1 rad m^{-2}, an order of magnitude higher than at 350 MHz. This allow us to map small column densities of magnetized plasma that are, in most cases, not possible to detect at higher radio frequencies. Interestingly, most of the observed structures at low-radio frequencies appear at Faraday depths $\Phi \leqslant 15$ rad m^{-2} and they are not correlated with structures in total intensity. This result will be relevant in later discussion of the polarization leakage in the cosmological 21 cm experiments (see Section 6.2.5).

Extragalactic polarized sources are not a big concern for the cosmological 21 cm experiments due to their sparsity in the sky. In the MWA 32 element prototype survey of 2400 deg^2 of the southern sky at 189 MHz only one polarized source was found (Bernardi et al. 2013). In a preliminary data release of the LOFAR Two-meter Sky Survey of the HETDEX field, covering an area of 570 square degrees, 92 polarized radio sources where found (Van Eck et al. 2018). This gives a lower limit to the polarized source surface density at 150 MHz of only 1 source per 6.2 square degrees. Somewhat higher value, 1 source per 2 degrees, was found based on

LOFAR observations of three 16 deg^2 fields (Jelić et al. 2015; Mulcahy et al. 2014; Van Eck et al. 2018).

6.1.4 Radio Frequency Interference

Terrestrially, radio frequency interference (RFI) from any human-made sources of radio transmission, such as wind turbines, leads to the necessary excision of frequency channels using a flagging technique (e.g., Offringa et al. 2012l Prasad & Chengalur 2012). The number of channels excised is significant, around 1% of channels of data for MWA and LOFAR (Offringa et al. 2019, 2015). Without careful mitigation in the calibration, imaging and diffuse foreground removal stages, RFI excision can result in an excess power that scales with the number of excised channels and does not integrate down with time, significantly dominating over the cosmological 21-cm signal by 1–2 magnitudes (Offringa et al. 2019).

6.2 Foreground Mitigation

The 21-cm signal emitted by high-redshift neutral hydrogen provides a window into the Epoch of Reionization (EoR), but it is a window that is obscured by layers of foregrounds. Extra-terrestrially, there exist a multitude of foregrounds which dominate all frequencies of observation and so more subtle methods than excision are required. This part of the chapter discusses the development and current use of Galactic and extragalactic foreground mitigation methods in Epoch of Reionization 21-cm experiments.

6.2.1 Foreground Mitigation in the Data Analysis Pipeline

6.2.1.1 Bright Source Removal

The first stage of foreground removal involves mitigating the effect of the very brightest sources on the sky: the point sources and extended sources. Bright source removal often comes under the umbrella of calibration as opposed to foreground mitigation however we will briefly summarize the process here. For example, the MWA real-time system (RTS) (Mitchell et al. 2008) carries out sequential bright source "peeling" on the visibilities, tracking a few hundred of the brightest sources and comparing to a sky model constructed from existing catalogues and MWA observations (Carroll et al. 2016). The gains are calibrated on the strongest source, before that source is peeled (subtracted) from the data, and the next strongest source is used to refine the calibration, and so on until it is deemed that enough bright sources have been removed, usually a few hundred to a thousand at most. The other MWA calibration pipeline, Fast Holographic Deconvolution (FHD) (Sullivan et al. 2012), uses the MWA extragalactic catalogue GLEAM (Hurley-Walker et al. 2017) to calibrate gains, modeling all sources out to 1% beam level in the primary lobe, amounting to approximately 50000 sources (Barry et al. 2019) and then removing a smaller population of them from the data. Similarly, LOFAR has built up a sky model over several years using the highest resolution LOFAR images and subtracts the sources in visibility space also (Yatawatta 2015; Yatawatta et al. 2013). As of 2017, the LOFAR EoR sky model contained around 20,800 unpolarized sources.

6.2.1.2 The EoR Window

It has previously been traditional when discussing diffuse foreground mitigation to assume that the previous stage of bright source subtraction has already been implemented perfectly. This is no longer seen to be a valid or safe assumption, as the chromaticity of the instrument, calibration errors and incorrect source sub-traction lead to significant bias in the EoR signal for all current and planned experiments (e.g., Ewall-Wice et al. 2017; Procopio et al. 2017; Barry et al. 2016; Patil et al. 2016; Datta et al. 2010; Liu et al. 2009), including redundant arrays (Byrne et al. 2019).

The spectral differences between the EoR signal and the bias introduced by the foregrounds and instrument lend themselves to a neat separation in $k_\perp - k_\parallel$ space, Figure 6.6. In this formalism, spectrally-smooth foregrounds live in a well-defined area of k-space, at the smallest $k_\parallel$ scales, equivalent to the red stripe at the bottom of Figure 6.6, excluding the wedge area. The assumption that the foregrounds would remain smooth and confined in a horizontal area at low $k_\parallel$ even after observation by a radio interferometer drove early foreground removal techniques such as those introduced in Section 6.2.3.1 but is now known to be an incorrect assumption. The chromaticity of the instrument results in a "mode-mixing" where power is trans-ferred from the angular to the frequency scales, throwing power upwards from the foreground area in the window into the larger $k_\parallel$ scales, with the effect increasing with larger $k_\perp$. This results in a wedge like structure, a structure that has been now extensively discussed and mathematically defined in the literature (e.g., Jensen et al. 2016; Dillon et al. 2014; Liu et al. 2014a, 2014b; Hazelton et al. 2013; Thyagarajan et al. 2013; Pober et al. 2013; Morales et al. 2012; Vedantham et al. 2012; Trott et al. 2012; Parsons et al. 2012a; Datta et al. 2010). Because the point sources reside on the largest $k_\perp$ scales they, or even their residuals when incorrectly calibrated, can overwhelm the EoR power in the frequency scales (e.g., Bowman et al. 2009 and immediately preceding references).

Now we have defined the problem, namely the overpowering magnitude and potential leakage of foregrounds onto the EoR signal, we can consider how to achieve our aim of making accurate statistical conclusions on the nature of the EoR using the data within this window. To proceed, we can consider two philosophies. The first, **foreground subtraction**, aims to remove foreground contamination on all scales. The benefit of this is that there are more k scales available for analysis. The drawback of foreground subtraction across all k-scales is that any failure in the method will potentially result in a foreground fitting bias across all scales of the window, providing another layer of contamination. One could instead avoid the foregrounds and therefore the need to remove them: **foreground avoidance**. This philosophy aims to then quantify the foregrounds and wedge such that any analysis occurs within a well-defined window free of contamination. The benefit of this is, as stated, the avoidance of foreground subtraction bias. The drawback is that any analysis is performed on a significantly reduced set of scales which can for example introduce its own bias into the spherically averaged power spectrum (Jensen et al. 2016). Additional to both philosophies, we can implement **foreground suppression**, which down-weights scales where the foregrounds or foreground removal residuals

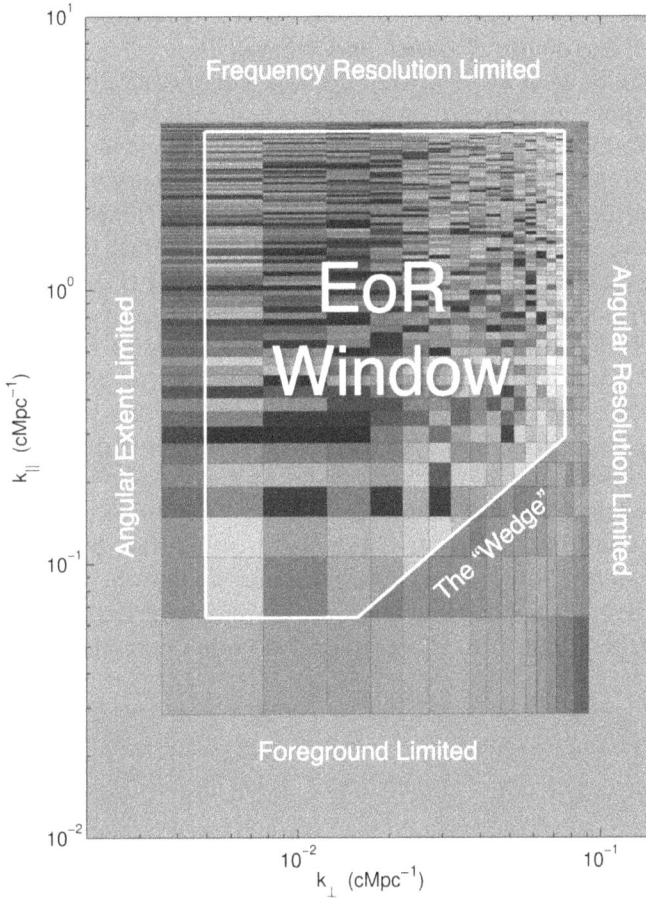

Figure 6.6. A schematic of the "EoR Window" in the cylindrical $k_\parallel$, $k_\perp$ Fourier plane, taken from Figure 1 of Dillon et al. (2014). In a perfect observation, with zero instrumental effects, the foregrounds would be entirely contained in the well defined horizontal band. In a realistic observation however, the chromaticity of the instrument results in a leakage of power up into the EoR window, into a region called the "wedge." Aside from these contaminated areas there should be a relatively clean area called the EoR window. Reprinted with permission from Dillon et al. (2014). Copyright 2014 by the American Physical Society.

are dominant. We will now discuss these approaches in further detail in the context of current EoR experiments.

6.2.2 Foreground Avoidance and Suppression

The Murchison Widefield Array (MWA) has two separate pipelines which differ in their application of foreground mitigation techniques and calibration methods, while mostly employing foreground avoidance. The way in which MWA is optimized for making images allows the option to directly subtract known foregrounds but in this case the direct foreground subtraction is primarily applied to get access to a cleaner EoR window, not to get access to within the wedge, as is the motivation of foreground subtraction in LOFAR.

The FHD (Sullivan et al. 2012) and ϵpsilon (Barry et al. 2019) pipeline builds a sky model of point sources based on a golden set of data, including all sources above a floor limit within the primary beam of the instrument, and those beyond the primary beam if they are above 1% of the maximum primary beam level. This point source model is used in calibration in a similar way to LOFAR, and contains about 7000 sources as of 2016 (Beardsley et al. 2016). In contrast to the RTS (Mitchell et al. 2008) and CHIPS (Trott et al. 2016) pipeline, the FHD-ϵpsilon pipeline also generates a diffuse foreground model by subtracting away the point source model from the observed data, and integrating over frequency to create a diffuse foreground model free of spectral information (Beardsley et al. 2016). They then subtract both the point source model and the diffuse model from the data to minimize the leakage from the wedge into the EoR window. In Figure 6.7 we see the effect of this foreground subtraction on the EoR window. The left image is the difference between the power spectrum of the MWA foreground model without diffuse foregrounds (i.e., just point sources) and with diffuse foregrounds. The plot shows that the diffuse foregrounds have power far up into the EoR window, due to non-uniform spectral sampling and the effect of windowing the data along frequency during the Fourier Transform. This figure if no other demonstrates the danger of

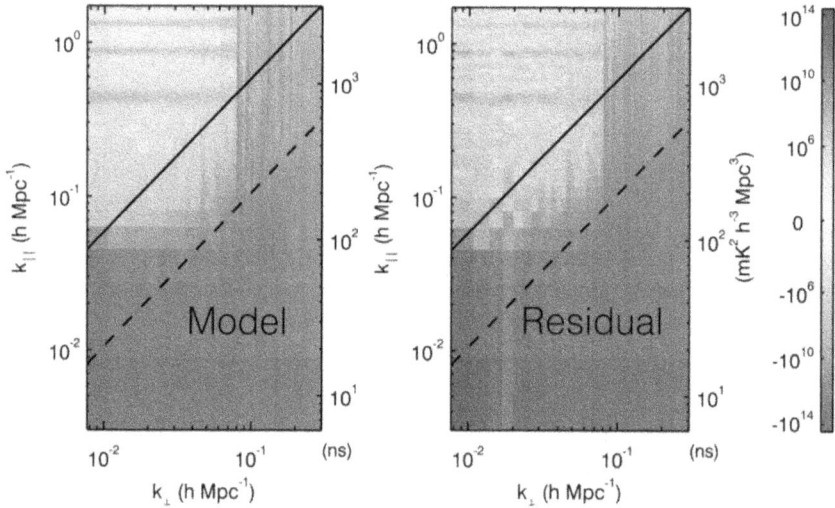

Figure 6.7. Left: The difference between the MWA foreground model without diffuse foregrounds (i.e., just point sources) and with diffuse foregrounds. Adding diffuse foregrounds into the model produces leakage far up into the EoR window and instrumental contamination can be seen in the horizontal lines throughout the EoR window. Right: the difference between the power spectrum of the residuals when only the point sources have been subtracted as described above, and the power spectrum of the residuals where the diffuse foregrounds have also been subtracted. There is a clear reduction in foreground residuals all along the wedge and the EoR window is noise-like, suggesting a lack of foreground contamination there. There is a 70% reduction in residual power of the foregrounds using this method. Reproduced from Beardsley et al. (2016). © 2016. The American Astronomical Society. All rights reserved.

assuming that the observed foreground signal is smooth and contained only at the smallest $k_\parallel$. Further instrumental complications can be seen in the horizontal lines throughout the EoR window, which is contamination due to the periodic frequency sampling function used by MWA Offringa et al. (2016). The right plot of Figure 6.7 shows the difference between the power spectrum of the residuals when only the point sources have been subtracted as described above, and the power spectrum of the residuals where the diffuse foregrounds have also been subtracted. There is a clear reduction in foreground residuals all along the wedge and the noise-like characteristic of the EoR window suggests a lack of foreground contamination there. Beardsley et al. (2016) report a 70% reduction in residual power of the foregrounds using this method.

The black lines in Figure 6.8 show the area of the EoR window used in the FHD-ϵpsilon pipeline, with the masks ensuring the avoidance of the horizontal contamination lines and the wedge.

The RTS-CHIPS pipeline subtracts significantly fewer sources, a few hundred to a thousand at most, and does so in visibility space. There is no diffuse foreground model in the subtraction stage and instead CHIPS down-weight modes with residual point source power. There is also the option of diffuse foreground weighting based on a simple foreground model where the covariances are known, though in practice this diffuse down-weighting is not currently utilized.

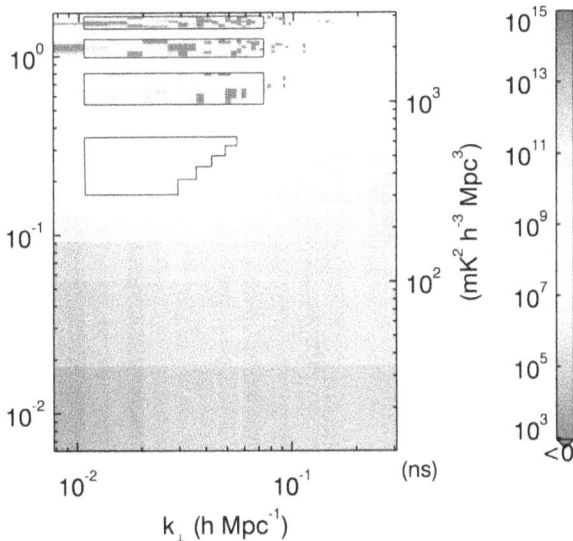

Figure 6.8. An example 2D cylindrical power spectrum of the first season MWA data after foreground mitigation. The data used for the upper limits can be seen bounded by black lines. The amount of data available for a power spectrum analysis has been severely reduced by the presence of foregrounds and instrumental contamination but the data within the bounded regions displays noise-like behavior indicative of successful foreground mitigation. Reproduced from Beardsley et al. (2016). © 2016. The American Astronomical Society. All rights reserved.

6.2.2.1 Delay Space Filtering

Delay space filtering is a method of foreground avoidance primarily adopted by the Donald C. Backer Precision Array for Probing the Epoch of Reionization (PAPER) Parsons et al. (2010). As with most foreground mitigation methods it requires the foregrounds to be reasonably smooth, even after instrumental effects. The wedge is the end-result of an instrument where the frequency-dependence of the instrument's sampling is dependent on the length of the baseline measuring the sky. Delay-space filtering exploits this relation by analyzing the data per baseline, circumventing the conspiracy of instrumental effects on the foregrounds and effectively isolating the foregrounds such that they are easily avoided. Figure 6.9 demonstrates that for a given baseline measurement the visibility sampled changes with frequency, with a steeper change for longer baselines. This results in the mode-mixing seen in the 2D cylindrical power spectrum and the wedge structure, where we see power thrown up into the EoR window increasingly on the largest $k_\perp$ scales, which are the scales sampled by the longest baselines. Delay space filtering aims to mitigate the mode mixing by performing a Fourier transform along the visibility sampled by a given baseline (a solid line in Figure 6.9), and not along the frequency direction (vertical axis of Figure 6.9) as is usual.

A delay transform takes a single time sample of a visibility from one baseline, for all observed frequencies (i.e., one of the solid lines on Figure 6.9, and Fourier transforms it to produce the delay spectrum (Parsons et al. 2012a, 2012b; Parsons & Backer 2009). The delay transform is:

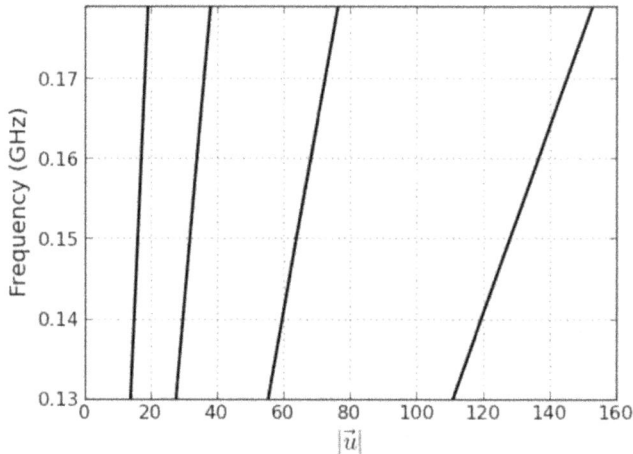

Figure 6.9. This figure demonstrates the frequency dependence of the wavemode sampled by baselines measuring 16, 32, 64, and 128 wavelengths at 150 MHz. For a given baseline measurement, the visibility sampled changes with frequency, with a steeper change for longer baselines. This results in the mode-mixing seen in the 2D cylindrical power spectrum and particularly "the wedge," where we see power thrown up into the EoR window increasingly on the smallest $k_\perp$ scales, which are the scales sampled by the longest baselines. Reproduced from Parsons et al. (2012). © 2012. The American Astronomical Society. All rights reserved.

$$\tilde{V}_{b}(\tau) = \int dl\ dm\ d\nu\ A(l, m, \nu)I(l, m, \nu)e^{-2\pi i \nu(\tau_g - \tau))} \qquad (6.6)$$

where l, m have their usual definition relating to angular coordinates on the sky (e.g., Richard Thompson et al. 2001). τ is the time-delay between the signal reaching both antennas and the geometric group delay associated with the projection of baseline $\vec{b} \equiv (b_x, b_y, b_z)$ in the direction $\hat{s} \equiv (l, m, \sqrt{1 - l^2 - m^2})$ is:

$$\tau_g \equiv \frac{\vec{b} \cdot \hat{s}}{c} \qquad (6.7)$$

For comparison, the usual equation where the Fourier transform is simply applied along the frequency axis is:

$$\tilde{V}(u, v, \eta) = \int dl\ dm\ d\nu\ A(l, m, \nu)I(l, m, \nu)e^{-2\pi i(ul + vm + \eta\nu)} \qquad (6.8)$$

where η is the Fourier transform of ν.

The delay transform transforms flat spectra sky emission into delta functions. Because the sky emission is not perfectly smooth, and the instrument adds in its own unsmoothing effects, this delta function is effectively convolved with a kernel, which broadens the delta function in delay space. For the smoother foregrounds, that kernel will be narrow, and confined within the "horizon limits," the geometric limit in delay space beyond which no flat spectra emission can enter the telescope. Spectrally unsmooth sky emission can enter beyond these horizon limits and emission such as the cosmological signal finds itself with a wide convolving kernel, spreading power well beyond the horizon limit where the foregrounds are theoretically confined. In Figure 6.10 we see the delay transform at 150 MHz for several spectrally smooth sources and how they remain confined within the horizon limits of

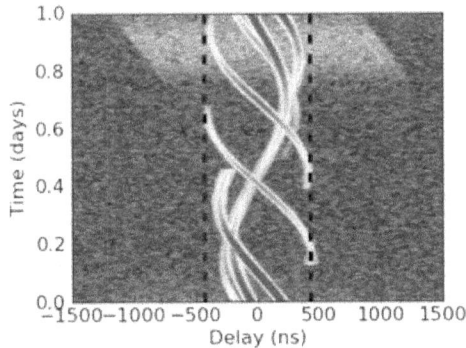

Figure 6.10. The delay spectra of several smooth-spectra sources, which remain largely confined within the geometric horizon limits. The broad 21-cm cosmological signal delay spectra in cyan demonstrates that unsmooth spectral signals have a much wider convolving kernal and produce a much wider delay spectra. If analysis is carried out outside of the horizon limits then the foregrounds can be avoided. Reproduced from Parsons et al. (2012). © 2012. The American Astronomical Society. All rights reserved.

the baseline (here 32 m). In contrast, the delay spectrum of spectrally unsmooth emission, such as the cosmological 21-cm signal, finds itself smeared to high delays. Full mathematical detail can be found in Parsons et al. (2012a, 2012b) and Parsons & Backer (2009).

By performing this delay space transform, we are effectively moving into the sidelobes of the 21-cm signal in delay space. The cosmological signal is scattered to high delays whereas the foregrounds are not, allowing the data analysis in that large delay space to be free of foregrounds and foreground removal bias. This method also removes the need for imaging in order to remove the foreground directly, making it suitable for a redundant array with little or no ability to image, but a high sensitivity to the 21-cm power spectrum (Parsons et al. 2012b).

PAPER is a radio interferometer with a highly redundant antenna layout, with multiple baselines of the same length and orientation. Because these multiple baselines all measure the same sky signal, any differences in the signal received would be due to instrumentation, allowing a quick calibration for multiple calibration parameters—"redundant calibration" (e.g., Joseph et al. 2018; Li et al. 2018; Dillon & Parsons 2016; Zheng et al. 2014; Wieringa 1992).

PAPER avoided the use of the delay modes dominated by foregrounds and downweighted residual foregrounds using inverse covariance weighting in order to form an upper limit power spectrum measurement (Ali et al. 2015). The latter method of inverse covariance weighting where the covariance is calculated based on the data itself has now been shown to carry the considerable risk of overfitting the EoR data (Cheng et al. 2018). To be clear, despite the retraction of the PAPER-64 results due to power spectrum estimation errors (Ali et al. 2018), the delay space filtering technique remains a promising approach to foreground mitigation.

6.2.3 Foreground Subtraction

Foreground subtraction methods all seek to find a model for the observed foregrounds and remove that model from the observed signal, leaving the cosmological signal, instrumental noise and any foreground fitting errors. Foreground removal is usually applied on all scales, meaning that it potentially allows access into the lowest $k_\parallel$ scales where foregrounds traditionally dominate. A caveat of this is that any foreground fitting bias has the potential to affect all scales in the window: foregrounds may remain within the wedge and cosmological signal may be erroneously fitted out within the previously clean EoR window. As an aside, there has been no method so far that can separate out the cosmological 21-cm signal entirely by itself, separate from instrumental noise. Currently when the foregrounds are subtracted or avoided the noise and cosmological signal are still mixed together in what are often termed the "residuals." The instrumental noise can be obtained from the data for example by the differencing of very fine bandwidth frequency channels, such that both the foregrounds and EoR signal are smooth. The noise power spectrum can then be removed from the residual power spectrum to form the recovered cosmological signal power spectrum. We will now introduce some of the main foreground subtraction techniques.

6.2.3.1 Polynomial Fitting and Global Experiments

As we have seen in the first half of this chapter, the astrophysical foregrounds are 3–5 magnitudes brighter than the cosmological 21-cm signal and so, by magnitude alone, appear to be the most ominous obstacle to the first detection. Despite, or perhaps because of, their overwhelming magnitude they are well constrained, following power laws with known indices and evolution. The sheer magnitude of the foregrounds means that purely spatial separation, i.e., separation based on only one frequency slice, is not possible: the 21-cm signal and foregrounds are not statistically different enough when only considering spatial scales (see left-hand panel of Figure 6.11) Santos et al. (2005), Di Matteo et al. (2004), Oh and Mack (2003), Di Matteo et al. (2002). While separation based purely on spatial scales is not feasible, the high frequency coherence of the foregrounds compared to both the instrumental noise and cosmological signal provides a way to separate out the two signals (foregrounds and both cosmological signal and noise) (see right-hand panel of Figure 6.11).

Santos et al. (2005) and Zaldarriaga et al. (2004) exploited the large cross-correlation of the foregrounds in slices at different frequencies to model and remove the foregrounds noting that the frequency coherence was also a useful tool for separation. Polynomial fitting went on to exploit the frequency coherence of the foregrounds across the bandwidth, removing the foregrounds along the line of sight without using any spatial correlation information (e.g., Bowman et al. 2009; Wang

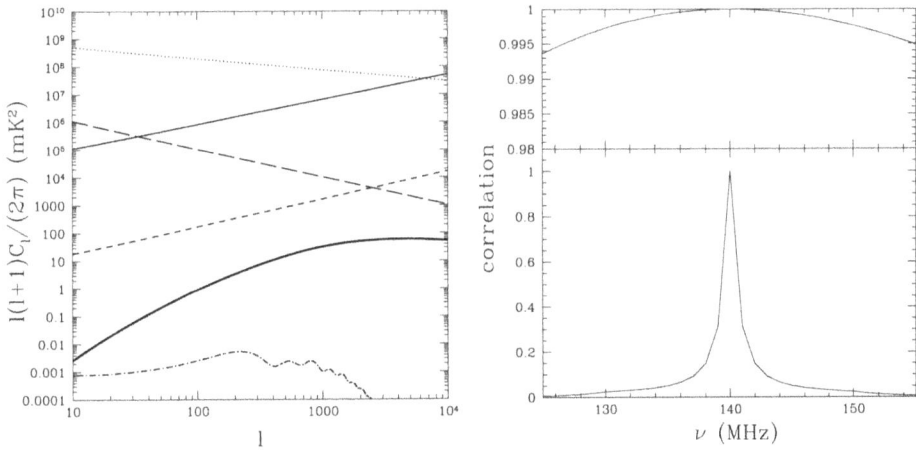

Figure 6.11. Left: The 2D power spectrum at 140 MHz for the cosmological signal (thick, solid), point sources (thin, solid), Galactic synchrotron (thin, dotted), extra-Galactic free–free (thin, dash), Galactic free–free (thin, long dash) and the CMB (dot–dash). The cosmological signal is dominated by foregrounds at all scales, such that separation based purely on spatial differences in not feasible. Figure taken from Figure 5 of Santos et al. (2005). Right: The simulated frequency correlations for the foregrounds (top) and cosmological signal (bottom). This plot shows how the correlation between frequency slices (with the comparison made to a slice at 140 MHz), drops off with increasing frequency separation. The foregrounds are highly frequency coherent, whereas the cosmological signal is significantly less so. Reproduced from Santos et al. (2005). © 2005. The American Astronomical Society. All rights reserved.

et al. 2006; McQuinn et al. 2006). In this method, the foregrounds are modeled by a polynomial function, for example in log-log space such as:

$$\log I = a_3 + a_2 \log \nu + a_1 (\log \nu)^2 + \cdots \tag{6.9}$$

where I is the brightness temperature of the data, ν is the frequency of observation and a_1, a_2, a_3 are the coefficients which are to be determined in the fit.

Polynomial fitting is a parametric foreground mitigation method. It uses knowledge from simulated foregrounds to tune the coefficients of the polynomial function (e.g., Jelić et al. 2008). There are two areas of concern when using this method. First, the effect of the instrument results in a signal which can differ significantly from the frequency-coherent theoretical foreground model (see Section 6.2.1.2). By incorporating weighting according to the amount of information in a particular uv cell, this could possibly be overcome (Liu et al. 2009; Bowman et al. 2009). The second area of concern was that the success of the method relies heavily on having an accurate model for the foreground signal. There are many more instrumental effects than the frequency dependence of the beams, for example polarization leakage (e.g., Nunhokee et al. 2017) and excess instrumental noise (Patil et al. 2016). Wang et al. (2013) demonstrated that polynomial removal across the EoR frequency band resulted in significant signal loss when using simulations of complex foregrounds, though they also showed that by fitting a polynomial simultaneously in smaller bandwidth segments this signal loss could be mitigated. Polynomial removal is now rarely used within the interferometric experiments with the exception of the upper limit from GMRT (Paciga et al. 2011) which used a similar philosophy to remove their foregrounds, albeit by applying a piecewise linear function, as opposed to a polynomial function.

Aside from interferometric experiments, polynomial fitting does have a prominent place in global EoR experiments (e.g., Singh et al. 2018; Bowman et al. 2018) which, due to the coherence of the 21-cm global signal over frequency, means that so far all the more sophisticated methods of foreground mitigation have been unworkable on global simulation and data. For example, the very small number of lines of sight observed by a single global experiment mean that there is not enough spatial information for some non-parametric methods to work.

The Experiment to Detect the Global EoR Signature (EDGES) detection (Bowman et al. 2018) used a five term polynomial based on the properties of the foregrounds and ionosphere, incorporating the actions of the instrument into their foreground model. The level of accuracy of this method has since questioned however, with the results showing dependence on the description of the foregrounds (Bradley et al. 2019; Hills et al. 2018). Overall, polynomial fitting correctly exploits the foreground coherence but it is vulnerable to unknown systematics and unexpected foreground signals. For global experiments there is currently no other option, but for interferometric experiments the methods in the following section provide an alternative.

6.2.3.2 Non-parametric Foreground Removal

The concern that the instrument might introduce complex spectral structure into the foreground signal has driven research into foreground mitigation methods which rely less on a strongly constrained foreground model. Wp smoothing (Harker et al. 2009a) fits a function along the line of sight while penalizing the "Wendepunkt," inflection points, that give the method its name. Unlike polynomial fitting, the function is permitted to be rough but inherently favors the more smooth models. Wp smoothing is applied along each line of sight individually and so spatial correlations of the foregrounds are not utilized in making the foreground fit. The current method employed by the LOFAR EoR pipeline, Gaussian Process Regression (GPR) Mertens et al. (2018) also relies purely on spectral information. GPR models the foregrounds, mode mixing components, 21-cm cosmological signal and noise by Gaussian Processes, allowing a clear separation and uncertainty estimation (see Figure 6.12). GPR does not require specification of a functional form for each component but instead allows the data to find its own model, while taking into account the covariance structure priors incorporated by the user. This allows a certain level of control, for example splitting the foreground covariance into a smooth intrinsic foreground model and an unsmooth mode mixing component, while still not imposing a strict level of smoothness or a parametric form on the data.

Blind Source Separation (BSS) methods have been used in Cosmic Microwave Background experiments (Planck Collaboration 2018a,b) and their application to EoR data is a natural evolution. BSS methods are used across a wide range of fields

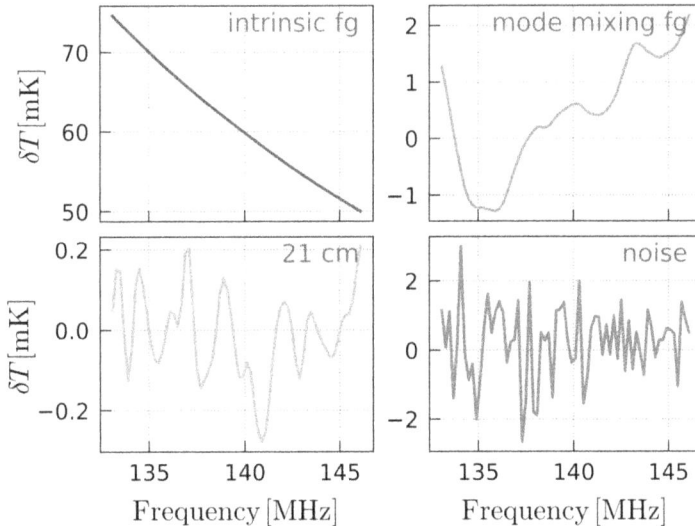

Figure 6.12. Simulated components of the observed signal, demonstrating that the smooth foreground signal is accompanied by an unsmooth mode mixing signal. GPR models each of these foreground components separately, making use of prior information about each component in the form of covariance functions. Reproduced from Mertens et al. (2018), by permission of Oxford University Press on behalf of the Royal Astronomical Society.

in order to separate mixed signals into independent components. The data can be expressed in terms of the mixing model:

$$\mathbf{X} = \mathbf{AS} + \mathbf{N} \tag{6.10}$$

where $\mathbf{X}$ is the observed signal, $\mathbf{S}$ are the independent components of that signal, $\mathbf{N}$ is the noise and $\mathbf{A}$ is a matrix determining how the components are mixed, the "mixing matrix." For an observation of m frequency channels each constituting t pixels and a foreground model of n independent foreground components, the dimensions of these quantities are $\mathbf{X}[m,t]$, $\mathbf{S}[n,t]$, $\mathbf{N}[m,t]$ and $\mathbf{A}[m,n]$.

When this framework is applied to EoR data, the foregrounds are contained within $\mathbf{S}[n,t]$ while the cosmological signal is contained along with the instrumental noise in $\mathbf{N}[m,t]$. The independent components of the foreground model are not directly related to the Galactic synchrotron, Galactic free–free and extragalactic foregrounds, but instead each independent component is potentially a mixture of all these physical foregrounds. This leaves the user without a physically motivated choice for the number of independent components, so that the number must be chosen empirically based on simulated data. Once a foreground model $\mathbf{AS}$ has been determined this can then be subtracted from the observed signal, leaving the residual data as with the other methods.

The two BSS methods introduced for use on EoR data differ by their definition of independence. FastICA (Chapman et al. 2012; Hyvärinen et al. 2004; Hyvarinen 1999) is a long-established independent component analysis technique which uses statistical independence to separate out the foreground components. FastICA constrains the different components by maximizing the negentropy of the signal components, utilizing central limit theorem which states that the more independent components a signal contains, the more Gaussian the probability distribution function of that signal will be. In contrast, GMCA (Bobin et al. 2016, 2015; Chapman et al. 2013; Bobin et al. 2008) is a method developed for use on CMB data that uses morphological diversity to separate out components. GMCA assumes that the data is represented in a sparse manner which can be achieved by a wavelet decomposition. With the independent components unlikely to have the same few non-zero basis coefficients in wavelet space, the method is able to separate out the components according to the differing sparse basis coefficient values. As with FastICA, we actually care little for the independent components individually, it is the combination of those as a whole which form the foreground model, with the method naturally separating out the decoherent noise and cosmological signal. In simulation both these methods have behaved well, opening up the EoR window into the lowest scales even when subjected to unsmooth foreground simulations, Figure 6.13. GMCA was used to achieve the current LOFAR upper-limit (Patil et al. 2017) but since then has not been able to remove the foregrounds down to the same level as, for example, GPR (Mertens et al. 2018). The reason for this remains unknown and a full comparative analysis is currently underway. Mertens et al. (2018) also expressed concern that because BSS methods are not based on defining the components in a statistical framework relating to the contributions from

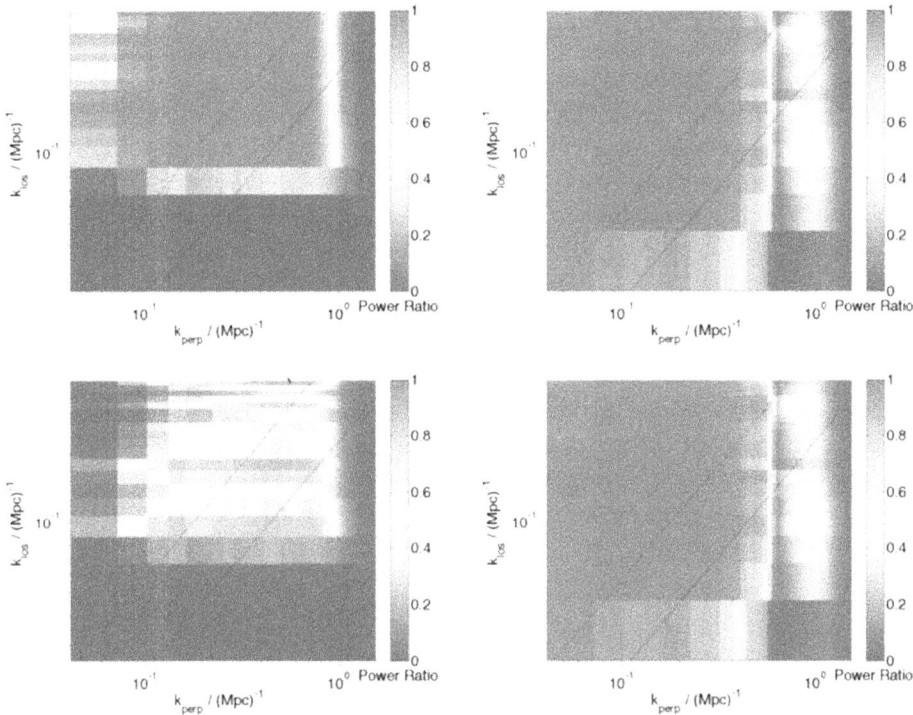

Figure 6.13. The left column shows the ratio of the simulated components, (cosmological signal/(cosmological signal + foregrounds)), demonstrating that the area of the window free from foreground contamination is small when the foregrounds are unsmooth. The top row is where the foreground model has a random wiggle along the line of sight equal in magnitude to 0.1% of the foreground signal. The bottom row shows a 1% wiggle. On the right is the same ratio but with foreground fitting errors after foreground removal by GMCA instead of the simulated foregrounds, demonstrating that the method can open up the EoR window significantly even when the smoothness of the foregrounds is under threat. Reproduced from Chapman et al. (2016), by permission of Oxford University Press on behalf of the Royal Astronomical Society.

foregrounds and mode-mixing, they are not easily assessed for uncertainty and physical meaning. The blind methods are very useful as a separate check on results from what are extremely complex experiments, with many unknown unknowns. There is scope to move these methods toward a more parametric framework, perhaps constraining the mixing matrix columns according to the first-hand knowledge about the instrumental effects and foregrounds we have built up from the pathfinder telescopes. This is a similar philosophy as introduced by Bonaldi & Brown (2015) in Correlated Component Analysis (CCA). While still based on a mixing matrix framework, CCA is a parametric method which constrains the mixing matrix to represent power law behavior over frequency, fixing the spectral index for a Galactic free–free contribution explicitly.

While Wp smoothing, GMCA, GPR and FastICA are all labeled non-parametric in the literature, it is important to note than none of them are fully blind or indeed fully non-parametric. Each of them require the selection of parameters to define the fit: whether it is the smoothing parameter in Wp smoothing, or the number of independent components in GMCA and FastICA. So far these parameters have been chosen based on minimizing the foreground fitting error on simulated data, where the foreground model is known. A more robust method is to implement a Bayesian model selection model, as GPR does already. In addition, Gleser et al. (2008) developed a method based on the Bayesian maximum *a posteriori* probability (MAP) formalism, assuming priors for the smoothness of the contaminating radiation and for the correlation properties of the cosmological signal and Zhang et al. (2016) introduced HIEMICA (HI Expectation Maximization Independent Component Analysis), an extension of ICA with a fully Bayesian inference of the foreground power spectra, allowing their separation from the cosmological signal power spectra. Machine learning has also been applied in an effort to seek a foreground model defined by the data itself (Li et al. 2019). There are now a multitude of non-parametric foreground subtraction methods available which have each proved their own principle on simulated, and in the case of GPR and GMCA, observed data. Now we know the constraints of the instrument much better, work on the relative advantages and disadvantages of all these approaches are a logical next step.

6.2.4 Residual Error Subtraction

The final stage of foreground mitigation is residual error subtraction (Morales et al. 2006; Morales & Hewitt 2004). The residual foreground mitigation errors from the previous two stages (bright source subtraction and diffuse foreground mitigation) produce distinct shapes in the spherical power spectrum, Figure 6.14. One can take the spherical power spectrum of the residual data and apply a multi-parameter fit according to the foreground residual and EoR template power spectrum. This allows a final cleaning of residual foreground contamination. Morales et al. (2006) also notes that "because the residual error subtraction relies on the statistical characteristics of the subtraction errors, the foreground removal steps become tightly linked

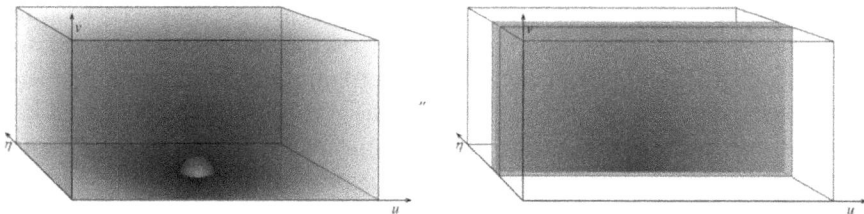

Figure 6.14. The 3D spherical power spectrum of the EoR signal (left), and an example residual foreground signal template (right), where zero is at the centre of the bottom face of the cuboid. The foreground signal displays a separable-axial symmetry while the EoR signal has a symmetric power spectrum. This contrast allows a further separation stage in order to clean the foreground fitting errors which have accumulated from the previous two stages of bright source subtraction and diffuse foreground mitigation. Reproduced from Morales et al. (2006). © 2006. The American Astronomical Society. All rights reserved.

and we must move from focusing on individual subtraction algorithms to the context of a complete foreground removal framework." This statement leads us neatly to the conclusion of this chapter.

6.2.5 Polarization Leakage

One of the challenges in calibration is to minimize leakage of polarization signals in total intensity. Otherwise, the polarization leakage can contaminate the cosmological 21-cm signal. A level of contamination depends strongly on characteristics of a radio telescope, its calibration strategy, and of polarized emission itself.

Antennas in the low-frequency radio telescopes are dipoles. Dipoles usually come in pairs. In each pair dipoles are orthogonal to each other and each dipole is sensitive to a certain polarization. Since antennas are also fixed to the ground, it is not possible to preform observations like with the traditional dish-like radio telescopes, where the tracking is done by steering the dish. Here, the sources are tracked by the beam-forming or simply the observation is done in a drift-scan mode. Depending on the position of the sources in the sky, the sources will see different projections of dipoles. If this geometrical projection is not corrected during the calibration, or the modeling of and correction for the beam polarization is not accurate, polarized signals can leak to total intensity and vice versa.

Since the polarized emission from the Milky Way can have a very complex frequency dependence, a leakage of this signal to the total intensity can contaminate the cosmological 21-cm signal, making extraction and analysis more demanding (Jelić et al. 2010; Moore et al. 2013; Spinelli et al. 2018 and see Figure 6.15). A number of studies addressed this problem for different low-frequency radio telescopes: LOFAR (Asad et al. 2015, 2016, 2018), MWA (Sutinjo et al. 2015) and PAPER (Kohn et al. 2016; Nunhokee et al. 2017). Although the assessed polarization leakage in these studies is not limiting current observations, it will become relevant once we reach a better sensitivity in the data. This will be especially the case for future 21 cm experiments, like HERA and SKA.

Most of the observed structures appear at Faraday depths $|\Phi| \lesssim 15 \, \text{rad m}^{-2}$, which measures the amount of Faraday rotation by intervening interstellar medium (see Section 6.1.3). Relatively small Faraday depths indicate polarized emission that fluctuates along frequency on scales larger than the expected cosmological 21-cm signal in total intensity (e.g., Moore et al. 2013). Thus, associated leaked signals can be in principle mitigated, as it was shown in the case of a simple and thin Faraday screen (Geil et al. 2011). On the contrary, polarized emission at Faraday depths $|\Phi| \gtrsim 15 \, \text{rad m}^{-2}$ can introduce frequency dependent signals, which if leaked can resemble the cosmological signal and make foreground mitigation difficult. Prior to the CD/EoR observations, it is therefore important to assess the properties of Galactic polarized emission in targeted region of the sky.

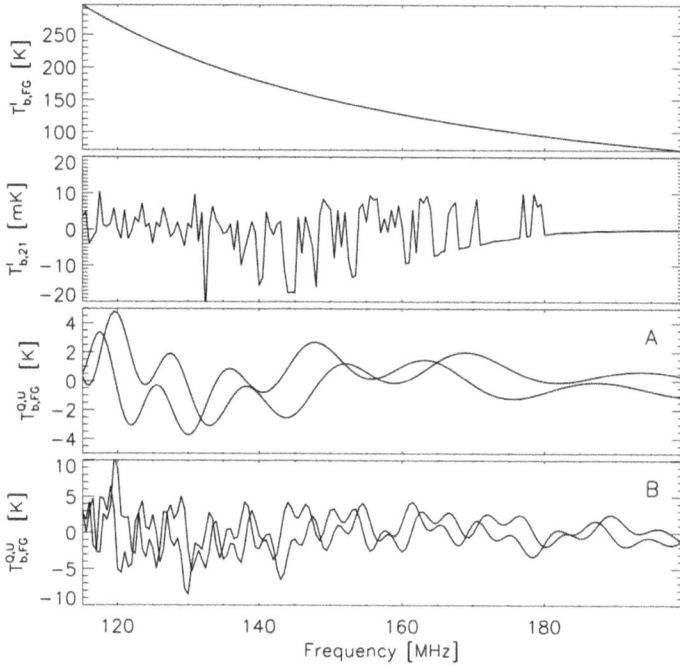

Figure 6.15. Galactic synchrotron emission given as a function of frequency in total intensity (Stokes I, $T^I_{b,FG}$) and polarization (Stokes Q and U, $T^{Q,U}_{b,FG}$). The spectra are generated by Jelić simulations (Jelić et al. 2010, 2008). Polarized emission can have a very complex frequency dependence compared to the plain power-law behavior in the total intensity, due to distinct Faraday rotation and depolarization at low-radio frequencies (see Section 6.1.3). If polarized emission consists of the multiple Faraday components and/or if some of the components are at $\Phi \gtrsim 15$ rad m^{-2} (example B versus A) this can create a leaked signal in total intensity that looks like the cosmological 21 cm signal ($T^I_{b,21}$, in this case generated by 21cm FAST Mesinger et al. 2011).

6.3 Conclusions

The foreground emission of our Galaxy and extragalactic radio sources dominates over the whole frequency range of the cosmological 21 cm experiments. In order to mitigate the foreground emission from the data we need to study and constrain its properties in great details. Thanks to the observations with LOFAR and MWA this is becoming possible.

The current EoR experiments are now modeled and constrained to an excellent degree but during that process there has been a blurring of boundaries between the analysis modules. The calibration stage, once assumed to mitigate foreground point sources only, can erroneously suppress diffuse foregrounds (Patil et al. 2016) and the mode-mixing of the instrument has required more complex modeling as wide-field effects have become apparent (Thyagarajan et al. 2015). There are a promising number of foreground mitigation techniques now available providing the necessary diversity of pipelines necessary for verifying the first detection. So far, there has not been a wide-reaching comparison of all of these methods or a complete assessment of their strengths and weaknesses for recovery of the different aspects of the EoR signal such as power spectra or images. Foreground subtraction, suppression and avoidance are now used in combination in the experimental pipelines and the further development of the best combination for these methods will provide an exciting area of research in the next decade.

References

Ali, Z. S., Parsons, A. R., Zheng, H., et al. 2015, ApJ, 809, 61

Ali, Z. S., Parsons, A. R., Zheng, H., et al. 2018, ApJ, 863, 201

Asad, K. M. B., Koopmans, L. V. E., Jelić, V., et al. 2018, MNRAS, 476, 3051

Asad, K. M. B., Koopmans, L. V. E., Jelić, V., et al. 2016, MNRAS, 462, 4482

Asad, K. M. B., Koopmans, L. V. E., Jelić, V., et al. 2015, MNRAS, 451, 3709

Barry, N., Beardsley, A. P., Byrne, R., et al. 2019, PASA, 36, e026

Barry, N., Hazelton, B., Sullivan, I., Morales, M. F., & Pober, J. C. 2016, MNRAS, 461, 3135

Beardsley, A. P., Hazelton, B. J., Sullivan, I. S., et al. 2016, ApJ, 833, 102

Bernardi, G., de Bruyn, A. G., Brentjens, M. A., et al. 2009, A&A, 500, 965

Bernardi, G., de Bruyn, A. G., Harker, G., et al. 2010, A&A, 522, A67

Bernardi, G., Greenhill, L. J., Mitchell, D. A., et al. 2013, ApJ, 771, 105

Bobin, J., Moudden, Y., Starck, J. L., Fadili, J., & Aghanim, N. 2008, StMet, 5, 307

Bobin, J., Sureau, F., & Starck, J. L. 2016, A&A, 591, A50

Bobin, J., Rapin, J., Larue, A., & Starck, J.-L. 2015, ITSP, 63, 1199

Bonaldi, A., Bonato, M., Galluzzi, V., et al. 2019, MNRAS, 482, 2

Bonaldi, A., & Brown, M. L. 2015, MNRAS, 447, 1973

Bondi, M., Ciliegi, P., Schinnerer, E., et al. 2008, ApJ, 681, 1129

Bowman, J. D., Morales, M. F., & Hewitt, J. N. 2009, ApJ, 695, 183

Bowman, J. D., Rogers, A. E. E., Monsalve, R. A., Mozdzen, T. J., & Mahesh, N. 2018, Natur, 555, 67

Bradley, R. F., Tauscher, K., Rapetti, D., & Burns, J. O. 2019, ApJ, 874, 153

Brentjens, M. A., & de Bruyn, A. G. 2005, A&A, 441, 1217

Byrne, R., Morales, M. F., Hazelton, B., et al. 2019, ApJ, 875, 70

Calistro, R. G., Williams, W. L., Hardcastle, M. J., et al. 2017, MNRAS, 469, 3468

Carroll, P. A., Line, J., Morales, M. F., et al. 2016, MNRAS, 461, 4151

Ceraj, L., Smolčić, V., Delvecchio, I., et al. 2018, A&A, 620, A192

Chapman, E., Bonaldi, A., Harker, G., et al. 2015, in Advancing Astrophysics with the Square Kilometre Array, PoS(AASKA14)005

Chapman, E., Abdalla, F. B., Bobin, J., et al. 2013, MNRAS, 429, 165

Chapman, E., Abdalla, F. B., Harker, G., et al. 2012, MNRAS, 423, 2518

Chapman, E., Zaroubi, S., Abdalla, F. B., et al. 2016, MNRAS, 458, 2928

Cheng, C., Parsons, A. R., Kolopanis, M., et al. 2018, ApJ, 868, 26

Condon, J. J. 1992, in AIP Conf. Proc. 254, Testing the AGN Paradigm, ed. S. H. Stephen, G. N. Susan, & C. M. Urry (Melville, NY: AIP), 629

Cooray, A., & Furlanetto, S. R. 2004, ApJ, 606, L5

Datta, A., Bowman, J. D., & Carilli, C. L. 2010, ApJ, 724, 526

de Oliveira-Costa, A., Tegmark, M., Gaensler, B. M., et al. 2008, MNRAS, 388, 247

Delvecchio, I., Smolčić, V., Zamorani, G., et al. 2017, A&A, 602, A3

Di Matteo, T., Ciardi, B., & Miniati, F. 2004, MNRAS, 355, 1053

Di Matteo, T., Perna, R., Abel, T., & Rees, M. J. 2002, ApJ, 564, 576

Dillon, J. S., Liu, A., Williams, C. L., et al. 2014, PhRvD, 89, 023002

Dillon, J. S., & Parsons, A. R. 2016, ApJ, 826, 181

Ewall-Wice, A., Dillon, J. S., Liu, A., & Hewitt, J. 2017, MNRAS, 470, 1849

Fanaroff, B. L., & Riley, J. M. 1974, MNRAS, 167, 31P

Geil, P. M., Gaensler, B. M., & Wyithe, J. S. B. 2011, MNRAS, 418, 516

Ghosh, A., Mertens, F. G., & Koopmans, L. V. E. 2018, MNRAS, 474, 4552

Gleser, L., Nusser, A., & Benson, A. J. 2008, MNRAS, 391, 383

Guzmán, A. E., May, J., Alvarez, H., & Maeda, K. 2011, A&A, 525, A138

Harker, G., Zaroubi, S., Bernardi, G., et al. 2010, MNRAS, 405, 2492

Harker, G., Zaroubi, S., Bernardi, G., et al. 2009a, MNRAS, 397, 1138

Harker, G. J. A., Zaroubi, S., Thomas, R. M., et al. 2009b, MNRAS, 393, 1449

Haslam, C. G. T., Klein, U., Salter, C. J., et al. 1981, A&A, 100, 209

Haslam, C. G. T., Salter, C. J., Stoffel, H., & Wilson, W. E. 1982, A&AS, 47, 1

Haverkorn, M., Boulanger, F., Enßlin, T., et al. 2019, Galaxies, 7, 17

Hazelton, B. J., Morales, M. F., & Sullivan, I. S. 2013, ApJ, 770, 156

Helou, G., Soifer, B. T., & Rowan-Robinson, M. 1985, ApJ, 298, L7

Heywood, I., Bannister, K. W., Marvil, J., et al. 2016, MNRAS, 457, 4160

Hills, R., Kulkarni, G., Daniel Meerburg, P., & Puchwein, E. 2018, Natur, 564, E32

Hopkins, A. M., Afonso, J., Chan, B., et al. 2003, AJ, 125, 465

Hurley-Walker, N., Callingham, J. R., Hancock, P. J., et al. 2017, MNRAS, 464, 1146

Hyvarinen, A. 1999, ITNN, 10, 626

Hyvärinen, A., Karhunen, J., & Oja, E. 2004, Independent Component Analysis, Vol. 46 (New York: Wiley)

Iacobelli, M., Haverkorn, M., Orrú, E., et al. 2013, A&A, 558, A72

Jarvis, M. J., Smith, D. J. B., Bonfield, D. G., et al. 2010, MNRAS, 409, 92

Jelić, V., de Bruyn, A. G., Mevius, M., et al. 2014, A&A, 568, A101

Jelić, V., de Bruyn, A. G., Pandey, V. N., et al. 2015, A&A, 583, A137

Jelić, V., Zaroubi, S., Labropoulos, P., et al. 2010, MNRAS, 409, 1647

Jelić, V., Zaroubi, S., Labropoulos, P., et al. 2008, MNRAS, 389, 1319

Jensen, H., Majumdar, S., Mellema, G., et al. 2016, MNRAS, 456, 66

Joseph, R. C., Trott, C. M., & Wayth, R. B. 2018, AJ, 156, 285

Kohn, S. A., Aguirre, J. E., Nunhokee, C. D., et al. 2016, ApJ, 823, 88

Landecker, T. L., & Wielebinski, R. 1970, AuJPA, 16, 1

Lawson, K. D., Mayer, C. J., Osborne, J. L., & Parkinson, M. L. 1987, MNRAS, 225, 307

Lenc, E., Anderson, C. S., Barry, N., et al. 2017, PASA, 34, e040

Lenc, E., Gaensler, B. M., Sun, X. H., et al. 2016, ApJ, 830, 38

Li, W., Pober, J. C., Hazelton, B. J., et al. 2018, ApJ, 863, 170

Li, W., Xu, H., Ma, Z., et al. 2019, MNRAS, 485, 2628

Liu, A., Parsons, A. R., & Trott, C. M. 2014b, PhRvD, 90, 023018

Liu, A., Parsons, A. R., & Trott, C. M. 2014a, PhRvD, 90, 023019

Liu, A., & Tegmark, M. 2012, MNRAS, 419, 3491

Liu, A., Tegmark, M., Bowman, J., Hewitt, J., & Zaldarriaga, M. 2009, MNRAS, 398, 401

Liu, A., Tegmark, M., & Zaldarriaga, M. 2009, MNRAS, 394, 1575

Longair, M. S. 2011, High Energy Astrophysics (Cambridge: Cambridge Univ. Press)

Mancuso, C., Lapi, A., Prandoni, I., et al. 2017, ApJ, 842, 95

McQuinn, M., Zahn, O., Zaldarriaga, M., Hernquist, L., & Furlanetto, S. R. 2006, ApJ, 653, 815

Mertens, F. G., Ghosh, A., & Koopmans, L. V. E. 2018, MNRAS, 478, 3640

Mesinger, A., Furlanetto, S., & Cen, R. 2011, MNRAS, 411, 955

Mitchell, D. A., Greenhill, L. J., Wayth, R. B., et al. 2008, ISTSP, 2, 707

Moore, D. F., Aguirre, J. E., Parsons, A. R., Jacobs, D. C., & Pober, J. C. 2013, ApJ, 769, 154

Morales, M. F., Bowman, J. D., & Hewitt, J. N. 2006, ApJ, 648, 767

Morales, M. F., Hazelton, B., Sullivan, I., & Beardsley, A. 2012, ApJ, 752, 137

Morales, M. F., & Hewitt, J. 2004, ApJ, 615, 7

Mozdzen, T. J., Bowman, J. D., Monsalve, R. A., & Rogers, A. E. E. 2017, MNRAS, 464, 4995

Mozdzen, T. J., Mahesh, N., Monsalve, R. A., Rogers, A. E. E., & Bowman, J. D. 2019, MNRAS, 483, 4411

Mulcahy, D. D., Horneffer, A., Beck, R., et al. 2014, A&A, 568, A74

Murray, S. G., Trott, C. M., & Jordan, C. H. 2017, ApJ, 845, 7

Murray, S. G., Trott, C. M., & Jordan, C. H. 2018, in IAU Symp. 333, Peering towards Cosmic Dawn, ed. V. Jelić, & T. van der Hulst (Cambridge: Cambridge Univ. Press), 199

Novak, M., Smolčić, V., Schinnerer, E., et al. 2018, A&A, 614, A47

Nunhokee, C. D., Bernardi, G., Kohn, S. A., et al. 2017, ApJ, 848, 47

Offringa, A. R., Mertens, F., & Koopmans, L. V. E. 2019, MNRAS, 484, 2866

Offringa, A. R., Trott, C. M., Hurley-Walker, N., et al. 2016, MNRAS, 458, 1057

Offringa, A. R., van de Gronde, J. J., & Roerdink, J. B. T. M. 2012, A&A, 539, A95

Offringa, A. R., Wayth, R. B., Hurley-Walker, N., et al. 2015, PASA, 32, e008

Oh, P. S., & Mack, K. J. 2003, MNRAS, 346, 871

Orlando, E., & Strong, A. 2013, MNRAS, 436, 2127

Pacholczyk, A. G. 1970, Radio Astrophysics. Nonthermal Processes in Galactic and Extragalactic Sources (San Francisco, CA: Freeman)

Paciga, G., Chang, T.-C., Gupta, Y., et al. 2011, MNRAS, 413, 1174

Parsons, A. R., & Backer, D. C. 2009, AJ, 138, 219

Parsons, A. R., Backer, D. C., Foster, G. S., et al. 2010, AJ, 139, 1468

Parsons, A. R., Pober, J. C., Aguirre, J. E., et al. 2012a, ApJ, 756, 165

Parsons, A., Pober, J., McQuinn, M., Jacobs, D., & Aguirre, J. 2012b, ApJ, 753, 81

Patil, A. H., Yatawatta, S., Koopmans, L. V. E., et al. 2017, ApJ, 838, 65

Patil, A. H., Yatawatta, S., Zaroubi, S., et al. 2016, MNRAS, 463, 4317

Pen, U.-L., Chang, T.-C., Hirata, C. M., et al. 2009, MNRAS, 399, 181

Petrovic, N., & Oh, P. S. 2011, MNRAS, 413, 2103

Planck Collaboration, 2018a, arXiv: 1807.06205

Planck Collaboration, 2018b, arXiv: 1807.06208

Platania, P., Bensadoun, M., Bersanelli, M., et al. 1998, ApJ, 505, 473

Pober, J. C., Parsons, A. R., Aguirre, J. E., et al. 2013, ApJ, 768, L36

Prandoni, I., Gregorini, L., Parma, P., et al. 2001, A&A, 365, 392

Prandoni, I., Guglielmino, G., Morganti, R., et al. 2018, MNRAS, 481, 4548

Prandoni, I. 2018, in IAU Symp. 333, Peering towards Cosmic Dawn, ed. V. Jelić, & T. van der Hulst (Cambridge: Cambridge Univ. Press), 175

Prasad, J., & Chengalur, J. 2012, ExA, 33, 157

Procopio, P., Wayth, R. B., Line, J., et al. 2017, PASA, 34, e033

Reich, P., & Reich, W. 1988, A&AS, 74, 7

Remazeilles, M., Dickinson, C., Banday, A. J., Bigot-Sazy, M. A., & Ghosh, T. 2015, MNRAS, 451, 4311

Rogers, A. E. E., & Bowman, J. D. 2008, AJ, 136, 641

Rybicki, G. B., & Lightman, A. P. 1986, Radiative Processes in Astrophysics (New York: Wiley)

Santos, M. G., Cooray, A., & Knox, L. 2005, ApJ, 625, 575

Shaver, P. A., Windhorst, R. A., Madau, P., & de Bruyn, A. G. 1999, A&A, 345, 380

Shimwell, T. W., Röttgering, H. J. A., Best, P. N., et al. 2017, A&A, 598, A104

Shimwell, T. W., Tasse, C., Hardcastle, M. J., et al. 2019, A&A, 622, A1

Singh, S., Subrahmanyan, R., Udaya Shankar, N., et al. 2018, ApJ, 858, 54

Smolčić, V., Delvecchio, I., Zamorani, G., et al. 2017a, A&A, 602, A2

Smolčić, V., Novak, M., Bondi, M., et al. 2017b, A&A, 602, A1

Smolčić, V., Novak, M., Delvecchio, I., et al. 2017c, A&A, 602, A6

Sokoloff, D. D., Bykov, A. A., Shukurov, A., et al. 1998, MNRAS, 299, 189

Spinelli, M., Bernardi, G., & Santos, M. G. 2018, MNRAS, 479, 275

Strong, A. W., Orlando, E., & Jaffe, T. R. 2011, A&A, 534, A54

Strong, A. W., Moskalenko, I. V., & Ptuskin, V. S. 2007, ARNPS, 57, 285

Sullivan, I. S., Morales, M. F., Hazelton, B. J., et al. 2012, ApJ, 759, 17

Sun, X. H., & Reich, W. 2009, A&A, 507, 1087

Sutinjo, A., O'Sullivan, J., Lenc, E., et al. 2015, RaSc, 50, 52

Richard Thompson, A., Moran, J. M., & Swenson, G. W. Jr. 2001, Interferometry and Synthesis in Radio Astronomy (2nd ed.; New York: Wiley)

Thyagarajan, N., Jacobs, D. C., Bowman, J. D., et al. 2015, ApJ, 807, L28

Thyagarajan, N., Udaya Shankar, N., Subrahmanyan, R., et al. 2013, ApJ, 776, 6

Trott, C. M., Pindor, B., Procopio, P., et al. 2016, ApJ, 818, 139

Trott, C. M., Wayth, R. B., & Tingay, S. J. 2012, ApJ, 757, 101

Van Eck, C. L., Haverkorn, M., Alves, M. I. R., et al. 2019, A&A, 623, A71

Van Eck, C. L., Haverkorn, M., Alves, M. I. R., et al. 2018, A&A, 613, A58

Van Eck, C. L., Haverkorn, M., Alves, M. I. R., et al. 2017, A&A, 597, A98

Vedantham, H., Udaya Shankar, N., & Subrahmanyan, R. 2012, ApJ, 745, 176

Waelkens, A., Jaffe, T., Reinecke, M., Kitaura, F. S., & Enßlin, T. A. 2009, A&A, 495, 697

Wang, J., Xu, H., An, T., et al. 2013, ApJ, 763, 90

Wang, X., Tegmark, M., Santos, M. G., & Knox, L. 2006, ApJ, 650, 529

White, R. L., Becker, R. H., Helfand, D. J., & Gregg, M. D. 1997, ApJ, 475, 479

Wieringa, M. H. 1992, ExA, 2, 203

Wilman, R. J., Jarvis, M. J., Mauch, T., Rawlings, S., & Hickey, S. 2010, MNRAS, 405, 447

Wilman, R. J., Miller, L., Jarvis, M. J., et al. 2008, MNRAS, 388, 1335

Yatawatta, S., de Bruyn, A. G., Brentjens, M. A., et al. 2013, A&A, 550, A136

Yatawatta, S. 2015, MNRAS, 449, 4506

Zaldarriaga, M., Furlanetto, S. R., & Hernquist, L. 2004, ApJ, 608, 622

Zhang, L., Bunn, E. F., Karakci, A., et al. 2016, ApJS, 222, 3

Zheng, H., Tegmark, M., Buza, V., et al. 2014, MNRAS, 445, 1084

Chapter 7

Global Signal Instrumentation

L J Greenhill and R Subrahmanyan

Experiments that seek to detect a global (zero-mode) H I signature during the Epoch of Reionization (EOR) and Cosmic Dawn (CD) use purpose-built meter-wave instrumentation. For the EOR, radiometry has contributed early constraints on models. For CD, the EDGES, SARAS, and LEDA efforts are active. Radiometry by EDGES has delivered a first claim of detection but independent confirmation is not yet in hand. This chapter presents the rudiments of radiometry instrumentation, discussion of concepts that bear on design, and challenges going forward.

7.1 Introduction

Two experiments have established limits on the predicted sky-averaged (zero-mode) signature from H I emission at redshifts (z) associated with the EOR. The Experiment to Detect the Global EOR Signature—EDGES (Bowman et al. 2018; Rogers & Bowman 2012)—provided the first constraint, excluding substantial change in neutral fraction over an interval narrower than $\Delta z \sim 0.06$, with 95% confidence, for $z \lesssim 11$. Also, the second-generation Shaped Antenna Measurement of the Background Radio Spectrum (SARAS-2) has excluded some model parameter combinations corresponding to late X-ray heating and rapid reionization, with 68%–95% (Patra et al. 2013; Singh et al. 2017, 2018).

Several experiments have also targeted setting constraints on parameters describing conditions during the CD through detection of predicted H I absorption against the cosmic microwave background (CMB), which may be more readily separated from foreground contamination than the EOR signal owing to the narrowness in redshift recognizable in many models. Exploiting techniques and radio-frequency (RF) electronics refined during preceding work at lower redshift, EDGES has claimed detection of a trough (Bowman et al. 2018), though with an unlikely fitted amplitude, breadth, and shape. As of this writing, much-needed independent confirmation is pending (Greenhill 2018; Hills et al. 2018; Bradley et al. 2019; Spinelli et al. 2019). The successor SARAS-3 experiment has collected data corresponding to $15 \lesssim z \lesssim 29$ with multiple antenna architectures and at widely separated sites. The Large-aperture Experiment to Detect the Dark Age—LEDA (Greenhill & Bernardi 2012; Price et al. 2018)—simultaneously uses several configurations of the antenna originally engineered by the Long Wavelength Array (LWA) project (Taylor et al. 2012) and embeds them in a dense interferometric array to make possible calibration techniques unavailable for standalone antennas. SCI-HI (Voytek et al. 2014) and the related Probing Radio Intensity at High z from Marion island (PRIZM) experiment (Philip et al. 2019) have acquired early data at two of the most radio-quiet sites used thus far (judged from the dearth of FM radio contamination), while following similar methodologies and instrumentation approaches. With sights set initially on system characterization and assessment of technique, the Broadband Instrument for Global Hydrogen Reionization Signal—BIGHORNS—has also presented early calibrated data for $z \lesssim 17$ (Sokolowski et al. 2015).

7.2 Radiometer Basics

Karl Jansky discovered that any sensor of electromagnetic fields placed beneath open sky samples at its terminals "cosmic radio noise"[1] (Jansky 1933). A typical channelized radiometer comprises an antenna, an amplifying receiver that includes a band-limiting filter, a digitizer, and a spectrometer. The filter defines the measurement bandwidth. The digitizer samples data at a rate of at least twice the bandwidth. The spectrometer instantiates Fourier techniques to transform time series into spectra.[2] For an Earth-bound antenna, received RF power will include contributions from the cosmos, the ground proximate to the antenna, self-generated noise from the electronics, and artificial terrestrial interference. For scale, we note that an antenna with unit gain in equilibrium with a 300 K blackbody delivers a noise power of $O(1)$ pW over a 100 MHz band.

7.2.1 Antenna

One of the simplest forms of an antenna comprises two oppositely directed conductors (a dipole). Incident waves induce currents in the conductors, and a voltage is developed across the inward-facing ends (the terminals). The spectrum of power measurable at the terminals corresponds to the incident waves, modified by the frequency-dependent electromagnetic coupling of the antenna and surrounding space (including the ground), antenna efficiency, and transfer function between the terminals and instrumentation downstream.

A dipole with wire-like arms, each of length one-quarter wavelength (λ_0), will have a sharply peaked resonance and maximum power transfer efficiency at

---

[1] The term "noise" is commonly used because RF radiation from atomic processes in the cosmos, the ionosphere and troposphere (lightning), and terrestrial thermal sources is spatially and temporally incoherent. The fields are generally described statistically as Gaussian random variables of zero mean, following from the central limit theorem.

[2] The theory and implementation of digital spectrometers may be found in Thompson et al. (2017) and references therein.

frequency ν_0 in a narrow band $\Delta\nu/\nu_0 \ll 1$. The antenna acts as a transformer from the impedance of free space, 377 Ω, on one side to that of transmission lines and RF electronics on the other. Over a narrow band, the impedances of the antenna and RF electronics can be readily matched, and all of the incident power enters the receiver (ignoring resistive losses).

However, the H I signal is inherently broadband, and instrumentation requires at least an octave of bandwidth ($\nu_{\mathrm{high}}/\nu_{\mathrm{low}} = 2$ or $0.66\bar{\nu} < \nu < 1.33\bar{\nu}$) in order to enable the predicted complex spectral structure due to the 21 cm transition to be distinguished from smoothly varying foregrounds. In general, the impedance of a dipole modified to achieve reasonable power transfer efficiency across such broad bandwidths varies considerably in amplitude and phase as a function of frequency. It can be impossible to match to the RF electronics "everywhere." The consequence is that power is reflected at the antenna–receiver interface and lost. Depending on experiment design, the best that can be achieved with a broadband dipole may be an upper limit on variations in impedance with frequency or an imposed functional form such that even in the face of calibration error, the variations cannot mimic the science signal.

Broadband dipoles (Figure 7.1) may be planar with arms comprising 2D shapes (e.g., plates in the case of EDGES), 3D structures comprising planes (e.g., triangles in the case of LEDA), or more complex structures. Linear dipoles such as these couple to a single linear polarization mode (with electric field, **E**, oriented along the dipole arms). Dipoles with spiral and helical arms couple to the circularly polarized mode propagating on axis and jointly to circular and linear modes off axis. Self-similar planar and conical spirals may have operational bandwidths that exceed an octave and maintain good impedance matching, but other metrics may suffer (e.g., frequency-variable side-lobe structure in gain patterns that generates "chromaticity"

Figure 7.1. (Top left) EDGES antenna and serrated 30 m × 30 m ground screen. (Top right) Close-up of the EDGES single-polarization dipole comprising two rectilinear metal sheets. (Bottom right) Close-up of the LEDA dual polarization pyramidal dipole. (Bottom center) LEDA antenna and serrated 20 m × 20 m ground screen implemented in 2019. (Bottom left) SARAS-2 antenna comprising a spherical element atop a 0.87 m diameter disk. The maximum gains for the EDGES and LEDA antennas are toward zenith. In contrast, the maximum for SARAS-2 traces a ring on the sky at 30° elevation, centered on zenith, where there is a null. Top right: adapted by permission from Springer Nature: Bowman et al. (2018); top left: adapted by permission of J. D. Bowman, Arizona State University School of Earth and Space Exploration LoCo Lab © 2019; bottom left: reprinted with permission from Singh et al. (2018), © Springer Science+Business Media B.V., part of Springer Nature 2018.

in antenna response). Experiments weigh trade-offs differently, typically depending on calibration strategy.

7.2.1.1 Ground Sensitivity

Owing to its symmetry, a planar dipole receives radiation incident from the sky and ground equally well. "Drooping" or raising the arms of a linear dipole (e.g., SCI-HI) or projecting a spiral onto a cone—as for BIGHORNS (Sokolowski et al. 2015)—are common techniques to break the symmetry, effectively narrowing the field of view and increasing antenna "directivity." However, substantive suppression of coupling to the ground is best achieved by covering it with a conducting plane. This "ground screen" acts as a reflector. The antenna senses the sky above and below, boosting sensitivity for suitable antenna–screen separations (e.g., the direct and reflected paths interfere constructively for $\lambda_0/4$ separation) over some bandwidth.

Ground screens may be soldered or welded wire mesh, with a minimum conductor spacing or hole size $\ll \lambda_0$, so as to minimize leakage of radiation across the plane. However, reflection and scattering off the discontinuity represented by the edge of the ground screen creates interference patterns that are functions of the direction and frequency of incident radiation, thereby modulating antenna gain, possibly enough that fluctuations in the received spectrum may be apparent even after calibration. Adding a random component to the geometry suppresses the effect to a degree. In telecommunications at high frequencies, fractal-like designs $O(\lambda)$ may be etched or milled into a solid metal substrate. However, at low frequencies, sculpting fractals with wire mesh is impractical. Instead, the implementation of serrated edges has been an effective tool (Figure 7.1).

7.2.2 Receiver

The purpose of an analog receiver is to amplify, over a desired frequency range, the signal coupled to the antenna and passed at its terminals. The fluctuating voltage at the terminals propagates along a transmission line to a (typically) modest gain amplifier that boosts signal-to-noise ratios relative to thermal processes in the electronics. This amplifier also adds noise to the incoming signal, and this is amplified by later stages in the signal path. There is always a practical trade-off between achieving high gain and low noise during amplification, and having the lowest additive noise relative to the sky signal in the first-stage amplification is usually paramount, as follows from Friis (1946)—see also Pozar (2011) for a discussion. Amplification is most often done in stages to achieve an aggregate gain sufficient for conversion to a digital signal downstream without introduction of excessive thermal noise from particularly high-gain amplifiers.

7.2.2.1 Filtering

Receivers typically include bandwidth-limiting elements to enhance the performance of particular amplifiers or downstream electronics, as for digital processing. In the former case, filtering is intended to exclude signals that are strong enough to substantively degrade the amplifier linearity, which is susceptible to saturation-like effects, or to create artifacts during amplification where the beating of signals at different frequencies generate products that may be detectable. This is also known as intermodulation (e.g., if broadcast signals above the "Cosmic Dawn band" pass through an amplifier, e.g., at 90, 100, and 110 MHz, artifacts may appear in the amplifier output at 10 and 20 MHz).

For digital processing, filtering serves to suppress the "aliasing" of unwanted signals, at frequencies outside the science band, into the band. This occurs when data are sampled at too low of a rate during digitization. A digital signal is a sequence of samples. Sampling with no loss of information requires a rate that is twice the uppermost frequency of interest, as established by the Nyquist–Shannon theorem (Thompson et al. 2017). Absent the application of a sharp low-pass filter corresponding to the top end of the science band, signals at higher frequencies appear to be reflected about the top end and superposed on the science signal. The unwanted power may be broadcast interference (narrowband) or continuum as from the sky (broadband).

Common filter topologies are Butterworth, which has a maximally flat, structure-free response, and Chebyshev type 1 and 2, which provide sharper transitions between the passed and rejected intervals but exhibit substantial ripple in one or the other (i.e., the frequency structure to be exiled to where it does the least damage, depending on application). A third topology, Elliptic, exhibits the same amplitude of structure in the passed and rejected bands, and promises the steepest possible transition from one to the other for a given maximum allowable ripple and magnitude of transition. Often, a combination of filters is used—one to provide excellent rejection but with relatively slower roll-off in response, and others with sharp transitions at the design edges of the passed band.

7.2.2.2 Reflection

Not limited to the antenna–receiver interface, the reflection of RF signals occurs at the interfaces between components with different impedances. A 1% mismatch in resistive impedance corresponds to a 0.5% reflection in voltage or 0.003% in power. The chain of components along a signal path in a receiver creates instances of multipath propagation due to numerous fractional reflections. For well-chosen components, some are negligible, but a complete vector analysis of amplitude and phase is required to understand what frequency structure a receiver may impose in the process of amplifying the input signal.

In this regard, pairings of filters and amplifiers deserve particular attention. As noted, filters present frequency structure at their outputs, depending on the selected topology. They also present frequency structure in reflection at the input. In the case of a low-noise amplifier in the first stage of a receiver, if it is followed by a poorly chosen filter, then it receives back a fraction of its output power but with a potentially complicated frequency-dependent structure imposed. This propagates upstream through the amplifier (with finite loss, a.k.a. isolation) and reflects off the imperfect antenna–receiver interface and arrives at the amplifier input, added to the cosmic signal and conceivably at detectable levels.

Analysis of multipath propagation among components applies to additive thermal noise as well as the amplified sky signal. Where noise and an attenuated, phase-shifted reflection are coadded during propagation, it develops frequency structure, even though for any given circuit temperature, the noise intrinsically varies slowly and smoothly with frequency. Because the antenna–receiver interface presents the largest mismatch along the signal path, this is especially important in the case of the first amplifier, from which noise propagates upstream toward the antenna as well as downstream toward later stages of amplification.

A primary engineering formalism describing the noise characteristics of linear devices involves a quartet of parameters, one for each frequency: minimum noise temperature, optimum voltage reflection coefficient (magnitude), the optimum voltage reflection coefficient (phase), and the equivalent noise resistance (referring to the spectral density of noise). These may be used to estimate the frequency structure imposed on noise emanating from amplifiers, as well as the dependence of the amplifier noise temperature on frequency and impedance. (An amplifier facing a resistive (real) or reactive (imaginary) impedance at its input exhibits different gains and noise temperatures.) Building off Hu & Weinreb (2004), Rogers & Bowman (2012) presented a simplified formalism for low-frequency systems referred to as "noise wave" analysis, which characterizes the propagation of noise in terms of correlated and uncorrelated components. As applied, this works well provided that the reactive component outside the amplifier is sufficiently small.

7.2.3 Digitzer

Analog-to-digital converters (ADCs) sample the receiver output at the "Nyquist rate," described above (e.g., 5 ns for a bandwidth of 100 MHz). The number of bits used to represent each sample determines the dynamic range achievable in each spectrum generated by a Fourier transform of every N samples (e.g., for $N = 4096$, the frequency resolution, R, in the above example is 24.4 kHz). The number of bits per sample must be sufficient to represent the range of sky brightness integrated over the antenna gain pattern and prescribed bandwidth. In particular, for a sky with a steep spectral index and/or antenna with a steep change of gain with frequency, there must be enough bits to represent power at both the highest and lowest frequencies. At present, transport of an aggregate data rate of $O(2)$ Gbit s^{-1} can be readily achieved, corresponding to 10 bits per sample and a 2^{10}:1 dynamic range for voltage and 2^{20}:1 for power. ADC hardware providing 8 to 16 bits at sample ranges O (100) MHz is readily available. However, the minimum acceptable bit depth for a given site often depends on the presence and characteristics of interference. Where peak band-averaged power due to continuous or impulsive interference exceeds that of the sky, representing both without saturation demands greater bit depth in sampling. Additional considerations arise in the uniformity of steps during the quantization of analog data, linearity over the full analog range, and calibration accuracy where sampling may be parallelized over multiple samplers (a.k.a. interleaving).

7.3 Challenges Facing Experiments

7.3.1 Antenna Radiation Efficiency

If an antenna is lossless and has no resistive elements, then when viewed as if it were a transmitter, all of the power fed to the antenna and not reflected back along the transmission line at the antenna terminals will emerge as radiation. However, for an antenna placed on bare ground, part of the radiated power may be absorbed. Low-frequency electromagnetic waves penetrate soil to substantial depths—several meters for dry soil. For antennas that are placed on an infinite conducting ground screen, all of the radiated power goes to the sky either directly or on reflection off the ground screen. Passive reciprocal antennas may be viewed conversely as receivers where the loss to resistive elements of the antenna and the loss to the ground both reduce antenna radiation efficiency while adding thermal noise.

However, ground loss has a role in mechanisms that create two additional challenges where redress may be more difficult. First, the additive component of ground emission has a complex imprint of the antenna radiation efficiency, making it difficult to separate from zero-mode 21 cm signals unless the antenna is designed so that the radiation efficiency itself has characteristics orthogonal to expectations for zero-mode signals. Second, the loss depends on ground characteristics, specifically conductivity and dielectric constant, which depend on soil characteristics and moisture content, and are functions of depth and time. A sudden change in dielectric constant or conductivity at depth creates an impedance discontinuity, as does the intrusion of bedrock into the strata. Such a discontinuity drives multipath propagation among ground layers and the instrumentation above—e.g., see Bradley et al. (2019). Mapping and tracking the RF characteristics of the ground can require extensive additional dedicated instrumentation, such as a network of dielectric impedance reflectometers, and tools to make use of the data are relatively crude at present.

Antenna radiation efficiency is also influenced by the environment of the antenna, not only the ground beneath but also features above, such as shrubs and man-made structures at distances up to several wavelengths. Conducting cables that supply power to the radiometer and conduct signals to receivers located some distance away may also influence the efficiency. In measurements of the reflection efficiency as a transmitting antenna, power transmitted by the antenna reflect and scatter off trees and structures in the environment and return to the antenna, as in a radar. Thus, measurements of Γ sample the environment as well as the antenna. Conversely, these environmental features will influence the receipt of cosmic radiation in reverse. Scattering off these objects may generate spectral structure that is an imprint of the environment. Thus, it is essential to have clear space above ground and homogeneous, if not also dry, soil below.

7.3.2 Antenna Transfer Efficiency

The antenna transfer or reflection efficiency, $(1 - \Gamma^2)$, which is related to the voltage reflection coefficient Γ at the antenna terminals, determines what fraction of cosmic noise received by the antenna propagates into the receiver chain. In this consideration, it is the impedance of the antenna at its terminals, which is effectively the free space impedance transformed by the antenna to its terminals, as compared to the impedance of the first low-noise amplifier encountered by the cosmic noise as transformed by the interconnecting transmission line to the antenna terminals, that decides the reflection coefficient Γ.

A design goal for zero-mode 21 cm is an antenna that has high reflection efficiency over the full observing band. However, because the foreground Galactic sky has a brightness temperature that is significantly greater than the noise temperatures of modern low-noise amplifiers operating in the 10–200 MHz band, it is sufficient that the total efficiency of the antenna provide an antenna temperature that well exceeds the receiver noise. In that case, the system temperature and hence the measurement noise for any integration time would be independent of the receiver noise, and improving the total efficiency would not improve detection sensitivity or reduce the required observing time.

What is probably of greater importance is that the reflection efficiency be a smooth function of low order so that the product of the relatively bright foreground sky with the reflection efficiency, to give the dominant unwanted component of the observed spectrum, does not confuse the desired zero-mode 21 cm signal, and is separable from the 21 cm signal. If the antenna structure is electrically long, as would be the case, for example, in frequency-independent spiral antennas with large structural bandwidth, the reflection efficiency would have fine structure in frequency. Therefore, from the viewpoint of designing the antenna element to have reflection efficiency that is exclusively of low order, it is advantageous to have electrically small antennas.

If the antenna does not have resistive elements and the radiation efficiency is unity, then a measurement of the antenna reflection efficiency Γ would be a useful method for correcting the data for antenna efficiency and translating the measured spectrum to a sky spectrum. In this case, it is desirable and useful to provide a switch at the antenna terminals, which might allow a one-port network analyzer to access the antenna terminals and make an accurate measurement of Γ. This is best done at the observing site, where the antenna environment is the same as for the zero-mode observing. Deriving the reflection efficiency and total efficiency also requires a measurement of Γ for the low-noise amplifier, but that may be done in the laboratory provided that the amplifier temperature and operating conditions are the same.

7.3.3 Gain Pattern

A critical challenge in antenna design is the suppression of "mode coupling." This arises when an antenna gain pattern is chromatic, i.e., for each direction, gain varies with frequency.[3] Even for a sky with a constant spectral index, the consequence of

---

[3] We note in passing that there is no general, compact, quantitative definition of chromaticity.

chromatic response is potentially a complex frequency structure everywhere on the sky.

Mode coupling can be a fundamental hurdle to detection of the 21 cm signal with any given antenna. The most certain means to suppress it is to adopt antennas that are achromatic, i.e., frequency independent in gain in all directions. Chromaticity may not be easily quantified, but spectra being limited by statistical noise rather than undulations in the spectral baseline can be an effective figure of merit. Servicing this, spectra simulated using calculated gain patterns and sky models may be used in tuning antenna designs to achieve a required level of achromaticity.

7.3.3.1 Measurement and Simulation

High-accuracy direct calibration of gain patterns in situ (thereby taking into account all details of coupling to the ground, structures, ground screens, etc.) is an unsolved problem. There are no suitable standard antennas in communications engineering. Lofting transmitters on drones has been developed (Jacobs et al. 2017), but cancellation of systematics intrinsic to this scheme (e.g., multipath and uncertainty in the gain pattern of the transmitting antenna) has not been demonstrated and confirmed. The most widely employed alternative is numerical simulation.

Arguably, most available electromagnetic simulation packages are not capable of providing solutions with the precision necessary to quantify antenna response, though cross-referencing of results enables assessment of systematics stemming from the various simulation techniques applied. Exacerbating the above difficulty is the need to include in simulations coupling to stochastic elements in the surroundings (vegetation), soil chemistry, and time-variable changes in water content. Antennas also couple to strata below ground with complex permittivities (i.e., nonzero conductivity) and infrastructure such as trenched cables. In general, the design path necessarily requires iteration to achieve exacting performance tolerances and cross-referencing results obtained with different packages.

7.3.4 Cosmic Foregrounds

At millimeter wavelengths, where the CMB peaks, the sky is dominated by the CMB. Galactic emission and the extragalactic background are subdominant. At frequencies relevant to the 21 cm transition at high z, the radio sky is qualitatively different, dominated by synchrotron emission from the Galactic plane, which extends to considerable latitudes, and diffuse off-plane components. Furthermore, in neither case is the structure readily modeled in angle or frequency, not least because position-resolved spectra of the sky at these low frequencies are known with accuracies of only $O(10\%)$. Moreover, there are no known fiducial tracers that may be used to establish external constraints.

The peak amplitude of the cosmic 21 cm signal is expected to be a few tens to a few hundred millikelvin, and the foreground is expected to between a few hundred to a few thousand Kelvin brightness temperature. This requires a dynamic range of at least 10^4; clean signal detection requires aiming for a dynamic range of 10^5. Because the algorithms for component separation, which depend on orthogonality between the zero-mode 21 cm signal and other unwanted additives and foreground, are usually limited and the models for the unwanted components would subsume a significant part of the 21 cm signal, the typical design goal for the 21 cm radiometers is therefore to achieve an artifact-free spectrum of $O(1)$ mK sensitivity, about 10^6 below the dominant foreground.

7.3.5 Ionosphere

The ionosphere has time-varying electron densities that is commonly characterized by the total electron content (TEC) along any line of sight. The ionosphere modifies spectra of background sources in several ways. It refracts rays, so that sources appear at higher elevations (Vedantham et al. 2014). The ionosphere also both absorbs the background and adds emission from populations of hot electrons (Rogers et al. 2015). These effects of the ionosphere are strongly wavelength dependent and are anticipated to predominantly modify radiation at $\nu \lesssim 100$ MHz.

In general, the accuracy and spacing of TEC measurements is as yet inadequate to support time-varying correction of measured sky spectra. TEC data are primarily of use in deciding the relative severity of ionospheric conditions and the order of magnitude of distortions to be anticipated in spectra—a coarse weather report. Consequently, analyses of radiometry data must include model nuisance parameters that describe ionosphere effects, requiring marginalization in order to extract the zero-mode signal from data.

7.3.6 Polarization

Galactic and extragalactic foregrounds comprise sources often with significant fractional linear polarization. The detected foreground spectrum will be polarization dependent. Complications arise due to Faraday rotation of differing degrees. The effect is frequency dependent, and for a linearly polarized source, the source intensity received by any single polarization will be frequency dependent. Hence, Faraday rotation results in spectral structure that may potentially confuse attempts to detect zero-mode 21 cm spectral structure. For this reason, it is desirable and a design goal for zero-mode radiometers to be a dual polarized pair of radiometers, with full polarization calibration that allows the derivation of the Stokes I component on the sky.

7.3.7 Interference

Virtually all frequencies at which a zero-mode signal corresponding to the EOR or CD could appear are allocated to terrestrial and space communications. Transmitters at these frequencies exist in most parts of the world, and propagation paths may be line of sight, reflected (e.g., off meteor trails and aircraft), or bent by diffraction around obstacles. Extremely remote sites exist where interference due to long-range propagation is rare (Voytek et al. 2014; Philip et al. 2019), but apart from these unusual cases, low-frequency radiometry data is corrupted intermittently, raising the possibility of weak interference contributing weak artifacts to spectra. These may be narrow band (e.g., FM transmission) or broad band (e.g., digital television, where each channel allocation is several megahertz wide). They may be recognizable as discrete spectral features or solely by deviations from Gaussian statistics in time series.

7.4 Précis of Design Requirements

1. **Radiometer bandwidth**: an octave or more, to enable separation of smooth-spectrum foregrounds and the distinctly not smooth 21 cm spectral signature.
2. **Antenna gain pattern in situ**: maximally smooth in angle and frequency with minimum chromaticity, and characterization from direct measurement or simulation if necessary.
3. **Ground screen**: large enough to isolate the radiometer from propagation in ground strata and buried infrastructure; sufficient geometric irregularity along the edges so as to suppress coherent patterns in scattering of incident radiation toward the antenna.
4. **Antenna polarization**: dual polarization to enable construction of Stokes I and cancel artifacts that can arise from Faraday rotation of foreground emission.
5. **Site**: a radiometry site with a clear horizon out to several wavelengths, and related to item no. 3, homogeneous dry strata below.

7.5 Outside the Box Architectures

7.5.1 Single-element Sensor Radiometer

The simplest form of a radio telescope that would be appropriate for detecting the zero-mode 21 cm signal is a single-elemental wideband antenna followed by a spectrometer. Such a radiometer would ideally have a frequency-independent antenna, a self-calibratable receiver that corrects for the bandpass, and switching schemes to cancel internal additives including receiver noise.

The considerations that drive the design of such a radiometer have been discussed above. Limitations and design challenges to the performance of such a single-element spectral radiometer are manifold. Therefore, there have been new concepts and design attempts to develop alternate schemes or configurations that might avoid some of the potential showstoppers.

7.5.2 Outriggers to Fourier Synthesis Telescopes

A key challenge in zero-mode 21 cm radiometers is knowing the antenna beam pattern, its chromaticity, and the bandpass of the antenna element. Antenna measurements at long wavelengths are exceedingly difficult because of the parasitic effects of environment and the ground, which influence both the device under test and also the test and measurement antenna. Switched calibration, using broadband noise sources, may serve to calibrate the bandpass of the receiver chain; however, this leaves the antenna bandpass, the radiation efficiency, uncalibrated. This leads to a situation where the antenna characteristics may have to be derived from electromagnetic simulations, which may not have the accuracy needed to correct the measurement data.

A solution to these issues is to deploy the single-element sensor based radiometer as an outrigger to an array of antennas, which operate in Fourier synthesis interferometer mode. The radiometer antenna and receiver chain then form another element of the array, which together observe the sky sources within the antenna primary beams. The measured spectral visibilities are then used to solve simultaneously for the sky model and also the instrument parameters, which include the bandpass and beam shape of the outrigger antenna. If all array elements are dual polarized, then full Stokes calibration is also possible, providing polarization calibration solutions as well for the outrigger antenna.

7.5.3 Interferometric Methods

It is not often appreciated that interferometers are not totally blind to the zero mode in the sky temperature distribution. If the sky were uniformly bright, then the electromagnetic fields at two points in space separated by less than a wavelength will show mutual coherence, which may be measured by an interferometer. A pair of half-wave dipoles placed adjacent to each other, in line, will respond to the zero mode of the sky temperature distribution. If placed parallel to each other, the interferometer response will contain the zero mode, and the response in this case will be greater. The interferometers may be thought of as sampling the zero mode in the direction in which the projected spacing between the elements is zero. Of course, if the spacing between the dipoles increases, the response falls off progressively.

The coupling of the interferometer response to the zero-mode signal depends on the mutual coupling between the aperture fields in the pair of antennas. This is relatively high for closely spaced elemental antennas; however, a pair of aperture antennas placed adjacent to each other would have very little response to the zero mode if their aperture illuminations have little overlap.

Interferometers have the advantage that when they deploy a pair of sensors of the electromagnetic field at two separated locations in space, and measure the mutual coherence in the fields by cross-correlating the received and amplified voltage waveforms, they are insensitive to the additive receiver noise in the two arms, which are uncorrelated.

Thus, interferometer measurements are blind to internal receiver additives, therefore scoring over single-element radiometers on this count.

7.5.4 Zero-spacing Interferometer

If a pair of wideband antennas are placed adjacent to each other, with mutual coupling between the antennas, then the interferometer response includes the zero-mode signal. This response may be enhanced by placing a vertical beam splitter in between the antennas, so that incident field from any side propagates to the antenna on the far side through the screen and to the near side in a direct path and also after reflection off the screen. If the beam splitter is reactive, and lossless, then the response to the zero mode on the two sides of the screen cancel. However, if the beam splitter is resistive, then the interferometer responses add with the same sign. Optimally, it can be shown that the sensitivity is a maximum if the beam splitter has a sheet resistance equal to the impedance of free space.

Interferometers made from elemental antennas placed in a close packed configuration will have a telescope filter function, which defines the interferometer response to zero-mode 21 cm signals, that is highly frequency dependent and challenging to calibrate. The advantage of the zero-spacing interferometer made from frequency-independent antennas is that the telescope filter function in this case is flat, at least over the frequency range in which the resistive screen is frequency independent.

7.5.5 Lunar Occultation

An alternate and interesting approach to the detection of the zero-mode 21 cm signal is via Fourier synthesis imaging of the Moon. The brightness measured toward the Moon, over frequency, would be a difference between the Moon brightness and the mean brightness of the radio sky. Thus, if the Moon were assumed to be of flat spectrum, or if the temperature spectrum of the Moon were known, then the differential measurement may be used to infer the zero-mode 21 cm signal.

The measurement is not without difficulties. First, it is not clear that the temperature spectrum of the Moon is flat or even smooth; the lunar regolith may have structure and layered in depth, which would give spectral structure in the emission brightness. Additionally, the Moon is reflective and hence the brightness of the Moon has a component that is a reflection of the radio sky. Finally, the synthetic beam of the Fourier synthesis telescope will have side lobes that are chromatic; thus, mode coupling will induce spectral structure whose removal from the data will be limited by the depth of the deconvolution.

References

Bowman, J. D., Rogers, A. E. E., Monsalve, R. A., Mozdzen, T. J., & Mahesh, N. 2018, Natur, 555, 67

Bradley, R. F., Tauscher, K., Rapetti, D., & Burns, J. O. 2019, ApJ, 874, 153

Friis, H. T. 1946, PIRE, 34, 254

Greenhill, L. 2018, Natur, 555, 38

Greenhill, L. J., & Bernardi, G. 2012, arXiv:1201.1700

Hills, R., Kulkarni, G., Meerburg, P. D., & Puchwein, E. 2018, Natur, 564, E32

Hu, R., & Weinreb, S. 2004, ITMTT, 52, 1498

Jacobs, D. C., Burba, J., Bowman, J. D., et al. 2017, PASP, 129, 035002

Jansky, K. G. 1933, Natur, 132, 66

Patra, N., Subrahmanyan, R., Raghunathan, A., & Udaya Shankar, N. 2013, ExA, 36, 319

Philip, L., Abdurashidova, Z., Chiang, H. C., et al. 2019, JAI, 8, 1950004

Pozar, D. M. 2011, Microwave Engineering (3rd ed.; New York: Wiley)

Price, D. C., Greenhill, L. J., Fialkov, A., et al. 2018, MNRAS, 478, 4193

Rogers, A. E. E., & Bowman, J. D. 2012, RaSc, 47, RS0K06

Rogers, A. E. E., Bowman, J. D., Vierinen, J., Monsalve, R., & Mozdzen, T. 2015, RaSc, 50, 130

Singh, S., Subrahmanyan, R., Udaya Shankar, N., et al. 2017, ApJ, 845, L12

Singh, S., Subrahmanyan, R., Udaya Shankar, N., et al. 2018, ExA, 45, 269

Sokolowski, M., Tremblay, S. E., Wayth, R. B., et al. 2015, PASA, 32, e004

Spinelli, M., Bernardi, G., & Santos, M. G. 2019, MNRAS, 489, 4007

Taylor, G. B., Ellingson, S. W., Kassim, N. E., et al. 2012, JAI, 1, 1250004

Thompson, A. R., Moran, J. M., & Swenson, G. W. Jr. 2017, Interferometry and Synthesis in Radio Astronomy (3rd ed.; Berlin: Springer)

Vedantham, H. K., Koopmans, L. V. E., de Bruyn, A. G., et al. 2014, MNRAS, 437, 1056

Voytek, T. C., Natarajan, A., Jáuregui García, J. M., Peterson, J. B., & López-Cruz, O. 2014, ApJ, 782, L9

Chapter 8

Status of 21 cm Interferometric Experiments

Cathryn M Trott and Jonathan Pober

Interferometric experiments of the reionization era offer the advantages of measuring power in spatial modes with increased sensitivity afforded by multiple independent sky measurements. Here we review early work to measure this signal, current experiments, and future opportunities, highlighting the lessons learned along the way that have shaped the research field and experimental design. In particular, this chapter discusses the history, progress, challenges and forecasts for detection and exploration of the spatial structure of the 21 cm brightness temperature signal in the Epoch of Reionization using interferometric experiments. We discuss GMRT, PAPER, LOFAR, MWA, and the future HERA and SKA.

8.1 Introduction

Because they provide both rapid mapping speed and good angular resolution, interferometers have become the preferred instrument for experiments looking to measure the expected spatial fluctuations in the 21 cm signal. The current instruments hosting such experiments include the Murchison Widefield Array, MWA[1] (Bowman et al. 2013; Tingay et al. 2013; Jacobs et al. 2016); the Precision Array for Probing the Epoch of Reionization, PAPER[2] (Parsons et al. 2010); the LOw Frequency ARray, LOFAR[3] (van Haarlem et al. 2013; Patil et al. 2016); and the Long Wavelength Array, LWA[4] (Ellingson et al. 2009). In the future, we expect the Hydrogen Epoch Reionization Array, HERA (DeBoer et al. 2017) and the Square Kilometre Array, SKA-Low (Koopmans et al. 2015). Sensitivity predictions for most of the current experiments (e.g., Beardsley et al. 2013; Pober et al. 2014) find that they will not be capable of achieving the necessary signal-to-noise to image the

---

[1] http://www.mwatelescope.org
[2] http://eor.berkeley.edu
[3] http://www.lofar.org
[4] http://lwa.unm.edu

21 cm signal directly (although see Zaroubi et al. 2012 for a study with LOFAR). As such, most of these experiments are targeting a detection of the 21 cm power spectrum, which can be constrained with higher signal-to-noise compared with an image because the isotropy and homogeneity of the Universe allows the 3D k-space power spectrum to be averaged over spherical shells of constant $|k|$. Even using the power spectrum, typical predictions for the requisite observing time are of order 1000 h (see Figure 8.1).

However, beyond the need to achieve the requisite sensitivity, experiments are faced with the daunting task of isolating the 21 cm signal from foregrounds that can be up to five orders of magnitude brighter. While the two can, in principle, be separated by their distinct spectral behavior, the inherently frequency-dependent response of radio interferometers complicates the picture significantly. In this chapter, we review the challenges faced by current interferometric 21 cm experiments as well as the progress to-date in overcoming them. The detailed structure of this chapter is as follows. In Section 8.2, we present the history of experiments and techniques that led to the design of current 21 cm experiments. In Section 8.3, we discuss the distinct approaches each experiment has developed to overcome the challenges associated with these observations, and in Section 8.4 we review the current published upper limits on the 21 cm signal strength from these experiments. In Section 8.5, we highlight the currently unsolved problems at the forefront of experimental 21 cm cosmology and conclude in Section 8.6 with a discussion of the potential for both current and future experiments to overcome them.

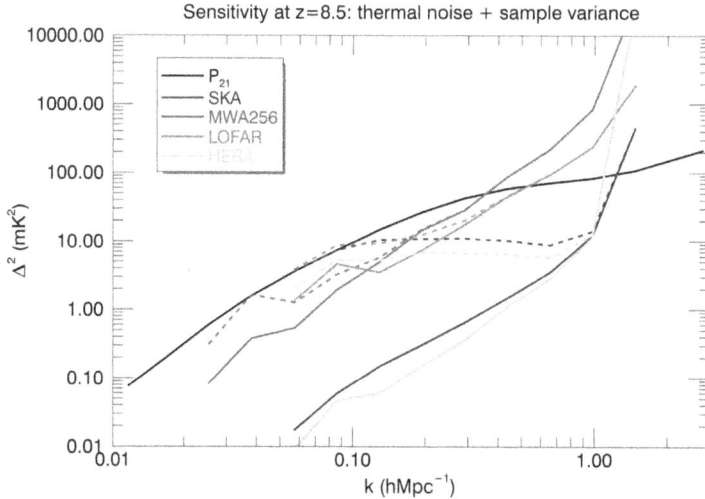

Figure 8.1. Estimated total uncertainty on the dimensionless power spectrum for a 1000 h observation with several experiments at $z = 8.5$ compared with a model input power spectrum (21cmFAST, Mesinger et al. 2011), and assuming that modes within the horizon are inaccessible. Solid lines are only thermal noise uncertainty, while dashed lines include sample variance. (Black) Model cosmological signal; (blue) SKA; (green) MWA256; (red) LOFAR; (orange) HERA.

8.2 Early Work

The origins of the approaches that current experiments are taking to detect the Epoch of Reionization power spectrum can be traced to the development of radio interferometry observational techniques and Cosmic Microwave Background (CMB) analysis methodology.

Radio interferometers measure the cross-correlation of voltages detected with two antennas, extracting the sky signal in a complex-valued dataset that encodes sky emission location and intensity, and as a function of antenna separation vector and frequency (Thompson et al. 1986). For small field-of-view instruments (large antenna aperture), the measured signal is well-approximated as the 2D Fourier Transform of the sky brightness, attenuated by the antenna response function (the primary beam).

Motivated by analysis of CMB datasets in the 1990s and 2000s, and the curved nature of full-sky imaging, early discussion of power spectrum estimators used spherical harmonic basis functions to describe the signal and extract optimal estimators (Tegmark 1997). CMB studies suffer from some of the challenges faced also by EoR experiments: wide fields-of-view, low sensitivity, limited angular resolution, and foreground contamination. Unlike EoR, which is an evolving signal in redshift space, CMB studies are single frequency experiments focused on angular statistics. As such, the foreground mitigation and treatment approaches of CMB studies are of limited use for EoR studies, which attempt to separate foregrounds from the 21 cm signal using the frequency axis. Nonetheless, the fundamental need to extract a weak signal from complex and highly-contaminated data is shared between the two fields, and Tegmark (1997) used this experience to apply CMB analysis techniques to early EoR methodology development. Since an interferometer natively measures in Fourier space, there was a transition from the natural basis of curved sky functions (spherical harmonics) to the interferometer measurement space (Fourier modes) in discussion of optimal estimators for EoR science (Liu & Tegmark 2011).

This work was supported by groundwork laid out for doing EoR power spectra with radio interferometers, including cosmological and unit conversions (Morales & Hewitt 2004; Parsons et al. 2012b) and noise considerations for astrophysical parameter estimation with specific future experiments (McQuinn et al. 2006). McQuinn et al. (2006) discussed a simple foreground model where fitting of a smooth spectral function could remove their effect cleanly, focusing on array sensitivity as the limiting factor for future experiments. However, the lack of any real-world experiments attempting the detection meant they failed to realise the extent of foreground spectral contamination.

More sophisticated approaches to foreground modeling and mitigation appeared in the mid-2000s, with Bowman et al. (2009) beginning a set of papers that explored the signature of smooth spectrum sources in the EoR power spectrum parameter space. Initially, low-order polynomials were explored to fit and remove these sources. However, lacking a physical motivation for this functional form to robustly separate foregrounds from cosmological signal, polynomials were replaced with more realistic functions. Ultimately, the likelihood of removing not only foregrounds but also cosmological signal when fitting and subtracting models, particularly considering the large difference in magnitude of the two signals, has steered the research field away from this approach to foreground mitigation.

As part of this better appreciation for the impact of foregrounds, particularly with the knowledge that they are used also for data calibration, Datta et al. (2010) explored the required accuracy of source models such that foregrounds may be subtracted to a level sufficient to detect the EoR. This work was the first to show the characteristic wedge in power spectrum parameter space, a triangular region in angular and line-of-sight wavenumber space representing the signature of smooth-spectrum sources observed with an interferometer.

8.3 Experimental Methodologies and Current Experiments

In this section we introduce the different instruments that have previously taken, or currently are taking and analyzing, data for an interferometric EoR experiment. We start by presenting the relevant parameters of the telescopes that these experiments use, highlighting and motivating the different observational and analysis approaches taken by each. Table 8.1 lists the location (including latitude), frequency (redshift) range, number of stations/antennas, station diameters, maximum baseline, and field-of-view at 150 MHz for the relevant instruments. Italicized telescopes are discussed in this Chapter. We also plot the full uncertainties (including sample variance) for a 1000 h observation at $z = 8.5$ (10 MHz bandwidth) for each experiment as a function of spatial wavenumber in Figure 8.1. We uniformly assume that the modes within the horizon are inaccessible due to foreground contamination, and note that this is a broad assumption that is not applicable to all experiments (see Chapter 5 for a discussion). Note also that MWA's and HERA's large fields-of-view gives them access to smaller wavenumbers. This figure also includes a nominal signal strength (black, 21cmFAST, Mesinger et al. 2011), but this level is highly uncertain because it depends on the unknown properties of high-z galaxies and the IGM (see Chapter 2). The proximity of the curves to this line highlights the difficulty with predicting the real sensitivity of experiments, particularly in light of the large number of observing

Table 8.1. General Parameters for the Telescopes Undertaking Interferometric Observations of the Cosmic Dawn and EoR

Facility	Location (Latitude)	Freq. [MHz] (z)	N_{ant}	Max. baseline	FOV$_{150}$
GMRT	India (19.1°N)	150–300 (3.7–8.5)	30	30 km	2.5°
MWA	Australia (26.5°S)	70–90, 135–195 (15–19, 6–10)	128	5 km	25°
LOFAR	Netherlands (52.9°N)	30–80, 120–190 (17–46, 6–11)	50–60	50 km	5°
PAPER	South Africa (30.6°S)	110–180 (7–12)	32–64	210 m	60°
LEDA[1]	USA (34°N)	45–88 (15–30)	256+	<10 km	70°
21CMA	China (42°N)	50–200 (6–27)	81	6 km	10°

Italicized telescopes are discussed in this chapter.
[1] LEDA is a total power experiment using interferometry for data calibration.

hours required to reach an expected detection. The parameters shown in the table, and the curves shown in Figure 8.1 motivate and frame the discussion of different experiments in the following sections. Experiments are forced to undertake different approaches to observations and data analysis, because the physical limitations of the systems promote different systematic errors into the forefront for each experiment. There is no silver bullet telescope for undertaking this experiment, however, and after reviewing the main experiments, we discuss the pros and cons of different features.

8.3.1 Giant Metrewave Radio Telescope—GMRT

The GMRT (Swarup et al. 1991) is a Y-shaped array of 30 45 m dishes spread over 25 km in western India. Operating between 50 MHz and 1420 MHz, its 153 MHz receiver has been most used for Reionization studies. Motivated by early work in the post-reionization era (325 MHz and 610 MHz receivers) to statistically detect 21 cm fluctuations and understand the foreground contamination of these data, the low frequency receiver opens the door to exploring the Reionization era. The methodology developed has focused on angular power spectra, measured at a range of frequencies, and pioneered much of the early work to use spectral correlation of foregrounds as a way of treating them. With a lack of short baselines and poor instantaneous *uv*-coverage (Figure 8.2), the array is suited to building high

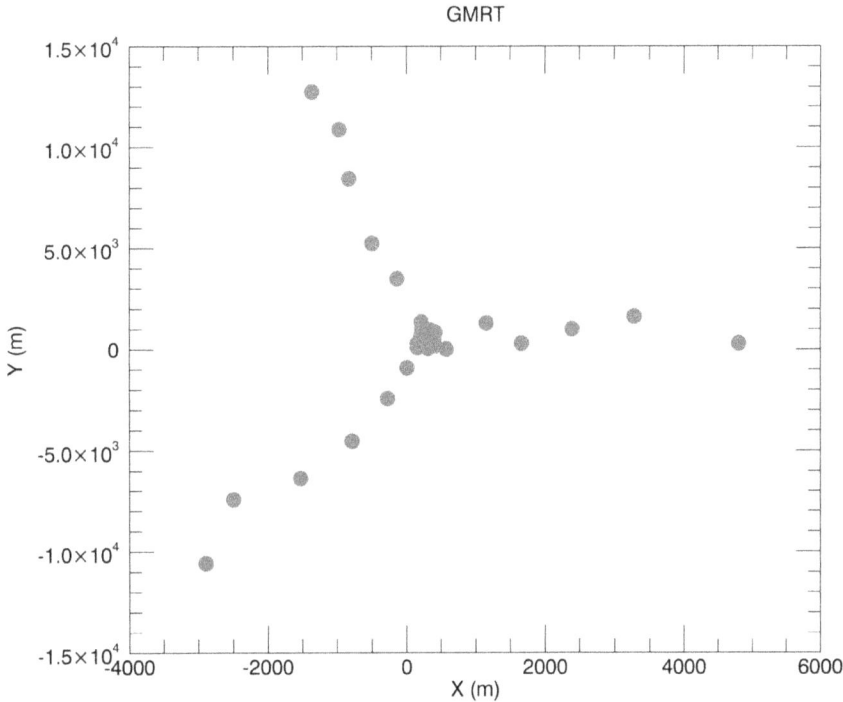

Figure 8.2. Array configuration for the GMRT: thirty 45 m dishes spread over 30 km.

resolution foreground models, and computing the foreground angular correlation function (Rana & Bagla 2019).

GMRT work has largely utilized the visibility correlation function, which cross-correlates visibilities to study the spectral and spatial structure of the sky. Visibility correlation functions were also explored for 21CMA analysis (Zheng et al. 2012). Unlike other experiments, which have cross-correlated interleaved time samples to remove noise power bias, GMRT has usually opted for cross-correlating visibilities from adjacent frequency channels. This has different systematics, with finer spectral resolution required to minimize visibility decorrelation. However, as is standard practise, time integration is used for reducing noise uncertainty.

During the 2000s, there was a series of papers developing a formalism for use of this visibility correlation function to measure angular modes. Bharadwaj & Sethi (2001) introduced a cross-visibility angular correlation function to measure HI fluctuations post-reionization. Bharadwaj & Ali (2005) then related the cross-visibility correlation function across baselines and frequencies to the power spectrum of brightness temperature fluctuations, presenting the full formalism and expected results in different epochs. They suggest that the cosmological signal is uncorrelated for frequency channel differences larger than 1 MHz, allowing signal to be "easily distinguished from the continuum sources of contamination". Ali et al. (2008) then extended this formalism to model the expected foreground continuum signatures in the cross-correlation visibility space, and compared with GMRT observations. Their results were hindered by calibration errors, which caused decorrelation of the signal over frequency, but presented the first application of this technique to data. Datta et al. (2007) provided an extension of the visibility cross-correlation approach to estimating power spectra to a multi-frequency angular power spectrum (MAPS), utilizing decorrelation of signals over frequency to extract information about bubble sizes and distributions as a function of redshift while suppressing the effects of foreground contamination.

Ghosh & Bharadwaj (2011a) published the first measurement of post-reionization neutral hydrogen fluctuations with GMRT (HI intensity mapping) at $z = 1.32$ (610 MHz) and using the MAPS formalism. They used a fourth-order polynomial to remove smooth foregrounds, in line with early attempts with many experiments to fit a parametric function without physical motivation. Ghosh & Bharadwaj (2011b) then demonstrated improved foreground removal for 610 MHz observations by tapering the primary beam function and reducing sidelobes; Ghosh et al. (2012) further extended the work to the reionization epoch using 150 MHz observations to characterize the foregrounds with the MAPS formalism.

Using an alternative analysis to the MAPS formalism, Paciga et al. (2011) analyzed 50 h of data at $z = 8.6$ with a simple piecewise linear foreground subtraction method and cross-correlation of foreground-subtracted images. The result was a reported upper limit on the 21 cm signal strength of $(70 \text{ mK})^2$. However, Paciga et al. (2013) re-analyzed the data with a more sophisticated foreground subtraction technique, including a calculation of signal loss due to foreground fitting. The result was an increase in the upper limit to $(248 \text{ mK})^2$, indicative of the degree to which signal loss can affect results.

More recently, Choudhuri et al. (2014) published a series of papers introducing and exploring the use of two new optimized power spectrum estimators using visibility correlations: the Tapered Gridded Estimator (TGE) and Bare Estimator (BE). The key concept for the TGE, which has been further discussed in the literature in subsequent papers (Choudhuri et al. 2016), is to use a Fourier beam gridding kernel that is larger than the physical beam kernel, thereby decorrelating sources at the edge of the field-of-view. Note that this approach is not a silver bullet to removing the effect of horizon sources, because their sidelobes remain in the data even if they have been attenuated. Originally developed as angular power spectra as a function of frequency, the TGE work has recently been expanded to use the line-of-sight spatial information (Bharadwaj et al. 2019). The BE directly squares adjacent visibilities to provide individual measurements of the power, but this has not been used further, possibly due to the large number of visibilities that are accumulated and stored.

Additionally to power spectra, Saiyad Ali (2006) predicted the amplitude of a bispectrum signal with GMRT using its shortest baselines by modeling non-linear clustering. They predicted the signal strength to be comparable to the power spectrum and detectable in 100 h but this project has not been explored observationally with this instrument.

8.3.2 Murchison Widefield Array—MWA

The Murchison Widefield Array (MWA) is a 256-element interferometer in the Western Australian desert. In Phase I of the array, operating from 2013–2016, it was composed of 128 tiles of 16 dual-polarization dipoles, spread over 3 km (Tingay et al. 2013). Phase II (2016–) expanded the array to 256 tiles, with longer baselines for improved survey science and sky model building (5 km), and two hexagonal sub-arrays of 36 tiles with short spacings available for redundant calibration and improved EoR sensitivity (Wayth et al. 2018). It operates in two distinct modes: Extended Array (128 tiles with long baselines), and Compact Array (128 tiles with short baselines including two 36-tile redundant subarrays in a hexagonal configuration). The Compact Array is principally used for EoR science (see Figure 8.3). The MWA is a general science telescope, with multiple science goals (Bowman et al. 2013). As such, it balances high surface brightness sensitivity on EoR scales, redundant and non-redundant elements, and longer baselines for good imaging capabilities. The instantaneous uv-coverage of the MWA is excellent, allowing for science-quality snapshot imaging (2 min). The MWA is also a wide-field instrument, with a field-of-view of 25 degrees at 150 MHz. This wide field-of-view, combined with the complex frequency-dependent shape of the aperture array primary beam, and analogue electronics, create challenges for data analysis. The two-stage analogue beamformer produces a frequency bandpass that contains 24 coarse channels over a 30.72 MHz bandwidth (chosen from the full bandwidth listed in Table 8.1), with missing regular channels between the coarse bands. This instrumental spectral structure provides a challenge to producing instrumentally-clean output EoR datasets.

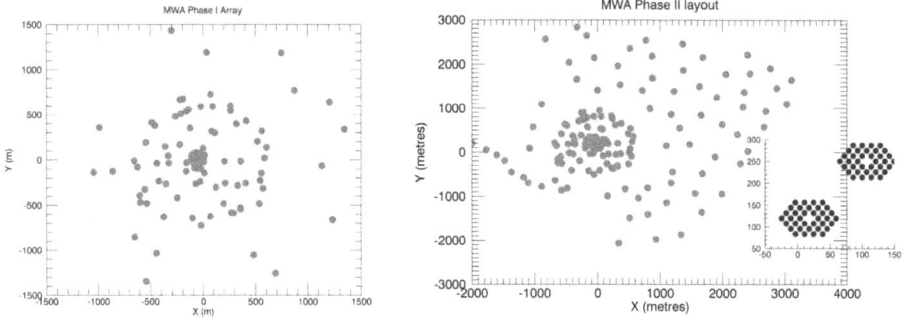

Figure 8.3. Array configurations for the MWA [Phase I, left; Phase II, right, including cutout of hexagonal subarrays (blue)]: 128 (256) 4.4 m aperture array tiles spread over 3 (5) km.

Early deployments of the array, with 32 tiles, were used for preliminary science, and to begin to survey the EoR fields (Williams et al. 2012). Upon completion of the 128 tiles, the MWA Commissioning Survey provided the first sky catalogue for use for calibration of EoR data (Hurley-Walker et al. 2014). This work paved the way for the GLEAM survey (Wayth et al. 2015) and catalogue (Hurley-Walker et al. 2017), yielding 300,000 sources in the southern sky. GLEAM provides the basis for the current sky model for point sources in EoR observations, augmented by individual models for extended sources.

In line with developments in concurrent experiments, prior to data acquisition the MWA EoR collaboration focused on relatively simplistic foreground fitting and removal methods, but with an increasing understanding of the signature of smooth-spectrum foregrounds in the wavenumber parameter space of an interferometer (Bowman et al. 2009; Datta et al. 2010; Trott et al. 2012). There are now two primary EoR data calibration and source subtraction pipelines used by the collaboration: the Real-Time System (RTS, Mitchell et al. 2008) and Fast Holographic Deconvolution (FHD, Sullivan et al. 2012). Both use underlying catalogues of sources that have been generated by cross-matching multiple low-frequency sky catalogues. PUMA (Line et al. 2017) generates an observation-specific sky model of point sources and double sources (Procopio et al. 2017), and includes shapelet-based and point source-based models for extended sources. The RTS calibrates the data in two steps, both of which rely on a weighted least-squares minimization: (1) overall direction-independent (flux density and phase) calibration on a full model of 5000 sources; (2) direction-dependent corrections along the line-of-sight to bright sources. The direction dependent corrections are then applied to sources in the region of the fit, and the 5000 source sky model is subtracted. FHD calibration (Sullivan et al. 2012) computationally optimizes the direction-dependent and wide-field imaging steps by pre-computing the mapping from Fourier to real space. FHD relies on an underlying point source sky model (Carroll et al. 2016), which generates an observation-specific calibration model for >10,000 sources based on the GLEAM catalogue and other cross-matched surveys.

Early developments of power spectrum pipelines stemmed from the inverse covariance quadratic estimator framework pioneered in CMB studies (Tegmark 1997), and applied to theoretical EoR datasets by Liu & Tegmark (2011). This work was further developed by Dillon in a series of papers that explored how to bridge some of the differences between the ideal estimator and a physical dataset (Dillon et al. 2015, 2013). In particular, Dillon discussed missing data, and large data volumes. An adapted approach was then applied to three hours of MWA data, showing promising results (Dillon et al. 2013).

One key feature of the optimal quadratic estimator formalism is the whitening of data according to the correlated covariance introduced by the uncertainty on residual foregrounds. This is effectively a down-weighting of data that are heavily affected by foregrounds, thereby improving signal-to-error. Subsequent analysis of a higher redshift dataset was used to estimate the principal eigenmodes of the data in spectral space, identifying these with bright foregrounds (Dillon et al. 2015). The covariances of these modes were then used in the estimator to down-weight and decorrelate data, yielding improved limits at $z = 6.8$. However, as with commensurate and subsequent work with PAPER that used this technique, it had the large potential to cause bias in the estimates. Re-use of the same dataset to empirically estimate the data covariance, and then fit for it, causes re-substitution bias, a well-known statistical effect where the performance of an estimator can appear much better than it actually is. In this work, Dillon was careful to estimate covariances empirically while omitting the uv cells in question, to avoid bias, however there was still limited information available in the remaining cells. Thus, although this work was careful to not try to subtract the foreground bias directly, use of the empirical covariance in the data weighting, and lack of a full end-to-end simulation to demonstrate no signal loss, makes this approach prone to large bias. It has not been used to analyze MWA data since the original analysis in Dillon et al. (2015).

In a later paper, describing the CHIPS estimator, Trott also developed an inverse covariance quadratic estimator formalism using a model foreground covariance (Trott et al. 2016). Unlike the empirical approach of earlier work, this does not use the data itself to form the foreground covariance, but a model for the expected spatial and spectral structure of point source foregrounds. However this approach can suffer from similar effects, whereby error in the covariance can propagate into the analysis. Therefore, this inverse foreground covariance has never been applied to data used in publication due to the output's sensitivity to the choice of foreground model.

A second principal power spectrum estimator for MWA EoR analysis, εppsilon, was independently developed from CHIPS (Barry et al. 2019). εppsilon prioritizes the propagation of thermal noise error from the visibilities (with estimates provided by FHD) through to the power spectrum while also providing a suite of diagnostics for assessing the performance of the estimator in a number of domains.

Both εppsilon and CHIPS (without the foreground covariance weighting) were used in the EoR limit paper led by Beardsley et al. (2016), which processed 32 h of MWA Phase I high-band data to power spectrum limits. At the time, these results were highly-competitive in the field, but the data were clearly still systematic-dominated.

At a similar time, Ewall-Wice published the first measurement of upper limit from the Cosmic Dawn (Epoch of X-ray heating, EoX) from 3 h of MWA data above $z = 15$ (Ewall-Wice et al. 2016).

One of the clear outcomes of the early upper limit publications from MWA (and other instruments, particularly LOFAR) was that the data were highly systematic-dominated in modes relevant for EoR, and accumulating more data into the power spectrum estimator would offer no advantage. With this realization, the MWA collaboration embarked on a two year program to prioritize understanding and treating systematics over processing large datasets, despite more than a thousand hours having been collected by the instrument. This work encompassed (1) improving the sky model (point, extended and multiple sources, Procopio et al. 2017; Trott & Wayth 2017); (2) understanding the impact of calibration choices on residuals and uncertainties (Barry et al. 2016; Trott & Wayth 2016; Trott et al. 2017; Ewall-Wice et al. 2017; Murray et al. 2017); (3) improving the primary beam modeling (Line et al. 2018); (4) developing data quality metrics for data triaging (RFI, ionospheric activity, Jordan et al. 2017; Trott et al. 2018; Wilensky et al. 2019), and (5) developing and refining redundant and hybrid calibration pipelines for Phase II (Li et al. 2018; Joseph et al. 2018; Byrne et al. 2019). A final important step was the development of a full end-to-end simulation to demonstrate that there was no signal loss in the chain from telescope to data product. The results of this work include upcoming EoR limits from re-analysis of Phase I data and new Phase II data, as well as exploration of new techniques for exploring the EoR (Trott et al. 2019).

A final, key insight from recent work helps to address the current questions in the EoR research field about robustness of any future claimed detection of cosmological signal. Along with confirmation by other telescopes, ability to detect the same signal in independent observing fields, where the foregrounds are different, is crucial. In Trott et al. (2019), MWA data from two observing fields was studied with a Kernel Density Estimator to understand the similarities and differences between the statistical structure of data from independent sky areas. This work can lead to a better understanding of robustly discriminating contamination from cosmological signal.

8.3.3 Low Frequency Array—LOFAR

LOFAR is a composite aperture array low-frequency radio interferometer. It has two primary station types; the High-Band Antennas (HBA, 120–190 MHz) and Low-Band Antennas (LBA, 30–90 MHz). Both station types have been used for EoR and Cosmic Dawn science. Athough LOFAR formally contains baselines of thousands of kilometres to the international stations, it is only the Dutch-based stations that are used for actual EoR measurements. Figure 8.4 shows the central stations (blue cut out) and the nearest remote stations (red).

The LOFAR latitude allows for circumpolar observations with long winter nights. As such, one of the primary observing fields is the North Celestial Pole, which can be observed for more than 12 h in the winter months. Early work with the

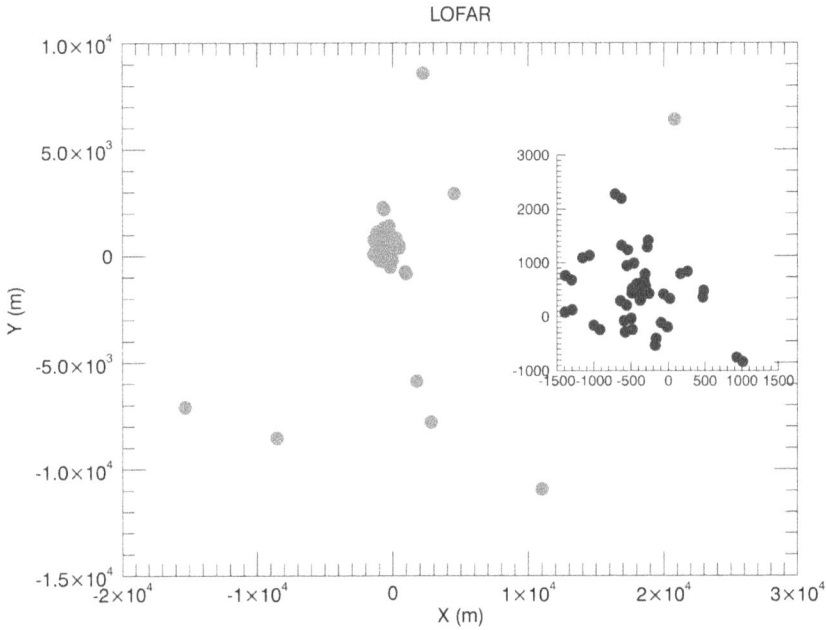

Figure 8.4. Array configuration for the central and inner remote stations of LOFAR (red), and subplot showing the central stations only (blue): 30–40 m aperture array dipole stations spread over tens of kilometres.

LOFAR EoR Key Science Project focused on foreground mitigation, choice of observing fields, and data analysis methodology. As with many of the published papers in this early epoch, foregrounds in Jelić et al. (2008) were modeled to be subtracted with a simple smooth fitting function. This work is notable because it provided realistic models for a range of different foreground components, and included discussion of the treatment of polarized foregrounds.

Jelić et al. (2010) extended the work from 2008 to focus on simulations of Faraday Rotation from polarized Galactic foregrounds. FR rotates the phase of the intrinsically-smooth foreground component yielding spectral structure that may mimic the EoR signal. In this work, Jelić shows the effect of inaccurate data calibration on polarized emission, which can imprint total intensity structure if the polarized instrumental response is incorrect.

The SAGE algorithm (Space Alternating Generalized Expectation Maximization) was first introduced in 2011 by Kazemi et al. (2011), and provides the basis for all calibration of LOFAR EoR datasets to the present day. Based on the well-known Expectation Maximization (EM) algorithm, which iteratively fits for calibration parameters (maximizes the likelihood with respect to a set of parameters and then alters the parameters to find a new likelihood) when the underlying system model contains unobserved variables, SAGE extends the traditional least-squares fitting to allow for more model flexibility, and improved convergence and efficiency. Part of the SAGE algorithm performs direction-dependent calibration toward clusters of sources on the sky, thereby allowing for ionospheric distortion of the sky model.

Detailed total intensity and polarized imaging of the LOFAR EoR fields were presented in Yatawatta et al. (2013) and Jelić et al. (2014). The North Celestial Pole field allows for deep, long winter nighttime observations, and was shown to be able to be calibrated over long integrations. Observations of low Faraday depth structures in the ELAIS-N1 field yielded structures that would be problematic for EoR science if the degree of polarization leakage into total intensity exceeded 1%. This quantification of the accuracy required of instrumental polarization models was the first of a set of papers that explored polarized signal and leakage for EoR science. Thus far, the LOFAR EoR collaboration has undertaken the most extensive work to quantify the impact of polarization leakage, while the MWA collaboration has made some observations of polarized emission in their data (Lenc et al. 2016, 2017; Bernardi et al. 2013). In Asad et al. (2015) and Asad et al. (2016), Asad and colleagues first studied the polarized emission in the 3C196 EoR field, finding them to be localized around a small Faraday depth, and quantified the leakage, and then studied the accuracy of the LOFAR polarized beam model to be able to limit leakage into total intensity. Given the level of polarized to total intensity, and a beam model accurate to 10% at the field centre, the leakage and subsequent spectral structure would be acceptable for EoR science. Finally, Asad et al. (2018) considered the more problematic impact of wide-field polarization leakage on the EoR power spectrum. Far from the field centre, the primary beam models are less accurate, and sources imprint additional spectral structure due to the chromaticity of an interferometer. In these cases, bias was found to persist in the EoR power spectrum.

In early work on fitting foregrounds, Harker et al. (2010) discussed the use of Wp smoothing as a non-parametric method for fitting a smooth function, based on limiting the number of inflection points in the fit. Further, they discuss the systematic errors introduced by the fitting routine, methods for estimating these, and for accounting for them in the final uncertainties. This approach is used again in the work of Mertens et al. (2018) for the Gaussian Process Regression fitting, and also is used generically in the PAPER analysis to try to understand signal loss.

In a series of papers, Chapman and collaborators explored novel approaches to fitting and removing foreground signal from image-based datacubes. In Chapman et al. (2012), they introduced the FastICA technique, as a non-parametric method that estimates independent foreground components and their mixing for each image pixel and frequency. The advantage of such methods is that they do not rely on any *a priori* knowledge of the signal, but instead only assume that the full signal can be represented by a small number of components (sparsity), thereby allowing for good estimation with a given dataset. The disadvantage lies in the sensitivity of results to the number of components the user chooses that the data should contain, and the potential for signal loss if the projection of the estimated components onto the EoR signal is non-negligible. ICA generically ignores stochastic components, thereby isolating smooth components of the data, and minimizing Gaussianity, thereby enforcing smoothness in the fitted components. Beyond ICA, a generalized method (GMCA; Generalized Morphological Component Analysis) was applied (Chapman et al. 2013, 2016) to use an underlying blind wavelet decomposition of the components, combined with the sparsity and mixing model methodology of the

ICA method. As with other methods when the underlying structure, spatial distribution and amplitude of the cosmological signal and foreground components is unknown, the potential for signal loss is present. The ICA and GMCA methods both assume that the cosmological signal has negligible amplitude and is absorbed in the noise. Structural deviations from this assumption can lead to signal loss.

In a new approach to foreground treatment, Mertens and colleagues discuss use of the well-known Gaussian Process Regression (GPR) technique to fit for foregrounds using only an understanding for the spectral data covariance of different components (Mertens et al. 2018). Unlike parametric methods that assume an underlying model, GPR (an extended version of kriging, which interpolates data based on their known covariance properties) relies only a statistical separation of foreground and cosmological signal via their spectral correlation lengths. This method also suffers from the potential for cosmological signal loss, but the authors attempt to capture the potential bias statistically through increased noise. The ultimate utility of this approach has yet to be demonstrated on a large dataset at the time of this writing.

Along with the power spectrum as a measure of the signal variance as a function of spatial scale, the variance statistic was explored with simulations in Patil et al. (2014). The variance of the brightness temperature, as the wavenumber integral over the power spectrum, quantifies the variability in the cosmological signal on the imaging scale (autocorrelation function). Although it provides limited cosmological information, detection of this variance can be theoretically obtained with fewer observing hours. Bayesian power spectrum extraction techniques were also explored in Ghosh et al. (2015), with a view to allowing for a spatially-smooth component to capture the unmodeled diffuse emission in the NCP field. Like other instruments, the data calibration and foreground models were limited to point and extended sources, with the complex diffuse emission difficult to measure and model. Increasingly, the impact of this incomplete sky model has become apparent.

In a landmark paper published by Patil and colleagues in 2016 (Patil et al. 2016) the source of "excess noise" and diffuse emission suppression in LOFAR data were studied. Excess noise is the identification of increased noise levels in the data post-calibration compared with expectations of thermal noise and Stokes V measurements. Ultimately, the lack of a diffuse model in the calibration sky model allowed for this signal to be absorbed into the gain calibration solutions, thereby yielding a direction-dependent bias and noise in the residual data. To address this problem, the short baselines containing the majority of the diffuse emission could be excluded, however this leads to increased noise on these scales (due to statistical leverage; effectively this amounts to additional flexibility in the gain solutions on these scales because they are not used in the modeling). This work was undertaken contemporaneously with that of Barry et al. (2016) and Trott & Wayth (2016), which both studied the effect of incomplete sky models and spectrally varying bandpass parameters on calibration and residual signal. The combined outcome of these studies is an understanding of the impact of sky model incompleteness, the need to enforce spectral correlation (e.g., regularization as in SAGECal or smooth model fitting) for calibration parameter fitting, and the approaches to calibration that can mitigate these.

Further exploration of the impact of calibration frameworks and data treatment were then explored in Mouri Sardarabadi & Koopmans (2019) and Offringa et al. (2019), with a view to having a complete understanding of the end-to-end data processing of LOFAR EoR data on the path to a detection. Unlike in the previous ten years before real observations were undertaken and thermal noise sensitivity was seen to be the major impediment for EoR detection, the field has come to appreciate the crucial roles of unbiased calibration, sky model completeness, and foreground treatment without cosmological signal loss.

The culmination of the lessons learned from statistical leverage and incomplete sky models was applied to two fluctuation upper limit papers published since 2017. In Patil et al. (2017), Patil and colleagues presented competitive results from a small set of data (~10 h) at $z = [9.6 - 10.6]$, with the best limit of $(59.6 \text{ mK})^2$. This work reported an excess variance, in line with previous discussions, and the use of Stokes V power to remove noise power. At higher redshifts (lower frequencies), Gehlot et al. (2019) reported upper limits above $z = 20$, with use of the Gaussian Process Regression foreground fitting technique introduced by Mertens et al. (2018) for EoR science.

Beyond the power spectrum, LOFAR has explored other tracers of the neutral hydrogen temperature field, namely the ability to produce low angular resolution images (Zaroubi et al. 2012) and the 21 cm Forest (Ciardi et al. 2013). LOFAR, like other current instruments, does not have the sensitivity to directly image neutral hydrogen bubbles at the instrumental resolution. However, by lowering the resolution of images (thereby improving the radiometric noise), Zaroubi and colleagues argue that the largest of bubbles may be detectable at low signal-to-noise ratio on the largest of scales late in reionization. The ability to detect the 21 cm Forest (absorption of continuum radio emission along the line-of-sight to high redshift AGN due to intervening neutral gas) remains a challenge and aim of many current interferometers. Ciardi and colleagues showed that LOFAR would have the ability to detect an absorption feature under ideal conditions. Unlike the statistical detection of the power spectrum of temperature fluctuations, the absorption signal amplitude is determined by the astrophysical conditions close to the gas, namely gas kinetic temperature. Cold gas is able to absorb light more readily than heated gas. The failure of this method to date is primarily due to the lack of any known high-redshift radio-loud AGN ($z > 6$). Given the sensitivity of current instruments, a source with flux density exceeding 10 mJy and cold neutral gas would be required. It is likely that the arrival of SKA will provide both the sensitivity and the detection (and confirmation) of high-redshift radio-loud AGN to be able to undertake this experiment. Of the current interferometric experiments, only LOFAR has sufficient sensitivity to be able to attempt this experiment at all.

The utility of extracting higher-order statistics of the 21 cm brightness temperature field were explored in a simulation study of foreground-subtracted image cubes by Harker et al. (2009). Again, the ability to smoothly treat and remove foregrounds placed the burden of detection on pure noise considerations.

8.3.4 Precision Array for Probing the Epoch of Reionization—PAPER

The PAPER experiment was designed as a testbed for developing novel 21 cm cosmology analysis techniques. PAPER antennas were chosen to be small, single dipoles on elevated ground-screens to enable reconfiguration of the array, and the system used a flexible digital correlator architecture that could scale as the number of antennas grew (Parsons et al. 2008). The small antenna size was also chosen to limit the frequency evolution of the antenna response over the instrument's 110–180 MHz of usable instantaneous bandwidth. The design and results from an initial 8-station deployment of PAPER in Green Bank, WV, USA were described in Parsons et al. (2010).

In its earlier stages, PAPER deployed its antennas in configurations designed for imaging, including a single-polarization 16-element 300 m diameter ring in Green Bank used for primary beam measurements in Pober et al. (2012). While the Green Bank array was upgraded to a single-polarization, 32-element array, all subsequent publications came using arrays deployed at the SKA-SA site in the Karoo, South Africa. Highlights of early PAPER studies include the creation of a 145 MHz Southern hemisphere sky-catalog using a single-polarization, 32-element array (Jacobs et al. 2011) and a study of the radio galaxy Centaurus A using a single-polarization, 64-element array (Stefan et al. 2013). In both of these cases, the elements were deployed in a randomized configuration over a circle of 300 m diameter to maximize *uv* coverage.

In 2012, however, members of the PAPER team developed what is now referred to as the "delay spectrum" approach for measuring the 21 cm power spectrum. In the delay spectrum approach, visibility spectra from individual baselines are Fourier transformed and cross-multiplied. Parsons et al. (2012a) demonstrated how these delay spectra can be used as estimates of the 21 cm power spectrum, without ever combining visibilities and making an image. Parsons et al. (2012a) also provided sensitivity estimates for the delay spectrum approach using a 128-element PAPER array. Parsons et al. (2012a) then demonstrated how 21 cm foregrounds isolate into what is now commonly referred to as "the wedge" and included the effects of foreground contamination in the sensitivity study. One consequence of the delay spectrum approach is a higher noise level than alternative approaches: power spectra estimated from individual baselines are averaged together, as opposed to coherently combining all the visibilities and forming a single power spectrum, so noise fluctuations average down more slowly. To make-up for this sensitivity sacrifice, Parsons et al. (2012a) proposed using a "maximum redundancy" configuration, in which antennas are arranged to create multiple copies of the same baseline spacing. These redundant baselines can then be averaged together before squaring, helping the noise level to integrate down faster. Although redundant layouts drastically reduce imaging fidelity, the delay spectrum approach does not requiring imaging and so is, in principle, not affected by this consequence.

The decision was made to reconfigure the PAPER array and test the delay spectrum technique in a maximum redundancy layout. However, a short dataset in a single-polarization, 64-element "minimum redundancy" (i.e., random layout) with a 300 m diameter was collected and used to make delay spectra from a range of baseline lengths and orientations in Pober et al. (2013). This analysis demonstrated good isolation of foreground emission to the wedge in 2D cosmological k-space, suggesting the promise of the delay spectrum technique.

Parsons et al. (2014) presented the first deep power spectrum limits from a dual-polarization, 32-element, maximum redundancy array (a grid of 8 columns and 4 rows, with a column spacing of 30 m and a row spacing of 4 m). Just over 1000 h of data were used in the analysis. In addition to the basic delay spectrum formalism, Parsons et al. (2014) introduced two additional analysis techniques: redundant calibration (Wieringa 1992; Liu et al. 2010), which was enabled by the redundant layout of the array, and a new technique for removing off-diagonal covariances between redundant baselines. These same techniques were applied to the same data over a range of redshifts in Jacobs et al. (2015).

The techniques of Parsons et al. (2014) were then applied to a new, 1000+ h, dual-polarization, 64-element PAPER dataset in Ali et al. (2015) (the layout of which is shown in Figure 8.5). This analysis improved upon the redundant calibration technique by using the OMNICAL package (Zheng et al. 2014), replaced the off-diagonal covariance removal technique with an inverse covariance weighting approach using empirically estimated covariance matrices (similar to Dillon et al. 2015) and applied a new technique known as "fringe rate filtering" (described in

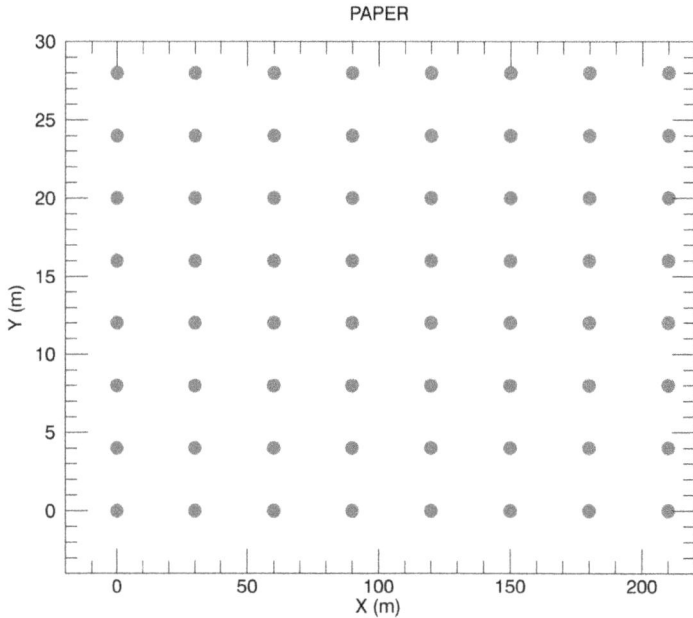

Figure 8.5. Array configuration for the PAPER-64 maximally-redundant array: 64 single dipoles spread over 210 m. Note the distinctly different scales on the x and y axes.

Parsons et al. 2016). At the time, the Ali et al. (2015) limits on the 21 cm power spectrum were believed to be the most stringent to be published and were followed by two separate publications using their measurement to constrain the temperature of the IGM at $z = 8.4$ (Pober et al. 2015; Greig et al. 2016).

However, re-analysis of the Ali et al. (2015) data by Cheng et al. (2018) revealed a critical error in the analysis: empirically estimated covariance matrices are correlated with the data, and weighting by them can bias the recovered signal low. (As described in Section 8.3.1, this bias has frequently been referred to as "signal loss"— the idea that an analysis technique can remove 21 cm signal along with foregrounds.) In practice, the signal loss in the PAPER analysis was *very* large (nearly four orders of magnitude of potential EoR signal was suppressed) due to the fringe-rate filtering technique that reduced the number of independent samples used to estimate the covariance matrix. Although the analysis in Ali et al. (2015) attempted to estimate signal loss using injection of mock EoR signals into the data, their method missed potential loss caused by data-signal cross terms in the covariance matrix and thus concluded that the original analysis was effectively lossless. Incorrect estimates of both the theoretical and observed noise levels in the data also contributed to the belief that the analysis of Ali et al. (2015) was sound.

In light of the analysis in Cheng et al. (2018), all of the PAPER results in Parsons et al. (2014), Jacobs et al. (2015), Ali et al. (2015) are considered to be invalid and do not place meaningful limits on the 21 cm signal.[5] A re-analysis of the full Ali et al. (2015) dataset using a lossless analysis is forthcoming, but the limits are not expected to be near the same level as Ali et al. (2015). The PAPER experiment also collected two years of data with a dual-polarization, 128-element array, but due to an increased amount of instrument systematics and failures in the aging system, these data are not expected to be published.

The delay spectrum approach does not allow for high accuracy polarization calibration, which needs to be performed in the image domain. Theoretical studies of the effect of Faraday rotated (i.e., frequency-dependent) polarized emission on the delay spectrum technique were presented in Moore et al. (2013) and Nunhokee et al. (2017) and studies using PAPER data were performed in Moore et al. (2017) and Kohn et al. (2016). Overall, the effect of polarized emission on the delay spectrum approach can be quite significant, but the overall amplitude is uncertain as there are few constraints on the polarization properties of the 150 MHz sky at the angular scales probed by PAPER. Ionospheric Faraday rotation can also attenuate the polarized signal in datasets averaged over many nights (Martinot et al. 2018).

8.4 Published Results

Here we collate the published best limits at each redshift from the current experiments (Table 8.2). PAPER measurements have been omitted. Despite the current

---

[5] Although Parsons et al. (2014) did not use the inverse covariance weighting that was the main source of the problem in Ali et al. (2015), its covariance removal technique has not been robustly vetted for signal loss and thus the results are considered suspect at best.

Table 8.2. Best Two Sigma Upper Limits on the EoR and Cosmic Dawn Power Spectrum for Each Experiment

Facility	z	k (h Mpc^{-1})	Upper Limit (mK)^2	Ref.
MWA	12.2	0.18	2.5×10^7	Ewall-Wice et al. (2016)
MWA	15.35	0.21	8.3×10^8	Ewall-Wice et al. (2016)
MWA	17.05	0.22	2.7×10^8	Ewall-Wice et al. (2016)
MWA	7.1	0.23	2.7×10^4	Beardsley et al. (2016)
MWA	6.8	0.24	3.0×10^4	Beardsley et al. (2016)
MWA	6.5	0.24	3.2×10^4	Beardsley et al. (2016)
MWA	9.5	0.05	6.8×10^4	Dillon et al. (2015)
LOFAR	10	0.053	6.3×10^3	Patil et al. (2016)
LOFAR	9	0.053	7.5×10^3	Patil et al. (2016)
LOFAR	8	0.053	1.7×10^4	Patil et al. (2016)
LOFAR	23	0.038	2.1×10^8	Gehlot et al. (2019)
GMRT	8.6	0.5	6.2×10^4	Paciga et al. (2013)

Only the lowest limits have been reproduced.

published values, there are publications in peer-review now for LOFAR, PAPER, and MWA improving on these results.

8.5 Current Challenges

21 cm experiments consist of many components, from the analog telescope design through to power spectrum estimation algorithms. One clear lesson from first generation experiments is that no one aspect of the system can provide the necessary 1-part-in-10^5 dynamic range required to detect the 21 cm signal; rather, the burden needs to be spread across the components of the experiment, alleviating the demands on each of the other components. In this section, we briefly review what we consider five key areas where 21 cm experiments continue to innovate: (1) analog instrument design; (2) data quality control; (3) calibration; (4) foreground mitigation and the associated potential for signal loss; and (5) end-to-end validation of analysis pipelines.

1. **Analog Instrument Design.** One of the major challenges for 21 cm cosmology experiments is to remove any spectral structure introduced by the instrument that might otherwise mix smooth spectrum foregrounds into the spectral modes occupied by the cosmological signal. One seemingly straightforward approach is to limit the amount of spectral structure in the instrument response through careful analog design. Initial specifications on the HERA system design were to limit spectral structure to a level that would enable the delay spectrum technique without any additional calibration or analysis requirements; however, further study has shown that the HERA design does not meet this stringent specification and will need data analysis algorithms to also model and remove spectral structure from the instrument (DeBoer et al. 2017). The push to larger bandwidths (e.g., 50–250 MHz for HERA and 50–350 MHz for the SKA) adds to the challenge of constructing a single instrument with a smooth

spectral response over a large range of wavelengths. Analysis with the MWA has also demonstrated how reflections in the analog system can contaminate modes of the EoR power spectrum, suggesting that more stringent specifications on impedance matches and cable lengths are necessary for future instruments (Barry et al. 2016; Ewall-Wice et al. 2017).

2. **Data Quality Control.** Given the extreme brightness of human-generated radio signals compared to the 21 cm signal, only a very small number of contaminated measurements are enough to significantly affect the analysis of a large dataset. The "gold standard" for identifying radio frequency interference, AOFlagger (Offringa et al. 2012), is used by both LOFAR and the MWA. However, additional quality metrics can still catch corrupted data that slips by this first round of flagging, including ultra-faint, broadband digital TV transmission (Wilensky et al. 2019) and effects due to ionospheric weather (Trott et al. 2018; Jordan et al. 2017). As interferometers grow in size, the large data rates may also require computationally faster algorithms for data quality checks (Kerrigan et al. 2019). Offringa et al. (2019) also demonstrate how even flagged RFI can affect power spectrum analysis if care is not taken.

3. **Calibration.** Instrument calibration is often regarded as the greatest challenge for existing and future 21 cm experiments. While both the analog design and the methodology used for power spectrum estimation can ease calibration requirements (Morales et al. 2019), experiments still need to control the spectral response of their telescopes over wide bandwidths at a level unprecedented in radio astronomy. Typically, antenna-based gain calibration is performed by forward-modeling visibilities and minimizing the difference with the observed data; however, Barry et al. (2016), Patil et al. (2016), and Trott & Wayth (2016) demonstrate that without additional constraints, calibration performed with an incomplete sky-model can lead to spurious spectral structure in the calibration solutions that can both overwhelm or remove the EoR signal. Redundancy based calibration has been viewed as a promising alternative because it does not reference a sky-model; however, recent work has shown that a sky model is still required to constrain the degeneracies inherent in redundant calibration, and that the same kind of contamination can affect the power spectrum as in sky-based calibration (Byrne et al. 2019; Li et al. 2018; Joseph et al. 2018). Calibration of the primary beam response of the instruments is also a major challenge, and several options have been explored, including sky-based calibration (Pober et al. 2012), using satellite broadcasts (Neben et al. 2015, 2016; Line et al. 2018), and with drones flying transmitters (Jacobs et al. 2017).

4. **Foreground Mitigation.** Fundamentally, the real challenges at the heart of 21 cm cosmology come from the intrinsic brightness of the foreground emission. While much of the work to date focuses on removing the instrument response from the foreground spectra, most current experiments use some form of foreground mitigation to help isolate or remove foregrounds. Many distinct approaches have been developed, which can be broadly

classified as either "foreground avoidance" and "foreground subtraction". Foreground avoidance methods attempt to isolate foregrounds into the wedge and minimize bleed into the EoR window; power spectra are then only estimated from within the EoR window. Examples of avoidance techniques includes the wide-band iterative deconvolution filter used in PAPER analyses (Kerrigan et al. 2018) and the inverse covariance weighting techniques also used by PAPER (Cheng et al. 2018). Foreground subtraction, on the other hand, attempts to remove specific models of the foregrounds—using either real sky catalogs or parametric models for their spectra—while leaving the 21 cm unaffected. Examples of foreground subtraction including the point-source forward modeling and subtraction performed by FHD (Barry et al. 2019) and the spectral based fitting methods used by LOFAR (Chapman et al. 2016; Mertens et al. 2018). It is worth stressing that while these techniques have historically been developed in the context of specific experiments, they are more generally applicable; see Kerrigan et al. (2018) for an example of PAPER-developed techniques applied to MWA data and MWA-developed techniques applied to PAPER data.

One of the greatest risks of foreground removal is the inadvertent removal of 21 cm signal, i.e., signal loss. Although many techniques have been developed using frameworks where signal loss is not expected, due to a presumed orthogonality of the foreground description and 21 cm signal basis, there are subtle challenges that arise when faced with a need to achieve five orders of magnitude of dynamic range. While cross-terms between the foreground and signal might have an expectation value of 0, there are still only a finite number of samples going into the analysis, and these cross terms will not have converged to their expectation value—as was the case in the PAPER analysis of Ali et al. (2015).

5. **Validation.** One of the last major challenges for current and future experiments is to rigorously test foreground removal and other analysis algorithms—ideally as part of a complete pipeline and not as an independent step—to confirm that 21 cm signal is not being biased or removed. And although foreground removal seems like the step most likely to cause signal loss, it is certainly not the only place that needs further scrutiny. In light of the PAPER retractions, 21 cm experiments are realizing the importance of simulation-based analysis vetting—ideally with independent, third-party simulations. Many inteferometric simulators exist, including CASA, PRISim (Thyagarajan et al. 2015), OSKAR, and pyuvsim (Lanman et al. 2019). In turn, it has become important to test the simulators against each other, to verify that they achieve the requisite precision for 21 cm cosmology. These validation efforts can be slow and painstaking, but as experiments push closer to a first detection of the 21 cm signal, they have become more vital than ever. The other avenue for verification is with other instruments and other pipelines providing independent analysis. Use of multiple observing fields can also show robustness to foreground treatment (Trott et al. 2019).

8.6 Prospects for the Future

8.6.1 Current Instruments

MWA and LOFAR are both currently pursuing deeper limits. Armed with new calibration and analysis, and critically, a deeper understanding of the effects of different processing approaches, the level of systematics in data are reduced, and more data can be processed to reduce noise. At this stage, it is difficult to predict whether systematics will remain at deeper levels, and if the fundamental limitations of the instrument will preclude a detection. While the reported detection of the Cosmic Dawn global signal from the EDGES experiment (Bowman et al. 2018) suggests that the spatial power spectrum amplitude may be larger than expected, this is highly uncertain, and the flexibility in possible strengths of the signal in the EoR emission part of the spectrum could help or hinder a detection by LOFAR and MWA. Pursuit of the Cosmic Dawn signal from 75–100 MHz observations with the MWA and LOFAR is also underway, but that introduces even greater challenges of large fields-of-view and poor extended source models at those frequencies.

8.6.2 Future Instruments

The SKA and HERA offer the future vision for EoR and Cosmic Dawn science. Like LOFAR and MWA, SKA is a general science instrument, being able to produce its own sky model and calibration framework, while needing to balance design with the other science aims of the observatory. HERA, like PAPER, is a custom EoR instrument, being able to design with a complete focus on EoR science, likely requiring external information to provide a full end-to-end calibration and source subtraction element.

The low-frequency telescope of the SKA Observatory, SKA-Low, will be centered at the Murchison Radioastronomy Observatory in Western Australia, on the same radio quiet site as MWA, ASKAP, EDGES and BiGHORNS (Koopmans et al. 2015; Dewdney et al. 2016). Despite being designed for 512 38 m stations (256 dual-polarization dipoles in each station) spread over >40 km, the core region will contain >200 stations within the central 1 km, with exceptional surface brightness sensitivity for EoR and CD science. With a frequency range available down to 50 MHz, the CD will be accessible to $z = 27$, with sub-stations able to be formed to produce the wider fields-of-view and shorter baselines required for early times. With its exceptional imaging capabilities, SKA-Low aims to pursue power spectrum, direct imaging (tomography) and 21 cm Forest studies. The prices to be paid for this highly-capable instrument are the complexity of the data and instrument, and the large data volumes that will be produced from the telescope, and is therefore faces a more severe version of the challenges currently experienced by multi-purpose dipole arrays such as MWA and LOFAR.

HERA (DeBoer et al. 2017) is a smaller instrument (although still significantly larger than any of the existing instruments) being constructed in South Africa. It comprises 350 14 m dipole elements spread over <1 km (331 in a 320 m core) for high EoR sensitivity and moderate imaging and calibration needs (19 outriggers). It will primarily pursue the statistical exploration of the EoR and CD using the delay spectrum technique, with some hope for imaging capability and alternate power spectrum analyses.

8.6.3 Future Analyses

Although the spatial power spectrum is the primary data product of most current EoR 21 cm experiments, there are other avenues of pursuit to explore this first billion years of the Universe, including an integrated product (the variance statistic, Patil et al. 2014). Direct imaging is beyond the capability of current instruments, demanding a high surface brightness sensitivity and thousands of hours. This will be pursued by the future SKA (Koopmans et al. 2015). However, there are other statistics that can be pursued through the 21 cm line, and also the opportunity for cross-correlating the signal with other tracers of early Universe evolution. The benefit of the latter approach is that the systematic errors may be different between the two tracers, offering an advantage over 21 cm alone.

At early times, the brightness temperature of the 21 cm line, relative to the CMB traces the matter power spectrum, and is highly Gaussian, but at later times the evolution of ionized bubbles dominates the spatial fluctuations and the signal is expected to have non-zero higher order terms (Furlanetto et al. 2006; McQuinn et al. 2006; Eisenstein & Hu 1999). The shape of the temperature distribution function evolves with time and spatial scale, and differs for different underlying models of the evolution of the Universe. As such, probing these non-Gaussian components can provide complementary information to the power spectrum, which, by design, only captures information in the second moment of the distribution (Wyithe & Morales 2007).

The bispectrum measures the three-point correlation function, and has been shown to encode non-Gaussianity. In early work to study the expected sensitivity of 21 cm experiments to the bispectrum, Yoshiura et al. (2015) computed theoretical expectations for a range of instruments, under the assumption of thermal noise only. More sophisticated recent work included the effects of calibration and foregrounds on the ability to detect the signal. In Trott et al. (2019), two bispectrum estimators were developed to take a practical approach to estimation with real data, and were applied to 20 h of data from Phase II of the MWA. This work discussed some of the advantages and challenges of doing such an experiment with real data.

Cross-correlation studies from the early Universe offer the potential for new astrophysical insight and reduced observational biases and errors. In the context of the MWA, Yoshiura et al. (2019) used data to explore the cross-correlation of the 21 cm image from the EoR-0 observing field, and the CMB field measured by Planck. An additional tracer that can be used is the population of high-redshift LAEs, which are observable in ionized regions (Yoshiura et al. 2018; Kubota et al. 2018; Hutter et al. 2017). The SKA's Synergy group is exploring the potential for multi-facility observations, including the exciting prospects available with WFIRST (Hutter et al. 2019).

References

Ali, S. S., Bharadwaj, S., & Chengalur, J. N. 2008, MNRAS, 385, 2166

Ali, Z. S., Parsons, A. R., Zheng, H., et al. 2015, ApJ, 809, 61

Asad, K. M. B., Koopmans, L. V. E., Jelić, V., et al. 2018, MNRAS, 476, 3051

Asad, K. M. B., Koopmans, L. V. E., Jelić, V., et al. 2016, MNRAS, 462, 4482

Asad, K. M. B., Koopmans, L. V. E., Jelić, V., et al. 2015, MNRAS, 451, 3709

Barry, N., Beardsley, A. P., Byrne, R., et al. 2019, PASA, 36, e026

Barry, N., Hazelton, B., Sullivan, I., Morales, M. F., & Pober, J. C. 2016, MNRAS, 461, 3135

Beardsley, A. P., Hazelton, B. J., Morales, M. F., et al. 2013, MNRAS, 429, L5

Beardsley, A. P., Hazelton, B. J., Sullivan, I. S., et al. 2016, ApJ, 833, 102

Bernardi, G., Greenhill, L. J., Mitchell, D. A., et al. 2013, ApJ, 771, 105

Bharadwaj, S., & Sethi, S. K. 2001, JApA, 22, 293

Bharadwaj, S., & Saiyad Ali, S. K. 2005, MNRAS, 356, 1519

Bharadwaj, S., Pal, S., Choudhuri, S., & Dutta, P. 2019, MNRAS, 483, 5694

Bowman, J. D., Cairns, I., Kaplan, D. L., et al. 2013, PASA, 30, 31

Bowman, J. D., Rogers, A. E. E., Monsalve, R. A., Mozdzen, T. J., & Mahesh, N. 2018, Natur, 555, 67

Bowman, J. D., Morales, M. F., & Hewitt, J. N. 2009, ApJ, 695, 183

Byrne, R., Morales, M. F., Hazelton, B., et al. 2019, ApJ, 875, 70

Carroll, P. A., Line, J., Morales, M. F., et al. 2016, MNRAS, 461, 4151

Chapman, E., Zaroubi, S., & Abdalla, F. B. 2016, MNRAS, 458, 2928

Chapman, E., Abdalla, F. B., Bobin, J., et al. 2013, MNRAS, 429, 165

Chapman, E., Abdalla, F. B., Harker, G., et al. 2012, MNRAS, 423, 2518

Cheng, C., Parsons, A. R., Kolopanis, M., et al. 2018, ApJ, 868, 26

Choudhuri, S., Bharadwaj, S., Ghosh, A., & Ali, S. S. 2014, MNRAS, 445, 4351

Choudhuri, S., Bharadwaj, S., Chatterjee, S., et al. 2016, MNRAS, 463, 4093

Ciardi, B., Labropoulos, P., Maselli, A., et al. 2013, MNRAS, 428, 1755

Datta, A., Bowman, J. D., & Carilli, C. L. 2010, ApJ, 724, 526

Datta, K. K., Roy Choudhury, T., & Bharadwaj, S. 2007, MNRAS, 378, 119

DeBoer, D. R., Parsons, A. R., Aguirre, J. E., et al. 2017, PASP, 129, 045001

Dewdney, P., Turner, W., Braun, R., et al. 2016, SKA System Baseline Design v2 SKA Document Series

Dillon, J. S., Liu, A., & Tegmark, M. 2013, PhRvD, 87, 043005

Dillon, J. S., Neben, A. R., Hewitt, J. N., et al. 2015, PhRvD, 91, 123011

Eisenstein, D. J., & Hu, W. 1999, ApJ, 511, 5

Ellingson, S. W., Clarke, T. E., Cohen, A., et al. 2009, IEEEP, 97, 1421

Ewall-Wice, A., Dillon, J. S., Hewitt, J. N., et al. 2016, MNRAS, 460, 4320

Ewall-Wice, A., Dillon, J. S., Liu, A., & Hewitt, J. 2017, MNRAS, 470, 1849

Furlanetto, S. R., Oh, S. P., & Briggs, F. H. 2006, PhR, 433, 181

Gehlot, B. K., Mertens, F. G., Koopmans, L. V. E., et al. 2019, MNRAS, 488, 4271

Ghosh, A., Bharadwaj, S., Saiyad Ali, S. K., & Chengalur, J. N. 2011a, MNRAS, 411, 2426

Ghosh, A., Bharadwaj, S., Saiyad Ali, S. K., & Chengalur, J. N. 2011b, MNRAS, 418, 2584

Ghosh, A., Koopmans, L. V. E., Chapman, E., & Jelić, V. 2015, MNRAS, 452, 1587

Ghosh, A., Prasad, J., Bharadwaj, S., Saiyad Ali, S. K., & Chengalur, J. N. 2012, MNRAS, 426, 3295

Greig, B., Mesinger, A., & Pober, J. C. 2016, MNRAS, 455, 4295

Harker, G., Zaroubi, S., Bernardi, G., et al. 2010, MNRAS, 405, 2492

Harker, G. J. A., Zaroubi, S., Thomas, R. M., et al. 2009, MNRAS, 393, 1449

Hurley-Walker, N., Callingham, J. R., Hancock, P. J., et al. 2017, MNRAS, 464, 1146

Hurley-Walker, N., Morgan, J., Wayth, R. B., et al. 2014, PASA, 31, e045

Hutter, A., Dayal, P., Malhotra, S., et al. 2019, BAAS, 51, 57

Hutter, A., Dayal, P., Müller, V., & Trott, C. M. 2017, ApJ, 836, 176

Jacobs, D. C., Hazelton, B. J., Trott, C. M., et al. 2016, ApJ, 825, 114

Jacobs, D. C., Aguirre, J. E., Parsons, A. R., et al. 2011, ApJ, 734, L34

Jacobs, D. C., Burba, J., Bowman, J. D., et al. 2017, PASP, 129, 035002

Jacobs, D. C., Pober, J. C., Parsons, A. R., et al. 2015, ApJ, 801, 51

Jelić, V., de Bruyn, A. G., Mevius, M., et al. 2014, A&A, 568, A101

Jelić, V., Zaroubi, S., Labropoulos, P., et al. 2008, MNRAS, 389, 1319

Jelić, V., Zaroubi, S., Labropoulos, P., et al. 2010, MNRAS, 409, 1647

Jordan, C. H., Murray, S., Trott, C. M., et al. 2017, MNRAS, 471, 3974

Joseph, R. C., Trott, C. M., & Wayth, R. B. 2018, AJ, 156, 285

Kazemi, S., Yatawatta, S., Zaroubi, S., et al. 2011, MNRAS, 414, 1656

Kerrigan, J., La Plante, P., Kohn, S., et al. 2019, MNRAS, 488, 2605

Kerrigan, J. R., Pober, J. C., Ali, Z. S., et al. 2018, ApJ, 864, 131

Kohn, S. A., Aguirre, J. E., Nunhokee, C. D., et al. 2016, ApJ, 823, 88

Koopmans, L., Pritchard, J., Mellema, G., et al. 2015, in Advancing Astrophysics with the Square Kilometre Array, PoS(AASKA14)001

Kubota, K., Yoshiura, S., Takahashi, K., et al. 2018, MNRAS, 479, 2754

Lanman, A., Hazelton, B., Jacobs, D., et al. 2019, JOSS, 4, 1234

Lenc, E., Anderson, C. S., Barry, N., et al. 2017, PASA, 34, e040

Lenc, E., Gaensler, B. M., Sun, X. H., et al. 2016, ApJ, 830, 38

Li, W., Pober, J. C., Hazelton, B. J., et al. 2018, ApJ, 863, 170

Line, J. L. B., McKinley, B., Rasti, J., et al. 2018, PASA, 35, 45

Line, J. L. B., Webster, R. L., Pindor, B., Mitchell, D. A., & Trott, C. M. 2017, PASA, 34, e003

Liu, A., & Tegmark, M. 2011, PhRvD, 83, 103006

Liu, A., Tegmark, M., Morrison, S., Lutomirski, A., & Zaldarriaga, M. 2010, MNRAS, 408, 1029

Martinot, Z. E., Aguirre, J. E., Kohn, S. A., & Washington, I. Q. 2018, ApJ, 869, 79

McQuinn, M., Zahn, O., Zaldarriaga, M., Hernquist, L., & Furlanetto, S. R. 2006, ApJ, 653, 815

Mertens, F. G., Ghosh, A., & Koopmans, L. V. E. 2018, MNRAS, 478, 3640

Mesinger, A., Furlanetto, S., & Cen, R. 2011, MNRAS, 411, 955

Mitchell, D. A., Greenhill, L. J., Wayth, R. B., et al. 2008, ISTSP, 2, 707

Moore, D. F., Aguirre, J. E., Parsons, A. R., Jacobs, D. C., & Pober, J. C. 2013, ApJ, 769, 154

Moore, D. F., Aguirre, J. E., Kohn, S. A., et al. 2017, ApJ, 836, 154

Morales, M. F., & Hewitt, J. 2004, ApJ, 615, 7

Morales, M. F., Beardsley, A., Pober, J., et al. 2019, MNRAS, 483, 2207

Mouri Sardarabadi, A., & Koopmans, L. V. E. 2019, MNRAS, 483, 5480

Murray, S. G., Trott, C. M., & Jordan, C. H. 2017, ApJ, 845, 7

Neben, A. R., Bradley, R. F., Hewitt, J. N., et al. 2015, RaSc, 50, 614

Neben, A. R., Bradley, R. F., Hewitt, J. N., et al. 2016, ApJ, 826, 199

Nunhokee, C. D., Bernardi, G., Kohn, S. A., et al. 2017, ApJ, 848, 47

Offringa, A. R., Mertens, F., & Koopmans, L. V. E. 2019, MNRAS, 484, 2866

Offringa, A. R., van de Gronde, J. J., & Roerdink, J. B. T. M. 2012, A&A, 539, A95

Paciga, G., Albert, J. G., Bandura, K., et al. 2013, MNRAS, 433, 639

Paciga, G., Chang, T.-C., Gupta, Y., et al. 2011, MNRAS, 413, 1174

Parsons, A., Pober, J., McQuinn, M., Jacobs, D., & Aguirre, J. 2012, ApJ, 753, 81

Parsons, A. R., Backer, D. C., Foster, G. S., et al. 2010, AJ, 139, 1468

Parsons, A. R., Liu, A., Aguirre, J. E., et al. 2014, ApJ, 788, 106

Parsons, A., Backer, D., Siemion, A., et al. 2008, PASP, 120, 1207

Parsons, A. R., Liu, A., Ali, Z. S., & Cheng, C. 2016, ApJ, 820, 51

Parsons, A. R., Pober, J. C., Aguirre, J. E., et al. 2012a, ApJ, 756, 165

Parsons, A. R., Pober, J. C., McQuinn, M., Jacobs, D., & Aguirre, J. 2012b, ApJ, 753, 81

Patil, A. H., Yatawatta, S., Koopmans, L. V. E., et al. 2017, ApJ, 838, 65

Patil, A. H., Yatawatta, S., Zaroubi, S., et al. 2016, MNRAS, 463, 4317

Patil, A. H., Zaroubi, S., Chapman, E., et al. 2014, MNRAS, 443, 1113

Pober, J. C., Liu, A., Dillon, J. S., et al. 2014, ApJ, 782, 66

Pober, J. C., Ali, Z. S., Parsons, A. R., et al. 2015, ApJ, 809, 62

Pober, J. C., Parsons, A. R., Aguirre, J. E., et al. 2013, ApJ, 768, L36

Pober, J. C., Parsons, A. R., Jacobs, D. C., et al. 2012, AJ, 143, 53

Procopio, P., Wayth, R. B., & Line, J. 2017, PASA, 34, e033

Rana, S., & Bagla, J. S. 2019, MNRAS, 485, 5891

Saiyad Ali, S. K., Bharadwaj, S., & Pandey, S. K. 2006, MNRAS, 366, 213

Stefan, I. I., Carilli, C. L., Green, D. A., et al. 2013, MNRAS, 432, 1285

Sullivan, I. S., Morales, M. F., & Hazelton, B. J. 2012, ApJ, 759, 17

Swarup, G., Ananthakrishnan, S., Kapahi, V. K., et al. 1991, CSci, 60, 95

Tegmark, M. 1997, PhRvD, 55, 5895

Thompson, A. R., Moran, J. M., & Swenson, G. W. 1986, Interferometry and Synthesis in Radio Astronomy (Berlin: Springer)

Thyagarajan, N., Jacobs, D. C., Bowman, J. D., et al. 2015, ApJ, 804, 14

Tingay, S. J., Goeke, R., Bowman, J. D., et al. 2013, PASA, 30, e007

Trott, C. M., Pindor, B., Procopio, P., et al. 2016, ApJ, 818, 139

Trott, C. M., & Wayth, R. B. 2016, PASA, 33, e019

Trott, C. M., Wayth, R. B., & Tingay, S. J. 2012, ApJ, 757, 101

Trott, C. M., de Lera Acedo, E., & Wayth, R. B. 2017, MNRAS, 470, 455

Trott, C. M., Fu, S. C., & Murray, S. G. 2019, MNRAS, 486, 5766

Trott, C. M., Jordan, C. H., & Murray, S. G. 2018, ApJ, 867, 15

Trott, C. M., Watkinson, C. A., & Jordan, C. H. 2019, PASA, 36, e023

Trott, C. M., & Wayth, R. B. 2017, PASA, 34, e061

van Haarlem, M. P., Wise, M. W., & Gunst, A. W. 2013, A&A, 556, A2

Wayth, R. B., Lenc, E., Bell, M. E., et al. 2015, PASA, 32, e025

Wayth, R. B., Tingay, S. J., Trott, C. M., et al. 2018, PASA, 35, e033

Wieringa, M. H. 1992, ExA, 2, 203

Wilensky, M. J., Morales, M. F., Hazelton, B. J., et al. 2019, PASP, 131, 114507

Williams, C. L., Hewitt, J. N., Levine, A. M., et al. 2012, ApJ, 755, 47

Wyithe, J. S. B., & Morales, M. F. 2007, MNRAS, 379, 1647

Yatawatta, S., de Bruyn, A. G., & Brentjens, M. A. 2013, A&A, 550, A136

Yoshiura, S., Ichiki, K., & Pindor, B. 2019, MNRAS, 483, 2697

Yoshiura, S., Line, J. L. B., Kubota, K., Hasegawa, K., & Takahashi, K. 2018, MNRAS, 479, 2767

Yoshiura, S., Shimabukuro, H., Takahashi, K., et al. 2015, MNRAS, 451, 266

Zaroubi, S., de Bruyn, A. G., Harker, G., et al. 2012, MNRAS, 425, 2964

Zheng, H., Tegmark, M., Buza, V., et al. 2014, MNRAS, 445, 1084

Zheng, Q., Wu, X.-P., Gu, J.-H., Wang, J., & Xu, H. 2012, ApJ, 758, L24

Chapter 9

Future Prospects

Léon V E Koopmans and Gianni Bernardi

This chapter addresses limitations to current 21 cm signal detection instruments, be it related to the instrument, environment, signal processing or science, and what lies beyond the current horizon for 21 cm science, especially in the 2030s and beyond. We address how to overcome current challenges and drive the field forward, not only approaching a detection of the 21 cm signal but to a full characterization of its parameter space, in particular, probing an increasingly larger volume of k modes (spatially and in redshifts). We also will shortly touch upon the kinds of questions that could drive such future endeavors.

9.1 What Drives Future 21 cm Signal Experiment?

The past two decades have witnessed exciting advancements in the field of 21 cm cosmology. Both theoretically and observationally, great progress has been made, although a convincing detection of the 21 cm signal still has to be achieved both for the globally averaged 21 cm signal as well as its spatial fluctuations. We have observed the construction and operation of a vast number of ground-based interferometers and single-element receivers, covering a wide range in terms of collecting area, core filling factor, field of view, frequency coverage, observational strategies (e.g., drift scan versus tracking), and receiver technology (phased-array versus dishes). Aside from these instrument and technology developments driving the field forward, enormous effort has been undertaken to develop much more refined flagging, calibration, imaging, foreground removal, and 21 cm signal extraction methodologies (e.g., Liu & Shaw 2019). These two tracks (instrument and signal processing) have gone hand in hand and have led to a steady progression and ever more stringent 21 cm signal limits (e.g., The Hydrogen Epoch of Reionization Array Collaboration 2019). The first confirmed 21 cm signal detection could be well in reach in the coming years.

One of the most exciting and hotly debated recent developments has been the announcement of the detection of the global 21 cm signal by the EDGES

doi:10.1088/2514-3433/ab4a73ch9

collaboration (Bowman et al. 2018). Although confirmation of this claim is still needed, it shows that astrophysical effects (e.g., bright polarized foregrounds, ionospheric refraction, radio-frequency interference (RFI) mitigation) and instrumental challenges (e.g., chromatic leakage, band-pass structure, multipath propagation, etc.) are controllable over nearly six orders of magnitude, and further improvements are still coming. These improvements in instrument design, layout, and interferometer versus single-receiver technology also inform each other, and hybrid systems that are currently under construction (e.g., LEDA, NenuFAR). Aside from these ongoing experiments and observational programs, the next generation of instruments, of which HERA and the Square Kilometre Array (SKA; Section 9.2.2 and 9.2.1) are the most significant proponents, is now underway. Whereas current instruments are limited to a "statistical detection" of the 21 cm signal power spectra (as opposed to the direct detection of the global signal; Mesinger et al. 2016), mostly limited to the Epoch of Reionization (EoR) due to the increasingly bright foregrounds and stronger ionospheric errors when moving to lower frequencies, these next-generation instruments aim not only to measure the 21 cm signal statistically but image it directly to the millikelvin level during the EoR and expand the redshift range to the Cosmic Dawn. This level of sensitivity requires a substantial increase in collecting area and filling factor (by a factor of about 10) over current instruments, thus stepping away from the experimental stage in which many active instruments find themselves. These new or upgraded systems are incorporating many of the lessons learned from past and ongoing efforts.

In this chapter, we will touch upon some of these ongoing and forthcoming developments, although we will not describe each system in extreme detail. Several are already ongoing in terms of extensions of current instruments. Finally, we briefly contemplate what lies beyond the current 2030 horizon, in particular instrumentation that can expand the currently envisioned science scope of the next-generation instruments and might require significantly larger collecting areas ($\gg 1$ km^2) and deployed in space (including the lunar environment or surface) to allow one to escape the limits set by the ionosphere and human-made interference at low frequencies. This would allow them to observe the holy grail of high-redshift 21 cm cosmology, the era called the Dark Ages, which allows a direct probe of questions posed by fundamental physics, inaccessible via any other way except through the 21 cm signal (Koopmans et al. 2019).

9.1.1 Limits of Current 21 cm Signal Observations

As shown in Mellema et al. (2013) and Koopmans et al. (2015), the core collecting area, its filling factor, and the field of view (FoV) of the instrument primarily drive the statistical sensitivity of an interferometer to the 21 cm signal power spectrum and its imaging capabilities. For direct imaging of the 21 cm signal or fluctuations on small scales, the FoV is less critical since cosmic variance might not be the driving factor. A limited FoV will increase the sample variance on the large scales, however, in power-spectrum measurements. Aside from the sensitivity to a given 21 cm signal mode in the presence of thermal noise, there are many other instrumental limitations. Some of these limitations can be kept under control to some extent but some cannot be, and thus have to be avoided (e.g., by choosing the proper location or controllable experimental setup) or mitigated (i.e., correct for errors in the data in real time or during post-processing). Below we summarize some of the issues that are currently considered as limiting 21 cm experiments from reaching the thermal noise:

- (a) Collecting area: although collecting area is one of the driving factors in sensitivity, it also drives hardware costs, especially if the receiver systems being correlated are small, and many are needed to reach the required collecting area. Lack of collecting area can also be compensated by increasing the FoV of a system such that more measurements are made by the same *uv* cells, and more *uv* cells are sampled, increasing power-spectrum sensitivity. This drives up correlator costs, however, so a careful balance needs to be struck between the difference requirements.

- (b) Filling factor: placing receivers in a smaller core area, even for a fixed collecting area and FoV, increases the sensitivity of the instrument for the simple reason that more independent visibilities are measured per *uv* cell (i.e., k mode). Because the 21 cm signal of interest is the same per visibility, whereas thermal noise is not, a higher filling factor rapidly increases the power-spectrum sensitivity.

- (c) Field of view: a larger FoV, in general, is related to a smaller receiver element and more elements to be correlated. For the aforementioned reasons, this increases the number of independent *uv* cells and independent 21 cm signal k modes, thereby decreasing power-spectrum errors.

- (d) Frequency coverage: whereas maximizing the frequency coverage from the EoR to the Cosmic Dawn and even into the Dark Ages (i.e., $z \sim 6\text{--}200$) would be optimal, in practice this frequency range is split up in smaller bands. These bands are generally limited in their spectral resolution: some channels are lost to RFI (e.g., FM band, DAB/DVB; see below) and in some cases are not even covered (e.g., below the ionospheric cutoff, which limits observations of the Dark Ages). These instrumental and environmental effects have led to the development of new wideband receivers, the deployment of instruments in remote locations to avoid RFI, or even in space, where they are not affected by the ionosphere (see below).

- (e) *uv* coverage: this has already been discussed above, and in general is driven by the number of correlated receivers and the density of the core. The limiting factor is often the cost of the electronics, the correlators, and data storage. However, in some cases longer baselines are necessary to calibrate the instruments in case it is a highly nonredundant array.

- (f) (Polarized) Foregrounds: Foreground emission is a significant complicating factor in a 21 cm signal experiment, not only because they are bright but also because they are partly polarized and have spatial structure. These two effects couple to the ionosphere and the chromatic and polarized nature of the instrument itself even in the absence of any errors, and cause leakage terms from the foregrounds into the 21 cm signal;

- (g) Instrumental effects: instruments are not perfect. Receivers need amplifiers to increase the weak electromagnetic signals into a measurable voltage that can be digitized. These low-noise amplifiers are not 100% stable and can cause both amplitude and phase errors in the visibility data after correlation. When various receivers and amplifiers take part in station beamforming, these errors become direction dependent. Second, cross-dipole receivers or receivers that measure circular polarization partly mix Stokes I, Q, U, and V power. The reason is that antennas see radiation coming from different projections of the sky, leading to instrumental polarization. So, even if the instrument is nearly perfect, these effects are impossible to avoid, and if the sky is partly polarized, this power contaminates the 21 cm power spectrum.
- (h) Signal processing: one final difficulty is inherent to the data processing. Because processing affects the data via RFI excision, gain calibration, fore-ground removal, etc., it can both remove and enhance signals of interest from the data, including the 21 cm signal itself. Processing, if not done very precisely and accurately, can thus lead to a severe 21 cm signal bias (in general suppression).

Aside from these limiting factors, mostly instrument related, two crucial factors are harder to control and will ultimately drive the designs of large-scale future low-frequency instruments toward space.

(1) **Radio frequency interference:** RFI is becoming an increasingly more critical problem for low-frequency telescopes, as human-made signals from, e.g., transmitters, vehicles, mobile phones, satellites, and airplanes are now occupying many of the frequency bands that were previously clean (Offringa et al. 2013, 2015). This increasing RFI occupancy motivates the next generation of instruments to be built in remote desert environments such as the Karoo in South Africa and Western Australia. Going to space is another, albeit expensive, option. Just above Earth, however, any receiver would see a much larger number of transmitters, making the problem worse. Even at the distance of the Moon, an RFI suppression of eight orders of magnitude in power would be needed to mitigate it to a level that the 21 cm signal from the Cosmic Dawn can be observed. In a lunar orbit, the Moon would shield the receivers from Earth and also from solar radiation for a fraction of the time, creating an "RFI-free" cone. A receiver on the far side of the Moon itself might be shielded even more. Future low-frequency instruments will likely be in space to at least partly mitigate the worsening RFI situation on Earth.

(2) **Ionosphere:** the ionosphere causes both phase and amplitude fluctuations in the received electromagnetic signal, which increase in strength toward lower frequencies and is maximal near the plasma frequency cutoff of the iono-sphere (around 5–10 MHz). The ionosphere has restricted most instruments to frequencies above 30–50 MHz, up to the Cosmic Dawn. Observations of the Dark Ages will be extremely hard, especially in the presence of very bright foreground that couple to the ionosphere (Vedantham & Koopmans 2015, 2016). Reaching the Dark Ages 21 cm signal, therefore, requires space-based instruments.

Whereas these technical and environmental reasons will ultimately drive instruments to become ever larger and be built on remote places, or even in space, what do we hope such future instruments will tell us?

9.1.2 What Will Drive Future 21 cm Experiments?

Future experiments will primarily be driven by increasing sensitivity in 21 cm signal regimes already explored, and also by exploring new regimes in redshift and spatial scales. The former has fundamentally driven the development of the high-sensitivity SKA and HERA telescopes, with their distinct approach in sensitivity increase in different parameter space regions (see Section 9.2). Below we will briefly discuss where future instruments will represent a leap forward with respect to current instruments (assuming the latter will detect the 21 cm signal in the coming years).

- **Smaller spatial scales:** Most current instruments are limited to a rather small range (less than one decade) of power-spectrum k modes, where sufficient signal-to-noise can be reached for a detection. The reason is that shorter baselines add coherently for a more extended integration time per uv cell, and these are most sensitive to the larger spatial cases (except in the frequency direction). To observe larger k modes (or smaller three-dimensional spatial scales), instruments in general need much more collecting area, one of the SKA drivers. HERA aims to reach higher sensitivity by increasing its collecting area, its FoV, and its the filling factor, but at the cost of having redundant uv sampling. The latter makes direct imaging much harder, especially on smaller spatial scales, also posing a limitation to calibration strategies.

- **Direct imaging:** Direct imaging of the 21 cm signal becomes possible with larger collecting area and higher sensitivity on all spatial scales. One of the highest-priority science drivers for the SKA is, therefore, the direct imaging of the 21 cm signal throughout the EoR on scales larger than 10' (see Section 9.2.1). Ionized bubbles can be imaged on even smaller scales as their contrast with the globally averaged 21 cm signal is very large (about 30 mK versus fluctuations of only a few millikelvin around the bubbles).

- **Higher redshifts:** A third driver behind the sensitivity increase is that it enables observations at even higher redshifts, e.g., in the Comic Dawn or even the Dark Ages. At higher redshift, the foregrounds are much brighter, increasing the overall system temperature and hence the thermal noise. Integration times therefore rapidly increase with redshift, and observations of the Cosmic Dawn are the territory of the upcoming interferometric arrays.

In short, whereas present-day instruments aim for the first detection at several (lower) redshifts and over a limited range of spatial scales, the next generation of instruments will aim to expand these parameter spaces, but also do direct imaging rather than summarizing the signal in some statistics (e.g., the power spectrum). In Section 9.2, these next-generation instruments will be discussed, where we not only limit ourselves to SKA and HERA but also briefly touch upon extensions to current instruments and their main science drivers.

9.2 Ground-based Interferometers

In this section, we review the status of those 21 cm ground-based interferometers that are under construction, have been upgraded, or will be constructed shortly.

9.2.1 The Square Kilometre Array—SKA1&2

The SKA is a global effort by a consortium of member-state countries[1] that will consist of at least two entirely different arrays: a midfrequency array (SKA-mid) to be built in the Karoo desert of South Africa, and a low-frequency array (SKA-low) to be built in Western Australia, both located in radio-quiet zones where human-made RFI is very limited. Because only SKA-low will cover the redshifts of the Cosmic Dawn and EoR (50–200 MHz, or $6 < z < 27.4$), here, we focus only on this instrument. SKA-low has a design similar to LOFAR (Mellema et al. 2013; Koopmans et al. 2015), but has some distinct differences as well, primarily related to the receiver design. SKA-low aims to have 512 stations in Phase 1 (denoted here by SKA1-low), having 256 log-periodic cross-dipole receivers that cover the full frequency band and are semirandomly spread inside a 40 m diameter circle. The requirement of a high-gain receiver over the full spectral band limits the FoV to a cone with an opening angle of about 90° centered on the zenith, but one that maximizes forward gain. The current receiver design also aims for a spectrally smooth band pass, something rather difficult to realize for a wide-field and wide-band receiver (de Lera Acedo et al. 2017, 2018). About 212 stations will be placed inside a central core of about 600 m (Figure 9.1), making it about eight times more sensitive than LOFAR. The remaining stations will be distributed along three arms that "spiral" outward up to about 65 km, in the current design. The long baselines will enable proper direction-dependent gain calibration of the system. The SKA observational and calibration strategy leverages minimal redundancy to reduce the point-spread function side lobes, improve imaging fidelity and sky-based calibration capabilities.

The sheer collecting area (0.4 km^2) and instantaneous (300 MHz, or 150 MHz when split in dual-beam mode) bandwidth make SKA1-low the premier instrument in the late 2020s to directly image the 21 cm signal during the EoR from $6 < z < 12$, covering the 21 cm signal power spectrum in the range of roughly $0.02 < k < 1$ cMpc^{-1} (depending on redshift), and push power-spectrum measurements over more limited k ranges deep into the Cosmic Dawn, up to $z \sim 20$ (or even more). Observations of individual ionized structures will enable cross-correlations with many other instruments that aim to look for the sources of reionization (e.g., *JWST*, ALMA, SPICA). Figure 9.2 shows an example of how well SKA-low will constrain reionization parameters (Greig et al. 2020). Currently, SKA is planned to be operational around 2028, although early science is foreseen several years before that. Finally, ideas for a far-future ($\gg$2030) upgrade to SKA2-low have already been developed (e.g., Koopmans et al. 2015), which nominally foresees an increase in collecting area by a

---

[1] http://www.skatelescope.org.

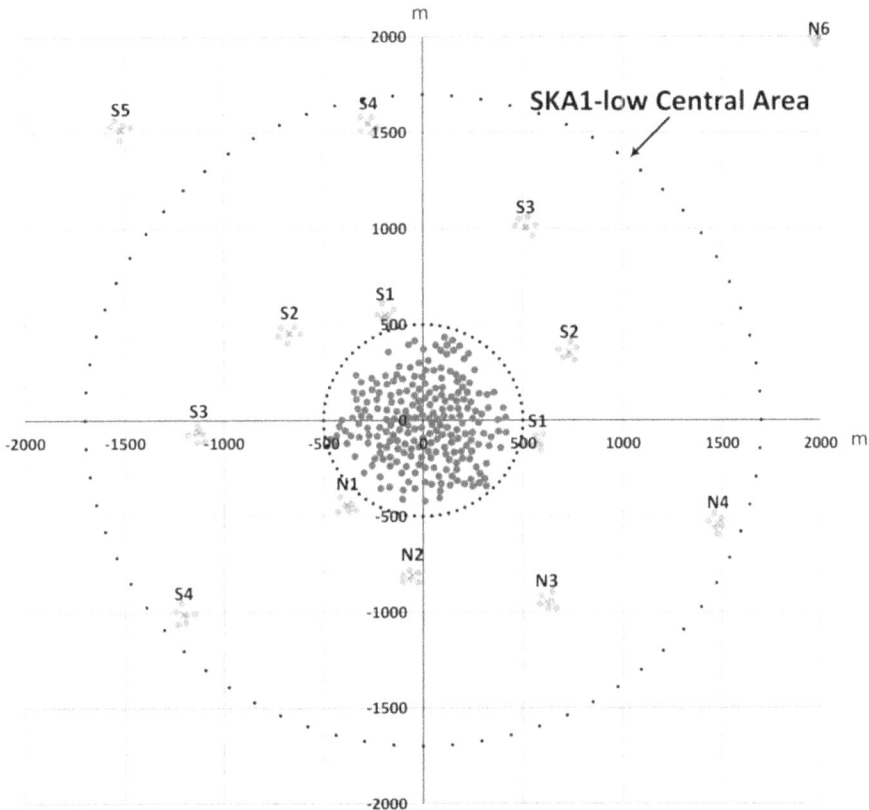

Figure 9.1. SKA1-low core layout, showing the station positions in the inner few kilometers, which are most sensitive to the 21 cm signal (Dewdney et al. 2009). Reproduced from a SKA internal memo. © 2016 SKA Organisation. All rights reserved.

factor of about four. This could allow for more detailed imaging due to lower thermal noise and increase the angular resolution also at higher redshifts.

9.2.2 The Hydrogen Epoch of Reionization Array—HERA

The Hydrogen Epoch of Reionization Array (HERA; DeBoer et al. 2017) is an array currently under construction in the Karoo reserve area in South Africa, following the decommissioning of the PAPER experiment (see Chapters 5 and 8 in this book for an overview of PAPER). HERA is built following the approach used for PAPER: a highly redundant array to maximize the sensitivity on a number of power-spectrum modes measured using the avoidance approach. In order to increase the sensitivity with respect to PAPER, it employs 14 m diameter nonsteerable dishes that, in the final configuration, will be densely packed in a highly redundant hexagonal array configuration of ~350 m diameter (see Figure 9.3). HERA's main goal is to measure the 21 cm power spectrum in the $6 < z < 12$ range with high significance in the $0.2 < k < 0.4$ Mpc^{-1} range (Pober et al. 2014), providing a full characterization of the

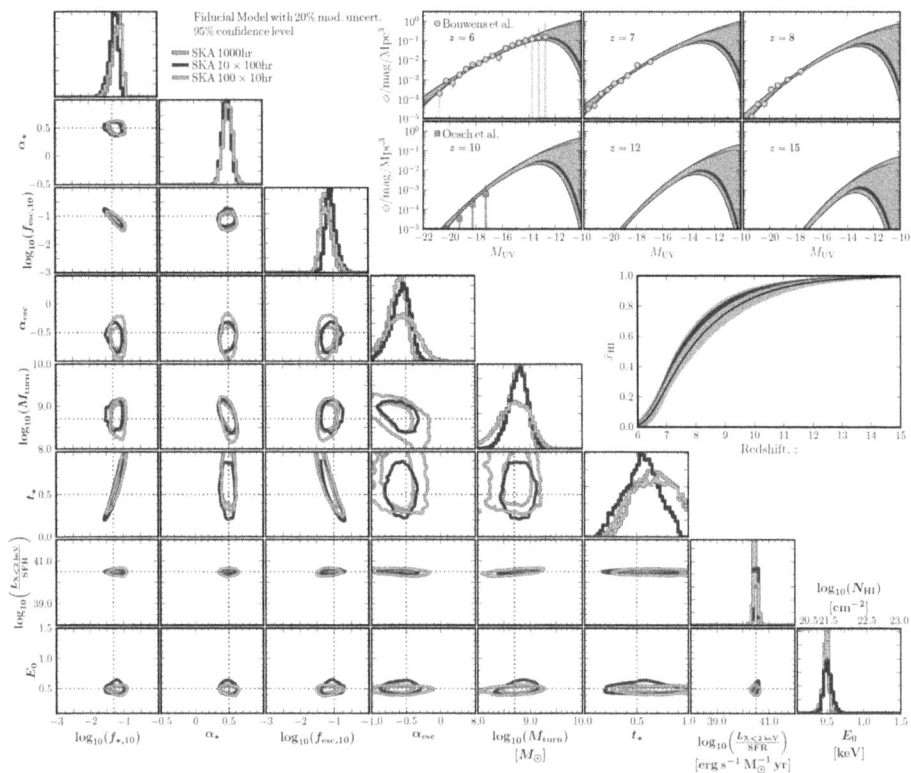

Figure 9.2. Astrophysical parameter constraints for a standard 21 cm signal model using 1000 hr of observations with SKA1-low (reproduced from Greig et al. 2020, by permission of Oxford University Press on behalf of the Royal Astronomical Society).

evolution of the neutral hydrogen fraction of the intergalactic medium (Figure 9.4). Given the highly redundant configuration, imaging tomography will remain challenging for HERA and likely the goal of a future-generation experiment. As foreground modeling and characterization will also be limited because of redundancy and the coarse angular resolution, significant effort was dedicated to keep the instrumental response from corrupting the intrinsically smooth foreground spectra and to accurately model it (Neben et al. 2016; Ewall-Wice et al. 2016; Thyagarajan et al. 2016; Patra et al. 2018). An alternative approach to redundant calibration is to apply foreground avoidance using closure phase quantities from antenna triads (Thyagarajan et al. 2018): closure phases are insensitive to errors in direction-independent interferometric calibration and, therefore, directly bypass the requirement of an accurate spectral calibration (see Chapter 5 in this book for an overview of calibration of 21 cm observations). A preliminary analysis of HERA closure phases seems to confirm these premises (Carilli et al. 2018). HERA is currently under construction, with more than 200 dishes deployed, and 21 cm observations are currently being analyzed. New feeds that extend the sensitivity to the 50–250 MHz range are currently deployed for testing in order to enable observations in the

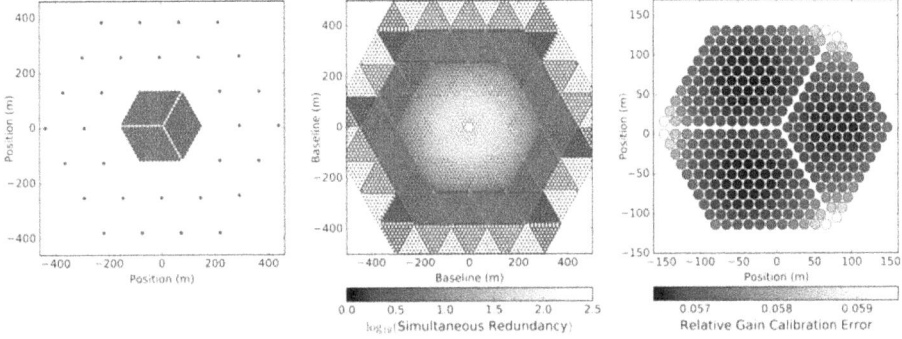

Figure 9.3. The HERA layout (left panel): 320 dishes are located in the hexagonal core and 30 more outrigger dishes are planned to be deployed out to a maximum baseline of ~800 m to improve angular resolution and imaging capabilities. The core is split into three sectors that are displaced from each other by a fraction of the dish diameter (see Dillon & Parsons 2016 for a detailed discussion). The split core provides a significantly improved instantaneous *uv* coverage (central panel) while retaining high redundancy. The right panel shows the expected relative antenna gain errors after using redundant calibration (from Dillon & Parsons 2016). Reproduced from DeBoer et al. (2017), © 2017. The Astronomical Society of the Pacific. All rights reserved. CC BY 3.0.

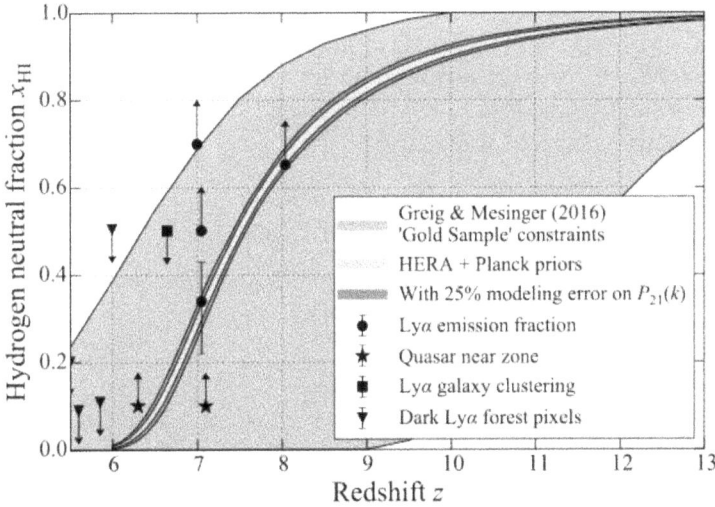

Figure 9.4. 95% confidence region on the hydrogen neutral fraction X_{HI} (gray; from Greig & Mesinger 2017). The inclusion of HERA measurements leads to a dramatic improvement in the constraints (red and pink areas; Liu & Parsons 2016). Constraints from other reionization probes are shown as well (see DeBoer et al. 2017 for a detailed description, © 2017. The Astronomical Society of the Pacific. All rights reserved. CC BY 3.0).

$12 < z < 35$ range (the Cosmic Dawn) and probe the nature of the first luminous sources and their impact on the thermal history of the intergalactic medium.

9.2.3 The Large Aperture Experiment to Detect the Dark Ages—LEDA

The Large aperture Experiment to detect the Dark Ages (LEDA; Bernardi et al. 2015; Kocz et al. 2015) is located at the Owens Valley Radio Observatory,

California. It operates in the 30–88 MHz frequency range, corresponding to $15 < z < 46$, seeking to detect the 21 cm signal from the Cosmic Dawn. The array layout consists of 251 dipoles pseudo-randomly deployed within a 200 m diameter core, and 23 dipoles are added out to a maximum 1.5 km baseline (see Figure 9.5). Five additional outrigger dipoles are custom-equipped to measure the global 21 cm signal via individual custom-built dipoles (see Section 9.3.2). The very dense core provides exceptional brightness sensitivity and a point-spread function with very low side lobes. The outrigger dipoles improve the angular resolution that helps to identify calibration sources and lower the confusion level. As the dipoles are individually correlated, visibilities have contributions from all-sky emission, particularly from Galactic diffuse emission—given the number of short baselines—and with significant ionospheric-induced refraction and scintillation. Despite these challenges, Eastwood et al. (2018) generated the first high-quality all-sky foreground maps. The LEDA approach to measure the 21 cm signal can be versatile, allowing foregrounds to be imaged and subtracted (Eastwood et al. 2018), but also to be avoided (similar to Beardsley et al. 2016). Eastwood et al. (2019) analyzed 20 hr of LEDA data calibrated using a compact source sky model and filtering foregrounds by using their statistical properties in a way similar to Dillon et al. (2014) and Trott et al. (2016). They reported an initial 10^8 mK^2 upper limit on the 21 cm power spectrum at $z = 18.4$. Several hundreds of hours of observations have been collected now and will be the focus of future analysis toward the detection of the power spectrum from the Cosmic Dawn and an independent confirmation of the reported detection by Bowman et al. (2018).

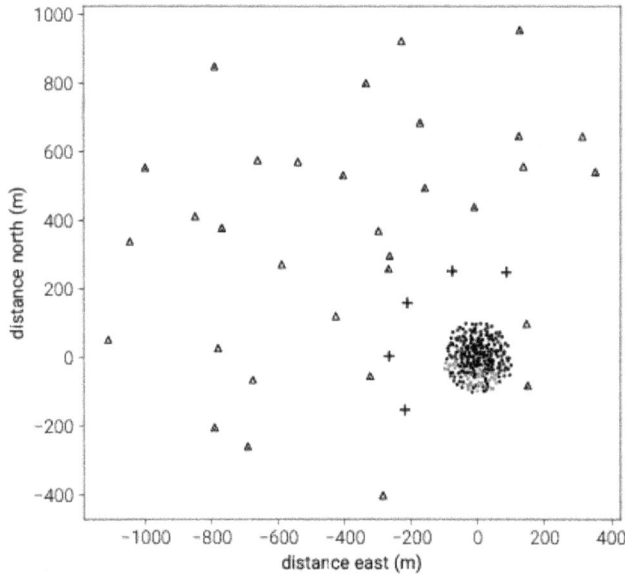

Figure 9.5. LEDA antenna layout: the dense core is surrounded by 32 dipoles in order to provide an exceptionally good instantaneous *uv* coverage (reproduced from Eastwood et al. 2018 © 2018. The American Astronomical Society. All rights reserved).

9.2.4 The Low Frequency Array 2.0—LOFAR2.0

LOFAR (van Haarlem et al. 2013) is already one of the most sensitive arrays to detect the 21 cm signal during the EoR, although its sensitivity is still limited in the Cosmic Dawn redshift/frequency range. This is due to the limited effective low-band antenna (LBA) collecting area and to the system temperature at low frequencies being much larger. Furthermore, the LBA dipoles are rather narrowband, and the gains sharply peak around 60 MHz, dropping rapidly at frequencies away from the resonance. LOFAR will undergo an upgrade in the coming years in order to improve its sensitivity and capability to calibrate the ionospheric distortions. Sensitivity will be increased by connecting the 48 dipoles in each LBA station that are not connected to the acquisition system because of budget limitations.

Moreover, LOFAR2 will enable simultaneous observations with the HBA and LBA systems, such that the more sensitive HBA system can be used to gain-calibrate the system, including (direction-dependent) ionospheric corrections. Finally, LOFAR2 will, in a later stage, also enable HBA observations with a dual analog tile-beam formation, enabling multiple target fields anywhere on the visible sky (not just limited to multiple beams inside the HBA tile beam, as is currently the case). Each of these upgrades improves the thermal-noise sensitivity to the 21 cm signal and the system calibratability. The first step in this process was recently taken with the installation of a GPU-based correlator.

- **Amsterdam-ASTRON Radio Transients Facility And Analysis Center— AARTFAAC**. Whereas LOFAR mainly operates in beam-formed mode, where dipoles or tiles are phased up in a given direction, the AARTFAAC system[2] currently enables all 576 LBA or HBA dipoles/tiles of the inner 12 stations to be cross-correlated, using two physical correlators, although currently only over a very limited 3.1 MHz bandwidth with 60 kHz resolution. This operational mode increases the FoV by a factor of about 25 for the HBA system and to all-sky for the LBA system (Figure 9.6), improving the power-spectrum sensitivity by a factor of about five or more per unit bandwidth (because both the collecting area and filling factor remain similar and long baselines do not add sensitivity to the 21 cm signal). AARTFAAC is currently already being used to target, for example, the 21 cm signal in the Cosmic Dawn with the goal to interferometrically confirm the recently reported global signal detection (Bowman et al. 2018). A system upgrade where all dipoles/tiles are fully correlated for 24 stations is envisioned over the full LBA and/or HBA bandwidth.

9.2.5 The Murchison Widefield Array Phase II

The Murchison Widefield Array (MWA) is located in Western Australia, operating in the 80–200 MHz range. In its initial phase (phase I; Tingay et al. 2013), it consists of 128 stations deployed in a pseudo-random configuration out to a ~3 km baseline. Each station includes 16 bow-tie dipoles arranged in a regular 4 × 4 square grid, 5 m

---

[2] http://www.aartfaac.org.

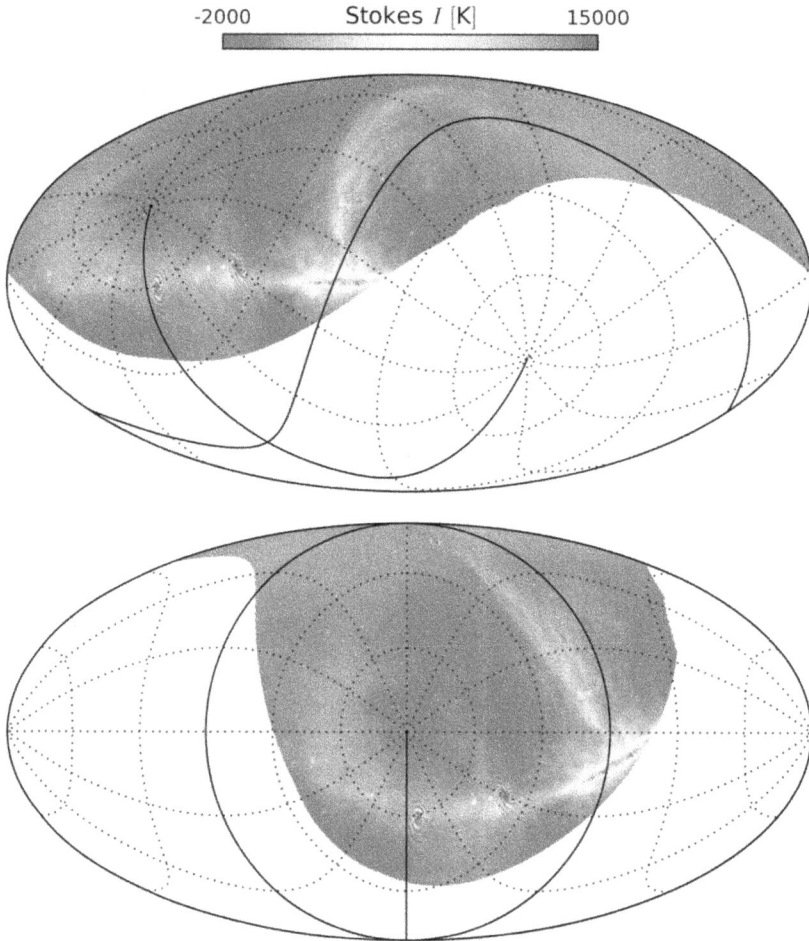

Figure 9.6. The first all-sky image of the northern sky taken with the AARTFAAC-LBA system, using 576 dipoles on the inner 12 LOFAR-LBA stations.

wide. Like LOFAR, MWA is a general purpose instrument, although one of its main science drivers is the measurement of the 21 cm signal from the EoR. It recently underwent an expansion (termed "phase II"; Wayth et al. 2018) where the number of stations was doubled: 72 new stations were placed in two highly redundant hexagons next to the array center and 56 stations outside the phase I array to extend the maximum baseline up ~5 km. Only 128 stations can still be instantaneously correlated, resulting in two different configurations: a compact configuration that includes the two redundant hexagons and the phase I compact core, and an extended configuration that excludes the two redundant hexagons and includes stations out to the longest baselines (see Chapter 7 in this book for an illustration of both configurations).

The MWA phase II array is therefore a fairly flexible instrument: its compact, redundant configuration is optimized for EoR power-spectrum observations

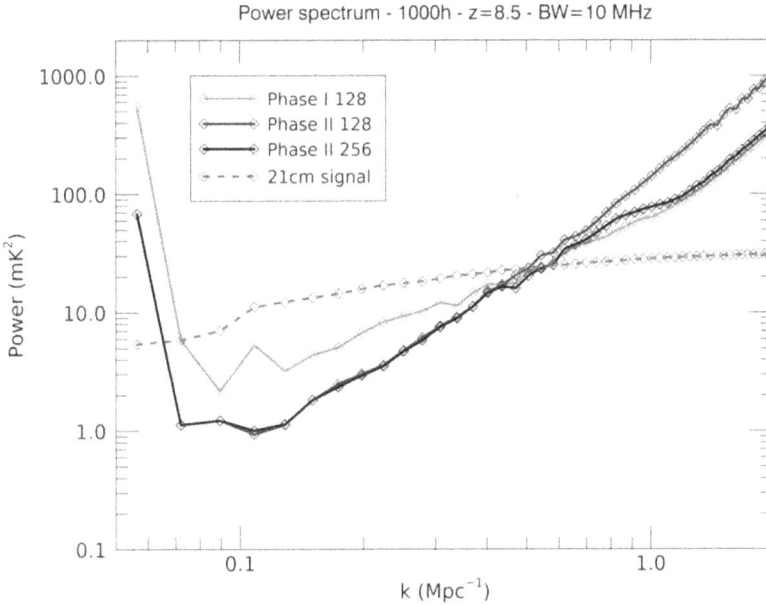

Figure 9.7. Fiducial 21 cm power-spectrum model at $z = 8.5$ with associated noise levels from the phase I and phase II arrays with a 1000 hr observation. "Phase II 256" shows the result from a future MWA upgrade where all 256 tiles are correlated simultaneously (reproduced with permission from Wayth et al. 2018).

following a strategy similar to HERA, i.e., leveraging upon redundant calibration schemes (Li et al. 2018) and improving the sensitivity by a factor of 4 with respect to phase I, leading to a $\sim 10\sigma$ detection of the fiducial 21 cm power spectrum at $k \sim 0.1$ Mpc^{-1} (Figure 9.7).

In its extended configuration, it has an exceptionally good instantaneous uv coverage (due to the high number of stations instantaneously correlated) with improved foreground modeling, thanks to the increased angular resolution from phase I.

9.2.6 New Extension in Nançay Upgrading LOFAR—NenuFAR

Another novel array currently in its rollout and early-science phase is NenuFAR (Figure 9.8; Acero et al. 2017). Whereas initially envisioned as an extremely sensitive beam-formed system operating in the (10)30–85 MHz band, the development of a relatively cheap GPU-based correlator for LOFAR will enable NenuFAR to correlate all envisioned 96 mini arrays, each consisting of 19 LEDA-like dipoles, over the full frequency band. The FoV of NenuFAR is about 20° at 60 MHz, and, inside the 400 m core, the filling factor reaches order unity at 35 MHz, and ~0.25 at 60 MHz, which makes it extremely sensitive to low-surface brightness structures. Currently, 56 stations are in place inside a core of about 400 m diameter. By late 2019, however, 80 stations will be in place, six of which will be placed farther out over a ~2.5 km diameter area. The correlator is already being installed and will also be operational by late 2019. The final goal is to have 96 mini arrays in place,

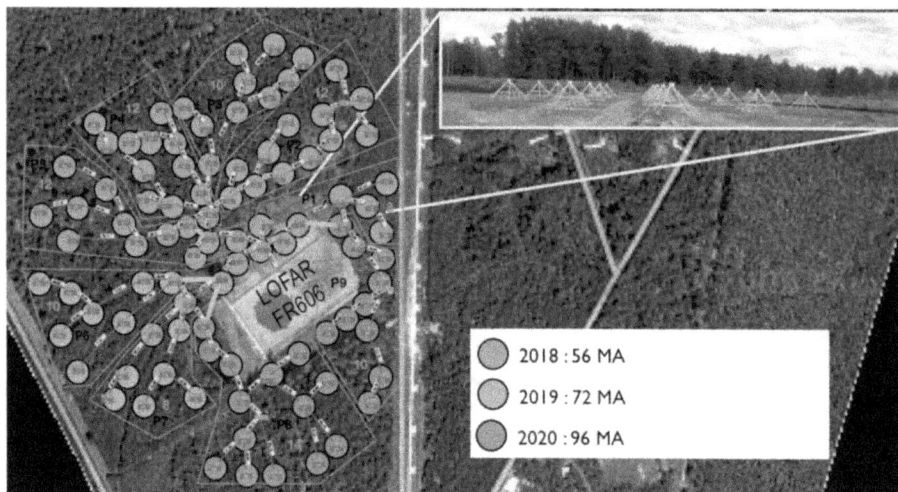

Figure 9.8. The current and planned layout of NenuFAR in Nançay, France. By late 2019, around 80 mini arrays (stations) will be in place and operating, to be expanded to the full 96 stations in 2020 (and beyond). A new GPU-based correlator will also be installed (Acero et al. 2017).

enabling maximum use of the system. This will make NenuFAR one of the most sensitive 21 cm signal arrays in the world, in principle able to reach the standard predicted 21 cm signal in the Cosmic Dawn redshift range (Mesinger et al. 2016).

9.3 Global Signal Experiments

In this section, we briefly review the status of ongoing global signal experiments.

9.3.1 The Experiment to Detect the Global EoR Signature—EDGES

The Experiment to Detect the Global EoR Signature (EDGES; Bowman et al. 2008) currently operates in two frequency bands: the 90–200 MHz (high) band in order to constrain the evolution of the neutral fraction throughout reionization, and the 50–100 MHz (low) band, in order to measure the expected heating of the intergalactic medium from primordial sources. The EDGES experiment has been pioneering techniques to accurately model all of the various instrumental components in order to carefully control systematic effects (Bowman et al. 2008; Monsalve et al. 2017a). Observations in the high band have constrained the duration of reionization Δz to be longer than $\Delta z > 1$ and started to constrain some properties of the first galaxies (Monsalve et al. 2017b, 2018). In the low band, Bowman et al. (2018) reported the surprising detection of an absorption trough that is twice deeper than the most extreme models, posing a serious challenge to its interpretation— assuming it is of cosmological origin. In light of this anomalous signal, the EDGES team is deploying a new dipole antenna tuned in size to simultaneously observe the 60–160 MHz range (i.e., $\sim$25% smaller than the low band antenna) and confirm the results in the low band. A further upgrade of the EDGES experiment with a more portable antenna that includes the electronics is under consideration for deployment in a quiet radio frequency environment in Oregon, USA.

9.3.2 The Large Aperture Experiment to Detect the Dark Ages—LEDA (Global Signal)

As mentioned in Section 9.2.3, LEDA includes a few custom-equipped dipoles to measure the global signal (Price et al. 2018). Initial observations were used to validate the end-to-end acquisition system and data analysis, leading to an 890 mK upper limit on the global signal amplitude in the $13.2 < z < 27.4$ range at the 95% confidence level (Bernardi et al. 2016). A series of upgrades has been implemented since the early system: filters with a sharper roll-off were installed in order to improve RFI rejection and extend the observing band up to 87.5 MHz; the noise diode stability was improved and a system to measure the ambient temperature was installed on the dipoles. The receiver seems to show the necessary stability to measure the global signal; however, other sources of systematic effects related to the antenna gain pattern remain less well known and are the subject of ongoing modeling and investigation. About 100 hr of observations were taken with the upgraded system and are currently being analyzed.

9.3.3 Shaped Antennas to Measure the Background Radio Spectrum—SARAS

The Shaped Antennas to measure the background RAdio Spectrum (SARAS) represent a progression of radiometers developed over the last decade at the Raman Research Institute and optimized to detect the global 21 cm signal in the 50–200 MHz range, i.e., in the Cosmic Dawn and the EoR. The SARAS antennas have been designed to provide nearly frequency-independent beams and avoid coupling of sky spatial structures into spectral structures in order to preserve the intrinsically smooth foreground spectrum (Sathyanarayana Rao et al. 2017). Initially, SARAS featured a fat-dipole antenna (Patra et al. 2013) that was later replaced by a shaped monopole antenna (SARAS 2, Singh et al. 2018). SARAS 2 was deployed in the radio-quiet Timbaktu Collective in Southern India, observing in the 110–200 MHz. SARAS 2 results disfavored models with inefficient heating of the intergalactic medium and rapid reionization (Singh et al. 2017, 2018). A new-generation experiment, SARAS 3, exploits a refined design to further reject spurious foreground structures and have control over systematics in order to target the 21 cm signal in the 50–100 MHz band.

9.4 Space-based Instruments

Whereas tremendous progress is being made from the ground to detect the globally averaged and spatially fluctuating 21 cm signal during the EoR and Cosmic Dawn, as discussed earlier, the stability of the system, RFI, the ionosphere, and even multipath propagation effects make ground-based observations hard and in some cases, such as a detection of the Dark Ages, even impossible. These motivations have been driving concepts and plans for space-based instrumentation.

9.4.1 The *Dark Ages Polarimetry Pathfinder—DAPPER*

The *Dark Ages Polarimetry Pathfinder* (*DAPPER*; Burns et al. 2019) is a space satellite that is intended to observe the global signal from a ~50,000 km lunar orbit, one of the quietest radio frequency environments, with an expected 26 month lifetime. Its goal is to observe the global signal absorption trough expected at $17 < \nu < 38$ ($83 > z > 36$), e.g., in the Dark Ages, well before the formation of the first luminous sources. In this epoch, the global signal is determined by linear perturbation theory, uncontaminated by complex astrophysical processes. *DAPPER* is expected to characterize the predicted signal, including any deviation that may be due to the additional cooling reported by Bowman et al. (2018). Its strategy includes the use of a polarimeter to measure the polarization induced by the anisotropic foregrounds; a large antenna beam to aid the foreground separation from the isotropic, unpolarized global signal (Nhan et al. 2017); and a pattern recognition data analysis that is trained on realistic smulations of observed foregrounds, instrument systematics, and the expected global signal (Tauscher et al. 2018). *DAPPER* is one of nine small satellite missions selected by NASA to be further studied for a possible launch in the next decade.

9.4.2 *Discovering the Sky at the Longest Wavelengths—DSL*

The *Discovering the Sky at the Longest Wavelengths* (*DSL*; Chen et al. 2019) is a mission concept that explores the possibility of deploying a constellation of micro-satellites circling the Moon on nearly identical orbits, performing interferometric observations of the sky below 30 MHz. Although its sensitivity is insufficient to detect 21 cm fluctuations from the Dark Ages, its goals will be to accurately image 21 cm foregrounds and to target the 21 cm global background using a calibrated single antenna. The current *DSL* concept includes a larger "mother" satellite that leads or trails five to eight smaller daughter satellites that carry out the radio observations and pass the data to the mother satellite through a microwave link. The mother performs the cross-correlation and handles communications with Earth. The *DSL* project is now undergoing a prototype study.

9.4.3 *Farside Array for Radio Science Investigations of the Dark Ages and Exoplanets*—FARSIDE

The *Farside Array for Radio Science Investigations of the Dark ages and Exoplanets* (*FARSIDE*; Burns et al. 2019) is a mission concept to place an interferometric array on the far side of the Moon, which offers complete isolation from terrestrial RFI and solar wind, allowing observations at sub-millihertz frequencies. The array would consist of 128 dual polarization antennas deployed across a 10 km area by a rover, observing in the 0.1–100 MHz (basically $z > 13$) range. *FARSIDE* would also include precision calibration of an individual antenna element via an orbiting beacon in order to attempt the detection of the global 21 cm from the Dark Ages ($50 < z < 100$). A NASA-funded design study, focused on the instrument, a deployment rover, the lander, and base station, delivered an architecture broadly consistent with the requirements for a probe mission (about 1.3 billion USD).

9.4.4 Netherlands–China Low Frequency Explorer—NCLE

The Netherlands–China Low frequency Explorer (NCLE) is a radio instrument payload on board on the Chinese *Queqiao* relay satellite that orbits behind the Moon. NCLE is designed, built, and tested by a Dutch consortium comprised of the Radboud University, ASTRON, and ISIS, in close collaboration with the National Astronomical Observatories of the Chinese Academy of Sciences. It is composed of three, 5 m long, carbon-fiber monopole antenna units that can be switched into dipole mode to observe in the 0.08–80 MHz range. Its main target is therefore the global 21 cm signal from the Dark Ages and the Cosmic Dawn, although it will also provide accurate, degree-scale foreground maps below 10 MHz and an extensive characterization of the RFI environment in the lunar far side. The *Queqiao* satellite was launched on 2018 May 21, and is currently behind the Moon, in the Earth–Moon second Lagrange point. NCLE is currently being commissioned, with first observations starting before the end of 2019.

9.5 The Far Future of 21 cm Cosmology

The 21 cm signal instruments and experiments described in this book and chapter often have continuously operated already for close to a decade (e.g., EDGES, LOFAR, MWA) and have made tremendous progress, possibly being on the verge of a detection or enabling a wide range of 21 cm signal models, either standard (Monsalve et al. 2017b; Singh et al. 2017; Monsalve et al. 2018; Singh et al. 2018) or "exotic" (Barkana 2018; Fialkov et al. 2018; Fialkov & Barkana 2019), to be excluded. Some have already been decommissioned and/or are being upgraded because they are reaching the end of their physical lifetime or are merely reaching the maximum of their capabilities (being thermal-noise or systematic-error limited). Entirely new ground-based instruments are also coming online or are being designed at the moment (e.g., NenuFAR, HERA, SKA) to push boundaries in various parameter spaces and which will likely dominate 21 cm signal science in the coming decade or two.

However, what lies beyond these instruments? What is still "left to do" when those future instruments have maximized their science return?

As touched upon earlier, aside from pushing the boundaries of parameter space in redshift, spatial scale, and signal-to-noise (e.g., imaging versus power-spectrum measurements), most ground-based instruments are running or will run against limits due to human-made RFI and ionospheric errors at very low frequencies.

The penultimate 21 cm signal instrument should, therefore, be in space, away from RFI and ionospheric errors, enabling not only extremely precise and accurate measurements of the 21 cm signal covering the redshifts of the EoR and Cosmic Dawn, but ultimately make a detection of 21 cm signal from the Dark Ages and also enable direct imaging of the Cosmic Dawn. The challenges that are facing ground-based instruments are impossible to overcome to reach those objectives (Koopmans et al. 2019). For example, at frequencies corresponding to Dark Ages redshifts, any radio signal from the sky is almost 100% distorted by the ionosphere on timescales of seconds with variable distortions over arcminute scales. It is therefore impossible to correct for these errors, and space instruments will be needed in order to observe the Dark Ages and image the Cosmic Dawn.

Space instruments are, however, complex and expensive, and developing light-weight, durable, and space-proof space technology is therefore critical. Moreover, space environment is not ubiquituosly RFI free, like Earth orbits. An optimal location is either in deep space where Earth becomes a faint radio source that can be dealt with using traditional excision techniques, but where the Sun remains a source of noise, the backside of the Moon (e.g., *FARSIDE*, Burns et al. 2019), or in lunar orbit (e.g., *DSL, DAPPER*; Chen et al. 2019; Burns et al. 2019), where it will be shielded from both Earth and the Sun for a fraction of the time. It should also be kept in mind that the lunar surface is partly charged due to solar radiation and cosmic rays and hence even on the lunar surface, effects of a lunar "ionosphere" are not completely absent. On the other hand, reflections of radio waves from the lunar surface back to any orbiter could lead to multipath propagation and also need mitigation because the dynamic range of the signal can be as high as 10^8 for the Dark Ages (a 10^5 K foreground sky versus a millikelvin signal).

Aside from these "environmental" effects, any space-based interferometer should have a collecting area far exceeding the area of upcoming instruments like HERA and the SKA. As the number of visibilities per spatial or *uv*-resolution element and their thermal noise determine the power-spectrum sensitivity, future instruments will feature an increased number of receiving elements that are cross-correlated (therefore measuring more modes) together with a larger collecting area (measuring each model with more sensitivity). Koopmans et al. (2019) suggest that the collecting area necessary to observe the Dark Ages will be as large as 10–100 km^2 in order to overcome the sky-dominated noise from foregrounds that are $\sim 10^5$ K bright. Despite these challenges, the information contained in the 21 cm signal during the Dark Ages and the early Cosmic Dawn will shed light on fundamental physics as well on the astrophysics of the infant universe, making these developments worth the effort.

References

Acero, F., Acquaviva, J. T., Adam, R., et al. 2017, arXiv:1712.06950

Barkana, R. 2018, Natur, 555, 71

Beardsley, A. P., Hazelton, B. J., Sullivan, I. S., et al. 2016, ApJ, 833, 102

Bernardi, G., McQuinn, M., & Greenhill, L. J. 2015, ApJ, 799, 90

Bernardi, G., Zwart, J. T. L., Price, D., et al. 2016, MNRAS, 461, 2847

Bowman, J. D., Rogers, A. E. E., & Hewitt, J. N. 2008, ApJ, 676, 1

Bowman, J. D., Rogers, A. E. E., Monsalve, R. A., Mozdzen, T. J., & Mahesh, N. 2018, Natur, 555, 67

Burns, J., Hallinan, G., Lux, J., et al. 2019, arXiv:1907.05407

Burns, J. O., Bale, S., & Bradley, R. F. 2019, AAS Meeting 234, 212.02

Carilli, C. L., Nikolic, B., Thyagarayan, N., & Gale-Sides, K. 2018, RaSc, 53, 845

Chen, X., Burns, J., Koopmans, L., et al. 2019, arXiv:1907.10853

DeBoer, D. R., Parsons, A. R., Aguirre, J. E., et al. 2017, PASP, 129, 045001 de Lera Acedo, E., Trott, C. M., Wayth, R. B., et al. 2017, MNRAS, 469, 2662

de Lera Acedo, E., Bolli, P., Paonessa, F., et al. 2018, ExA, 45, 1

Dewdney, P. E., Hall, P. J., Schilizzi, R. T., & Lazio, T. J. L. W. 2009, IEEEP, 97, 1482

Dillon, J. S., Liu, A., Williams, C. L., et al. 2014, PhRvD, 89, 023002

Dillon, J. S., & Parsons, A. R. 2016, ApJ, 826, 181

Eastwood, M. W., Anderson, M. M., Monroe, R. M., et al. 2018, AJ, 156, 32

Eastwood, M. W., Anderson, M. M., Monroe, R. M., et al. 2019, AJ, 158, 84

Ewall-Wice, A., Bradley, R., Deboer, D., et al. 2016, ApJ, 831, 196

Fialkov, A., & Barkana, R. 2019, MNRAS, 486, 1763

Fialkov, A., Barkana, R., & Cohen, A. 2018, PhRvL, 121, 011101

Greig, B., & Mesinger, A. 2017, MNRAS, 465, 4838

Greig, B., Mesinger, A., & Koopmans, L. V. E. 2020, MNRAS, 491, 1398

Kocz, J., Greenhill, L. J., Barsdell, B. R., et al. 2015, JAI, 4, 1550003

Koopmans, L., Barkana, R., Bentum, M., et al. 2019, arXiv:1908.04296

Koopmans, L., Pritchard, J., Mellema, G., et al. 2015, in Advancing Astrophysics with the Square Kilometre Array, PoS(AASKA14)001

Li, W., Pober, J. C., Hazelton, B. J., et al. 2018, ApJ, 863, 170

Liu, A., & Parsons, A. R. 2016, MNRAS, 457, 1864

Liu, A., & Shaw, J. R. 2019, arXiv:1907.08211

Mellema, G., Koopmans, L. V. E., Abdalla, F. A., et al. 2013, ExA, 36, 235

Mesinger, A., Greig, B., & Sobacchi, E. 2016, MNRAS, 459, 2342

Monsalve, R. A., Greig, B., Bowman, J. D., et al. 2018, ApJ, 863, 11

Monsalve, R. A., Rogers, A. E. E., Bowman, J. D., & Mozdzen, T. J. 2017a, ApJ, 835, 49

Monsalve, R. A., Rogers, A. E. E., Bowman, J. D., & Mozdzen, T. J. 2017b, ApJ, 847, 64

Neben, A. R., Bradley, R. F., Hewitt, J. N., et al. 2016, ApJ, 826, 199

Nhan, B. D., Bradley, R. F., & Burns, J. O. 2017, ApJ, 836, 90

Offringa, A. R., de Bruyn, A. G., Zaroubi, S., et al. 2013, A&A, 549, A11

Offringa, A. R., Wayth, R. B., Hurley-Walker, N., et al. 2015, PASA, 32, e008

Patra, N., Parsons, A. R., DeBoer, D. R., et al. 2018, ExA, 45, 177

Patra, N., Subrahmanyan, R., Raghunathan, A., & Udaya Shankar, N. 2013, ExA, 36, 319

Pober, J. C., Liu, A., Dillon, J. S., et al. 2014, ApJ, 782, 66

Price, D. C., Greenhill, L. J., Fialkov, A., et al. 2018, MNRAS, 478, 4193

Sathyanarayana Rao, M., Subrahmanyan, R., Udaya Shankar, N., & Chluba, J. 2017, ApJ, 840, 33

Singh, S., Subrahmanyan, R., Shankar, N. U., et al. 2018, ExA, 45, 269

Singh, S., Subrahmanyan, R., Udaya Shankar, N., et al. 2017, ApJ, 845, L12

Singh, S., Subrahmanyan, R., Udaya Shankar, N., et al. 2018, ApJ, 858, 54

Tauscher, K., Rapetti, D., Burns, J. O., & Switzer, E. 2018, ApJ, 853, 187

The Hydrogen Epoch of Reionization Array Collaboration 2019, arXiv:1907.06440

Thyagarajan, N., Carilli, C. L., & Nikolic, B. 2018, PhRvL, 120, 251301

Thyagarajan, N., Parsons, A. R., DeBoer, D. R., et al. 2016, ApJ, 825, 9

Tingay, S. J., Goeke, R., Bowman, J. D., et al. 2013, PASA, 30, e007

Trott, C. M., Pindor, B., Procopio, P., et al. 2016, ApJ, 818, 139

van Haarlem, M. P., Wise, M. W., Gunst, A. W., et al. 2013, A&A, 556, A2

Vedantham, H. K., & Koopmans, L. V. E. 2015, MNRAS, 453, 925

Vedantham, H. K., & Koopmans, L. V. E. 2016, MNRAS, 458, 3099

Wayth, R. B., Tingay, S. J., Trott, C. M., et al. 2018, PASA, 35, e033

www.ingramcontent.com/pod-product-compliance
Lightning Source LLC
Chambersburg PA
CBHW080530220326
41599CB00032B/6259